POLYMERIC AND NANOSTRUCTURED MATERIALS

Synthesis, Properties, and Advanced Applications

POLYMERIC AND NANOSTRUCTURED MATERIALS

Synthesis, Properties, and Advanced Applications

Edited by

Aparna Thankappan, PhD
Nandakumar Kalarikkal, PhD
Sabu Thomas, PhD
Aneesa Padinjakkara

Apple Academic Press Inc.
3333 Mistwell Crescent
Oakville, ON L6L 0A2
Canada

Apple Academic Press Inc.
9 Spinnaker Way
Waretown, NJ 08758
USA

© 2019 by Apple Academic Press, Inc.
Exclusive worldwide distribution by CRC Press, a member of Taylor & Francis Group
No claim to original U.S. Government works

International Standard Book Number-13: 978-1-77188-644-4 (Hardcover)
International Standard Book Number-13: 978-1-315-14749-9 (eBook)

All rights reserved. No part of this work may be reprinted or reproduced or utilized in any form or by any electric, mechanical or other means, now known or hereafter invented, including photocopying and recording, or in any information storage or retrieval system, without permission in writing from the publisher or its distributor, except in the case of brief excerpts or quotations for use in reviews or critical articles.

This book contains information obtained from authentic and highly regarded sources. Reprinted material is quoted with permission and sources are indicated. Copyright for individual articles remains with the authors as indicated. A wide variety of references are listed. Reasonable efforts have been made to publish reliable data and information, but the authors, editors, and the publisher cannot assume responsibility for the validity of all materials or the consequences of their use. The authors, editors, and the publisher have attempted to trace the copyright holders of all material reproduced in this publication and apologize to copyright holders if permission to publish in this form has not been obtained. If any copyright material has not been acknowledged, please write and let us know so we may rectify in any future reprint.

Trademark Notice: Registered trademark of products or corporate names are used only for explanation and identification without intent to infringe.

Library and Archives Canada Cataloguing in Publication
Polymeric and nanostructured materials : synthesis, properties, and advanced applications / edited by Aparna Thankappan, PhD, Nandakumar Kalarikkal, PhD, Sabu Thomas, PhD, Aneesa Padinjakkara.
Includes bibliographical references and index. Issued in print and electronic formats. ISBN 978-1-77188-644-4 (hardcover).--ISBN 978-1-315-14749-9 (PDF)
1. Nanocomposites (Materials). 2. Polymers. 3. Electrochemistry. I. Thankappan, Aparna, editor II. Kalarikkal, Nandakumar, editor III. Thomas, Sabu, editor IV. Padinjakkara, Aneesa, editor
TA418.9.N35P65 2018   620.1'15   C2018-904418-7   C2018-904419-5

Library of Congress Cataloging-in-Publication Data
Names: Thankappan, Aparna, editor.	Kalarikkal, Nandakumar, editor.	Thomas, Sabu, editor.	Padinjakkara, Aneesa, editor.
Title: Polymeric and nanostructured materials : synthesis, properties, and advanced applications / Editors, Aparna Thankappan, Nandakumar Kalarikkal, Sabu Thomas, Aneesa Padinjakkara.
Description: Oakville, ON ; Waretown, NJ : Apple Academic Press, 2019.	Includes bibliographical references and index.
Identifiers: LCCN 2018035500 (print)	LCCN 2018036483 (ebook)	ISBN 9781315147499 (ebook)	ISBN 9781771886444 (hardcover : alk. paper)
Subjects:	MESH: Nanocomposites	Polymers	Electrochemistry
Classification: LCC QP801.B69 (ebook)	LCC QP801.B69 (print)	NLM QT 36.5	DDC 572/.33--dc23
LC record available at https://lccn.loc.gov/2018035500

Apple Academic Press also publishes its books in a variety of electronic formats. Some content that appears in print may not be available in electronic format. For information about Apple Academic Press products, visit our website at **www.appleacademicpress.com** and the CRC Press website at **www.crcpress.com**

ABOUT THE EDITORS

Aparna Thankappan, PhD

Dr. Aparna Thankappan is currently an assistant professor, Baselius college, Kottayam, Kerala and was a postdoctoral fellow from the International and Inter University Centre for Nanoscience and Nanotechnology, Mahatma Gandhi University, Kottayam, India. She has published over a dozen articles in professional journals and has presented at several conferences as well. She has co-authored several book chapters. She received her PhD in photonics from Cochin University of Science and Technology, India. Her research interests include nonlinear optics, photonics, synthesis of nanomaterials, solar cell, and metal organic frameworks.

Nandakumar Kalarikkal, PhD

Dr. Nandakumar Kalarikkal, Director, International and Inter University Centre for Nanoscience and Nanotechnology, Mahatma Gandhi University, India. He received his PhD in Semiconductor Physics from Cochin University of Science and Technology, Kerala, India. Dr. Kalarikkal's research group is specialized in areas of nanomultiferroics, nanosemicondustors and nanophosphors, nanocomposites, nanoferroelectrics, nanoferrites, nanomedicine, nanosenors, ion beam radiation effects, phase transitions, etc. Dr. Kalarikkal's research group has extensive exchange programs with different industries and research and academic institutions all over the world and is performing world-class collaborative research in various fields. Dr. Kalarikkal's center is equipped with various sophisticated instruments and has established state-of-the-art experimental facilities that cater to the needs of researchers within the country and abroad.

Sabu Thomas

Professor Sabu Thomas is currently Pro-Vice Chancellor of Mahatma Gandhi University and the Founder Director and Professor of the International and Inter University Centre for Nanoscience and Nanotechnology. He is also a full-time professor of Polymer Science and Engineering at the School of Chemical Sciences of Mahatma Gandhi University, Kottayam, Kerala, India. Professor Thomas is an outstanding leader with sustained international acclaim for his work in nanoscience, polymer science and engineering, polymer nanocomposites, elastomers, polymer blends, interpenetrating polymer networks, polymer membranes, green composites and nanocomposites, and nanomedicine and green nanotechnology. Dr. Thomas's groundbreaking inventions in polymer nanocomposites, polymer blends, green bionanotechnological, and nano-biomedical sciences have made transformative differences in the development of new materials for automotive, space, housing, and in the biomedical fields. In collaboration with India's premier tire company, Apollo Tyres, Professor Thomas's group invented new high-performance barrier rubber nanocomposite membranes for inner tubes and inner liners for tires.

Professor Thomas has received a number of national and international awards, which include: Fellowship of the Royal Society of Chemistry, London (FRSC); Distinguished Professorship from the Josef Stefan Institute, Slovenia; Distinguished Faculty Award; Dr. APJ Abdul Kalam Award for Scientific Excellence—2016; Award for Outstanding Contribution in November 2016 from Mahatma Gandhi University; Lifetime Achievement Award from the Malaysian Polymer Group; Indian Nano Biologists Award—2017; Sukumar Maithy Award for the best polymer researcher in the country; and medals from the Materials Research Society of India, Chemical Research Society of India, and others. He is one of the most productive researchers in India and holds the fifth position. Recently, in acknowledgment of his outstanding contributions to the field of nanoscience and polymer science and engineering, Professor Thomas has been conferred with an Honoris Causa (DSc) Doctorate by the University of South Brittany, Lorient, France, and University of Lorraine, Nancy, France. He was also recently awarded a Senior Fulbright Fellowship to visit 20 universities in the United States.

Professor Thomas has published over 800 peer-reviewed research papers, reviews, and book chapters. He has co-edited over 100 books published by Royal Society, Wiley, Woodhead, Elsevier, CRC Press, Apple Academic Press, Springer, Nova, etc. The H index of Prof. Thomas is 90, and his work has been cited more than 36,000 times. He has delivered over 300 plenary/inaugural and invited lectures at national/international meetings in over 30 countries. He has established a state-of-the-art laboratory at Mahatma Gandhi University in the area of polymer science and engineering and nanoscience and nanotechnology through external funding from various organizations, such as DST, CSIR, TWAS, UGC, DBT, DRDO, AICTE, ISRO, DIT, KSCSTE, BRNS, UGC-DAE, Du Pont (USA), General Cables (USA), Surface Treat (Czech Republic), MRF Tyres, and Apollo Tyres. Professor Thomas has worked on several international collaborative projects with many countries. He also holds six patents.

Aneesa Padinjakkara

Aneesa Padinjakkara is a Junior Research Fellow at the International and Inter University Centre for Nanoscience and Nanotechnology at Mahatma Gandhi University, Kerala, India. She is currently working on a project involving "Engineering of Nanostructured High Performance Epoxy/Liquid Rubber Blends with Controlled Nanoparticle Localization," under the supervision of Dr. Sabu Thomas.

CONTENTS

Contributors .... *xiii*

Abbreviations .... *xvii*

Introduction .... *xix*

PART I: Research on Nanocomposites 1

1. **Thermal Lens Technique: An Investigation on Rhodamine 6G Incorporated in Zinc Oxide Low-Dimensional Structures** 3
   Aparna Thankappan and V. P. N. Nampoori

2. **Thermal Properties of Polypropylene Hybrid Composites** 15
   D. Purnima and Sagarika Talla

3. **Inverse Relaxation in Polymeric Materials: Special Reference to Textiles** 23
   P. K. Mandhyan, N. Shanmugam, P. G. Patil, R. P. Nachane, and S. K. Dey

4. **Synthesis and Characterization of Nanocrystalline ZnO Thin Film Prepared by Atomic Layer Deposition** 37
   T. V. Lidiya and K. Rajeev Kumar

5. **Vibrational Properties of a Glass Fabric/Cashew Nut Shell Liquid Resin Composite** 49
   Sai Naga Sri Harsha Ch., K. Padmanabhan, and R. Murugan

6. **A Comparative Study of Effect of Dye Structure on Polyelectrolyte-Induced Metachromasy** 55
   R. Nandini and B. Vishalakshi

7. **Nanostructured Epoxy/Block Copolymer Blends: Characterization of Micro- and Nanostructure by Atomic Force Microscopy, Scanning Electron Microscopy, and Transmission Electron Microscopy** 71
   Raghvendrakumar Mishra, Remya V. R., Jayesh Cherusseri, Nandakumar Kalarikkal, and Sabu Thomas

8. **Xanthene Dye-Doped PVA-Based Thin-Film Optical Filter Characteristics and Its Green Laser Beam Blocking** 127
   R. Renjini, George Mitty, V. P. N. Nampoori, and S. Mathew

9. **Fabrication and Fracture Toughness Properties of Cashew Nut Shell Liquid Resin-Based Glass Fabric Composites** 135
   Sai Naga Sri Harsha Ch. and K. Padmanabhan

10. **Phytosynthesis of Cu and Fe Nanoparticles Using Aqueous Plant Extracts** 141
    Subramanian L., Obey Koshy, and Sabu Thomas

11. **Effect of Concentration and Temperature on ZnO Nanoparticles Prepared by Reflux Method** 153
    Vaibhav Koutu, Najidha S., Lokesh Shastri, and M. M. Malik

12. **Graphene-Based Nanocomposite for Remediation of Inorganic Pollutants in Water** 159
    Vimlesh Chandra

PART II: Research on Polymer Technology 195

13. **Cellulose: The Potential Biopolymer** 197
    Aneesa Padinjakkara and Sabu Thomas

14. **Bioprocessing: A Sustainable Tool in the Pretreatment of Lycra/Cotton Weft-Knitted Fabrics** 203
    Shanthi Radhakrishnan

15. **Halloysite Bionanocomposites** 215
    P. Santhana Gopala Krishnan, P. Manju, and S. K. Nayak

16. **Flame Spraying of Polymers: Distinctive Features of the Equipment and Coating Applications** 235
    Yury Korobov and Marat Belotserkovskiy

PART III: Research on Electrochemistry 253

17. **Graphene-Based Hybrids for Energy Storage and Energy Conversion Applications** 255
    Anju K. Nair, Sabu Thomas, Kala M. S., and Nandakumar Kalarikkal

18. **Piezoelectric Polymer Nanocomposites for Energy-Scavenging Applications** 273
    Anshida Mayeen and Nandakumar Kalarikkal

19. **Optimization of Betanin Dye for Solar Cell Applications** 293
    Aparna Thankappan, V. P. N. Nampoori, and Sabu Thomas

20. **Polymer Electrolyte Membrane-Based Electrochemical Conversion of Carbon Dioxide from Aqueous Solutions** 307
    P. Suresh, K. Ramya, and K. S. Dhathathreyan

21. **Graphene-Based Nanostructured Materials for Advanced Electrochemical Water/Wastewater Treatment** 321
    Emmanuel Mousset and Minghua Zhou

Index.. *359*

CONTRIBUTORS

Marat Belotserkovskiy
JIME NSA Belarus, Academic Street 12, Minsk 220072, Republic of Belarus.
E-mail: mbelotser@gmail.com

Vimlesh Chandra
Department of Chemistry, Dr. Harisingh Gour Central University, Sagar 470003, Madhya Pradesh, India. E-mail: vchandg@gmail.com

Jayesh Cherusseri
Materials Science Programme, Indian Institute of Technology Kanpur, Kanpur 208016, Uttar Pradesh, India

Dr. S. K. Dey
ICAR-Central Institute for Research on Cotton Technology, Mumbai, Maharashtra, India

K. S. Dhathathreyan
Centre for Fuel Cell Technology, International Advanced Research Centre for Powder Metallurgy and New Materials (ARCI), 2nd Floor, IIT-M Research Park, No. 6, Kanagam Road, Taramani, Chennai 600113, India

P. Santhana Gopala Krishnan
Department of Plastics Technology, Central Institute of Plastics Engineering & Technology, Guindy, Chennai 600032, Tamil Nadu, India
Department of Plastics Engineering, Central Institute of Plastics Engineering & Technology, Patia, Bhubaneswar 751024, Odisha, India. E-mail: psgkrishnan@hotmail.com

Nandakumar Kalarikkal
International and Inter University Centre for Nanoscience and Nanotechnology, Mahatma Gandhi University, Kottayam 686560, Kerala, India
School of Pure and Applied Physics, Mahatma Gandhi University, Kottayam 686560, Kerala, India.
E-mail: nkkalarikkal@mgu.ac.in

Yury Korobov
UrFU, Mira St. 19, Ekaterinburg 620002, Russia. E-mail: yukorobov@gmail.com

Obey Koshy
International and Inter University Centre for Nanoscience and Nanotechnology, Mahatma Gandhi University, P.D. Hills, Kottayam 686560, Kerala, India. E-mail: Obey.Koshy@gmail.com

Vaibhav Koutu
Department of Physics, Nanoscience and Engineering Center, MANIT, Bhopal, Madhya Pradesh, India

Subramanian L.
Department of Chemistry, Amrita School of Arts and Sciences, Amrita Vishwa Vidyapeetham, Amrita University, Clappana PO Kollam 690525, Kerala, India

T. V. Lidiya
Department of Instrumentation, Cochin University of Science and Technology, Cochin, Kerala, India

Kala M. S.
Department of Physics, St. Teresa's College, Ernakulam 682011, Kerala, India

M. M. Malik
Department of Physics, Nanoscience and Engineering Center, MANIT, Bhopal, Madhya Pradesh, India

P. K. Mandhyan
ICAR-Central Institute for Research on Cotton Technology, Mumbai, Maharashtra, India

P. Manju
Department of Plastics Technology, Central Institute of Plastics Engineering & Technology, Guindy, Chennai 600032, Tamil Nadu, India

S. Mathew
International School of Photonics, Cochin University of Science and Technology, Kochi, India

Anshida Mayeen
School of Pure and Applied Physics, Mahatma Gandhi University, Kottayam, Kerala, India

Raghvendrakumar Mishra
International and Inter University Centre for Nanoscience and Nanotechnology, Mahatma Gandhi University, Kottayam, Kerala, India. E-mail: raghvendramishra4489@gmail.com

George Mitty
International School of Photonics, Cochin University of Science and Technology, Kochi, India

Emmanuel Mousset
Laboratoire Réactions et Génie des Procédés, UMR CNRS 7274, Université de Lorraine, 1 rue Grandville BP 20451, 54001 Nancy Cedex, France. E-mail: emmanuel.mousset@univ-lorraine.fr

R. Murugan
Department of Mechanical Engineering, Sri Venkateswara College of Engineering, Sriperumbudur 602117, India. E-mail: muruga@svce.ac.in

Dr. R. P. Nachane
ICAR-Central Institute for Research on Cotton Technology, Mumbai, Maharashtra, India

Anju K. Nair
International and Inter University Centre for Nanoscience and Nanotechnology,
Mahatma Gandhi University, Kottayam 686560, Kerala, India
Department of Physics, St. Teresa's College, Ernakulam 682011, Kerala, India.
E-mail: anjuillathu@gmail.com

V. P. N. Nampoori
International School of Photonics, Cochin University of Science and Technology, Kochi, India

R. Nandini
Department of Chemistry, MITE, Moodabidri 574226 (DK), Karnataka, India

S. K. Nayak
Department of Plastics Technology, Central Institute of Plastics Engineering & Technology, Guindy, Chennai 600032, Tamil Nadu, India

Aneesa Padinjakkara
International and Inter University Centre for Nanoscience and Nanotechnology, Mahatma Gandhi University, Kottayam 686560, Kerala, India. E-mail: anee18p@gmail.com

K. Padmanabhan
Centre for Excellence in Nano-Composites, School of Mechanical and Building Sciences,
VIT University, Vellore, India. E-mail: padmanabhan.k@vit.ac.in

Contributors

Dr. P. G. Patil
ICAR-Central Institute for Research on Cotton Technology, Mumbai, Maharashtra, India

D. Purnima
Department of Chemical Engineering, BITS-Pilani, Hyderabad, Telangana, India

Shanthi Radhakrishnan
Department of Fashion Technology, Kumaraguru College of Technology, Coimbatore 641049, India.
E-mail: shanradkri@gmail.com

K. Rajeev Kumar
Department of Instrumentation, Cochin University of Science and Technology, Cochin, Kerala, India

K. Ramya
Centre for Fuel Cell Technology, International Advanced Research Centre for Powder Metallurgy and New Materials (ARCI), 2nd Floor, IIT-M Research Park, No. 6, Kanagam Road, Taramani, Chennai 600113, India. E-mail: ramya.k.krishnan@gmail.com

R. Renjini
International School of Photonics, Cochin University of Science and Technology, Kochi, India

Najidha S.
Department of Physics, BJM Government College, Kollam, Kerala, India

Dr. N. Shanmugam
ICAR-Central Institute for Research on Cotton Technology, Mumbai, Maharashtra, India

Lokesh Shastri
Department of Physics, Nanoscience and Engineering Center, MANIT, Bhopal, Madhya Pradesh, India

Sai Naga Sri Harsha Ch.
Srujana-Innovation Center, L. V. Prasad Eye Institute, Hyderabad, India.
E-mail: sainagasriharsha@gmail.com

P. Suresh
Centre for Fuel Cell Technology, International Advanced Research Centre for Powder Metallurgy and New Materials (ARCI), 2nd Floor, IIT-M Research Park, No. 6, Kanagam Road, Taramani, Chennai 600113, India

Sagarika Talla
Department of Chemical Engineering, BITS-Pilani, Hyderabad, Telangana, India

Aparna Thankappan
International and Inter University Centre for Nanoscience and Nanotechnology, Mahatma Gandhi University, Kottayam, India. E-mail: aparna.subhash@gmail.com

Sabu Thomas
International and Inter University Centre for Nanoscience and Nanotechnology, Mahatma Gandhi University, Kottayam 686560, Kerala, India
School of Chemical Sciences, Mahatma Gandhi University, Kottayam 686560, Kerala, India

Remya V. R.
International and Inter University Centre for Nanoscience and Nanotechnology, Mahatma Gandhi University, Kottayam, Kerala, India

B. Vishalakshi
Department of Chemistry, Mangalore University, Mangalagangotri, Karnataka, India

Minghua Zhou
Key Laboratory of Pollution Process and Environmental Criteria, Ministry of Education, College of Environmental Science and Engineering, Nankai University, Tianjin 300071, China
Tianjin Key Laboratory of Urban Ecology Environmental Remediation and Pollution Control, College of Environmental Science and Engineering, Nankai University, Tianjin 300071, China

ABBREVIATIONS

AAS	atomic absorption spectroscopy
AFM	atomic force microscopy
ALD	atomic layer deposition
BET	Brunauer–Emmett–Teller
BOD	biological oxygen demand
CEC	cation exchange capacity
CNSL	cashew nut shell liquid
CNT	carbon nanotubes
COD	chemical oxygen demand
CTBN	carboxyl-terminated butadiene-acrylonitrile
CVD	chemical vapor deposition
DEG	diethylene glycol
DMBA	*N,N*-dimethylbenzylamine
DSC	differential scanning calorimetry
DSSC	dye-sensitized solar cells
EAOPs	electrochemical advanced oxidation processes
EDLCs	electrochemical double layer capacitors
EEG	electrochemically exfoliated graphene
EPA	Environmental Protection Agency
FFT	force frequency time
FRF	force response function
GA	graphene aerogel
GC	glass–ceramic
GO	graphene oxide
HDPE	high-density polyethylene
HMT	hexamethylenetetramine
HNTs	halloysite nanotubes
IR	inverse relaxation
JCPDS	Joint Committee on Powder Diffraction Standard
MB	methylene blue
NaHep	sodium heparinate
NaOH	sodium hydroxide
OD	optical density
ODT	order–disorder transition

ORR	oxygen reduction reaction
PA	phthalic anhydride
PA	polyamide
PBS	poly(butylene succinate)
PE	polyethylene
PEO–PEP	poly(ethylene oxide)–poly(ethylene-*alt*-propylene)
PET	polyethylene terephthalate
PLA	poly(lactic acid)
POPs	persistent organic pollutants
PP	polypropylene
PTCA	3,4,9,10-perylene tetracarboxylic acid
PVOH	poly(vinyl alcohol)
PZT	lead zirconium titanate
rGO	reduced graphene oxide
SB	styrene–butadiene
SBS	styrene–butadiene–styrene
SEM	scanning electron microscope
SEM	scanning electron microscopy
SIS	styrene–isoprene–styrene
TB	toluidine blue
TCO	transparent conducting oxide
TDS	total dissolved solids
TEA	triethanolamine
TGA	thermogravimetric analysis
THPA	tetrahydrophthalic anhydride
TL	thermal lens
TSS	total suspended solids
TTIP	titanium tetra-isopropoxide
UV	ultraviolet
VDCN	vinylidene cyanide
VR	viscoelastic recovery
WHO	World Health Organization
XRD	X-ray diffraction

INTRODUCTION

The world's growing awareness toward health and safety stresses the need for monitoring all aspects of the environment in real time. This awareness has led to the intense study of the development of environmentally safe devices.

However, recent hurdles have given scientists new tools to understand and benefit from the phenomena that occur naturally when matter is organized at the nanoscale. In essence, these phenomena are based on "quantum confinement." With the rapid development of the science and technology into the nanoscale regime, one needs to synthesize low-dimensional structures to address their peculiar physical, chemical, and optical properties related to their dimensionality with an emphasis on exploring their possible applications, such as in medicine, imaging, computing, printing, chemical catalysis, materials synthesis, and many other fields.

For nanomaterials, their morphological properties, such as size, shape, and surface properties, determine their characteristics and can affect the chemical properties, reactivity, catalytic activities of a substance as well as the energetic properties and their confinement. With the combination of different synthesis techniques and materials of nanometer dimensions, nanoparticles decoration has been used to enhance the properties of photonic devices.

Green energy and green technology are quoted often in the context of modern science and technology. Technology that is close to nature is necessary in the modern world, which is haunted by global warming and climatic alterations. Proper utilization of solar energy is one of the goals of the green energy movement. Looking at the 21st century, the nano/microsciences will be the chief contributor to scientific and technological development. As nanotechnology progresses and complex nanosystems are fabricated, a growing impetus is being given to the development of multifunctional and size-dependent materials. The control of the morphology, from the nano- to the micrometer scales, associated with the incorporation of several functionalities, can yield entirely new smart hybrid materials. They are a special class of materials that provide a new method for the improvement of the environmental stability of the material with interesting optical properties and opening a land of opportunities for applications in the field of photonics.

Organic–inorganic hybrid materials represent the natural interface between two worlds of chemistry, each with major contributions to the field of material science. The organic–inorganic composites can be broadly defined as nanocomposites that found a compromise between different properties or functions such as mechanics, density, permeability, and color. The structure–function relationship of organic–inorganic hybrid structure exhibits an interface where there is a synergistic interaction between the organic and inorganic components. It has influence on the biomineralization, hydrogen bonding, and hydrophilic–hydrophobic interactions.

Nanomaterials are an exciting subject in the field of both fundamental study and applied science. Recent research has progressed to nanocomposite materials with various structures and functions. Among the corporative candidates for constructing nanocomposites, polymers are the best choice, as they have many complementary functions such as elasticity, viscosity, and plasticity that inorganic nanocrystals lack. Photonic nano/microcomposites are generally constructed by embedding optically functional guest materials at nano/microscale into an optically transparent host matrix. The assembly of nanoparticles in matrices is of major interest in several optical and sensor applications, especially those which require large area coating. Polymeric materials are the perfect choice for such an integrated platform due to their lightweight and often ductile nature. They generally have a low-cost room-temperature fabrication process. However, they have some drawbacks, such as low modulus and strength compared to metal and ceramics. Polymers can be synthesized with customer-defined optical characteristics such as selective transparency bands in different spectral ranges, variable refractive indices, low birefringence, etc. Polymer matrices reinforced with micro- and nanoparticles possess properties superior to those of the starting materials and can enhance their properties of the polymers due to the strong interfacial adhesion or interaction between the organic polymer and the inorganic particles. They exhibit a high optical damage threshold, microhardness, good thermal stability, etc., and enable the transformation of radiations over a wide spectral range with better efficiency than polymers and dyes separately.

Thus, nano/microcomposites provide a new method to enhance the processability and stability of materials with interesting optical properties. Organic dye–polymer nanocomposites have been generating significant research attention by way of their high degree of compatibility, easy mode of preparation, flexibility in molecular design strategy, versatility, and diverse functionality.

Without any doubt, hybrid materials will soon generate smart materials and play a major role in the development of advanced functional materials. The goal of this book is to coordinate many smart nanomaterials and their recent developments in science and technology.

PART I
Research on Nanocomposites

CHAPTER 1

THERMAL LENS TECHNIQUE: AN INVESTIGATION ON RHODAMINE 6G INCORPORATED IN ZINC OXIDE LOW-DIMENSIONAL STRUCTURES

APARNA THANKAPPAN[1*] and V. P. N. NAMPOORI[2]

[1]*International and Inter University Centre for Nanoscience and Nanotechnology, Mahatma Gandhi University, Kottayam, India*

[2]*International School of Photonics, Cochin University of Science and Technology, Kochi, India*

*Corresponding author. E-mail: aparna.subhash@gmail.com

ABSTRACT

This chapter reports the use of the dual-beam thermal lens technique as a quantitative method to determine thermal diffusivity and quenching of fluorescence emission from ZnO nano/microcolloidal solution incorporated with rhodamine 6G. It is a sensitive technique that has been applied for measurements of low absorption values in materials. The investigation shows that the size- and structure-dependent interfacial interaction between the ZnO clusters and mainly the Brownian motion influenced the thermal transport properties.

1.1 INTRODUCTION

In recent years, wide-band-gap semiconductor composites with design flexibility have attracted a great deal of attention because of the intense commercial interest in developing photonic materials for many application fields from optoelectronics to energy conversion, gas sensing, and photocatalysis.[1] ZnO is a promising material for these applications because of its large room temperature band gap of 3.37 eV and also the fact that the exciton-binding energy of 60 meV should ensure excitonic survival well above room temperature.[2] The properties of ZnO make it of great commercial and scientific interest due to the enhancement of light-induced linear and nonlinear processes within the nanoscopic volume of the media surrounding nanoparticles. Laser-induced absorption of ZnO nanostructures has drawn great attention in recent years due to their potential device applications such as ultraviolet photosensors,[3] light-emitting diodes,[4] gas sensors,[5] dye-sensitized solar cells,[6,7] and piezo-nanogenerators[8] and is characterized by an increase in the absorption coefficient with light intensity. It can be induced by illuminating material with strong pump beam at one wavelength and observed by monitoring the sample of a weak probe beam at different wavelength.[9] Among the various methods for the thermal characterization of materials, such as photo-acoustic spectroscopy, photothermal lensing spectroscopy, and photothermal deflection spectroscopy, thermal lensing is a versatile and viable technique that has proved to be a powerful technique for determining the thermophysical parameters. In this method, the signal amplitude directly relates to the temperature rise in the light-illuminated material and to the amount of heat generated via optical absorption and subsequent nonradiative relaxation; a spatial distribution on the refractive index is generated in the absorbing medium. Therefore, the medium acts as lens-like optical element, called thermal lens (TL). The TL affects the propagation of the probe beam by distorting its wave front. The modified phase front produces additional convergence or divergence of the probe beam, the magnitude and dynamics of which can be correlated to the thermal conductivity and absorption coefficient. So, photothermal methods provide a higher sensitivity for the measurement of low absorption values in the materials; absorption-induced heating is detected directly.

According to the physical properties of the material, temperature profile, and incident optical beam, it's the focal length that will change. Hence, instead of directly measuring the optical spectrum, the derivative with respect to some parameter is evaluated. The effect of nano/microsized particles on the effective thermal conductivity of liquids has recently attracted considerable interest. Here, the synthesis of nano/microstructures is typically carried out using a wet chemical method. Among the numerous nano/micromaterial preparation methods, wet chemical process is more favorable, as the growth temperature can be as low as 80–100°C.

In this work, we focus on the investigation of photothermal studies of chemically prepared ZnO nano/micro-low-dimensional structures and are well dispersed in diethylene glycol (DEG), a base fluid. The objective is to study the nature of the thermal diffusion of nano/microfluids and rhodamine 6G with various volume fractions and probe into the energy transfer mechanisms on the overall cooling process by the nonradiative transition. Here, we also present the fluorescence measurements of the dye in the ZnO colloidal environment. The possibilities of composite materials with laser dyes like rhodamine to study the energy transfer processes also open up.

1.2 EXPERIMENTAL DETAILS

1.2.1 EXPERIMENTAL SETUP

The experimental architecture of the dual-beam TL technique employed in the present investigation is shown in Figure 1.1. A 150-mW diode-pumped solid-state laser with wavelength of 532 nm is used as the excitation source. Radiation of wavelength 632.8 nm from a low-power (2 mW) intensity-stabilized He–Ne laser source was used as the probe is arranged to be collinear with the pump and made to pass through the sample solution

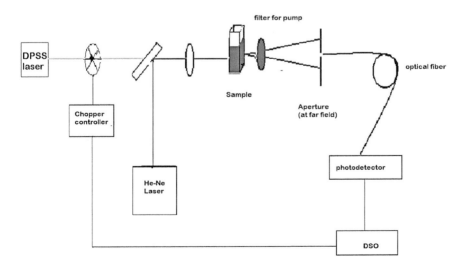

FIGURE 1.1 Schematic representation of the experimental setup.

by the use of a dichroic mirror and a convex lens. Sample was taken in cuvettes of 1 cm^2 path lengths for various sets of measurements. The pump beam was intensity modulated at 3 Hz using a mechanical chopper. The formation of a TL causes the probe beam to expand, and it is detectable at the far field. The beam is carefully positioned using mirrors and coupled to the photodetector–digital storage oscilloscope system using an optical fiber. A filter of 532 nm was used before the detector to remove the residual pump.

1.2.2 PREPARATION OF ZnO SAMPLES

The ZnO DB nanorods and microrods were synthesized by wet chemical method using zinc nitrate hexahydrate [$Zn(NO_3)_2 6H_2O$] and hexamethylenetetramine (HMT) [$(CH_2)_6 N_4$] on different reaction times. A quantity of 1 mmol zinc nitrate hexahydrate and HMT solutions were prepared in deionized water. The hexamine solution was added to the zinc nitrate solution drop wise while stirring. Finally, the solution was kept undisturbed at 80°C for 6 and 22 h.

ZnO nanoparticles/microspheres were prepared from precursor solutions of zinc acetate dehydrate ($Zn(CH_3COO)_2 2(H_2O)$) (3.35 mmol) in methanol (30 mL), and another solution of potassium hydroxide (KOH) was prepared by dissolving KOH (5.7 mmol) in methanol (16 mL). The KOH solution was added drop wise to the zinc acetate solution at 60°C under vigorous stirring. After a few times, the nanoparticles started to precipitate and the solution became turbid. ZnO nanoparticles were thus formed, and for the ZnO microsphere, the stirrer was removed after 1/2 h and the solution was capped and allowed to react till the solution evaporated off and concentrated to 10% of its original volume. The ZnO microspheres settled at the bottom and the excess mother liquor was removed and the precipitate was washed twice with methanol. The precipitate was then dispersed in 5 mL methanol. After the growth, the ZnO structures were dispersed in DEG and small amount of dye was added to it, which helps in improving light absorption.

The surface morphology of samples was studied by scanning electron microscopy (SEM, JEOL/EO, and JSM6390) (shown in Fig. 1.2) and fluorescence emission was recorded using a Cary Eclipse fluorescence spectrometer (Varian).

Thermal Lens Technique: An Investigation

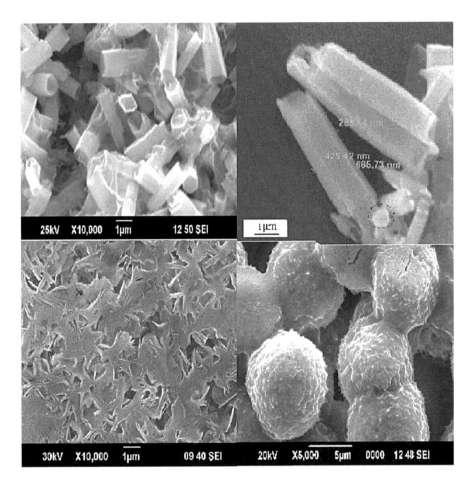

FIGURE 1.2 Scanning electron microscopic images of ZnO low-dimensional structures. (a) DB microrods, (b) microrods, (c) nanoparticles, and (d) microsphere.

1.3 RESULTS AND DISCUSSION

A TL is produced illuminating a sample with a laser beam with a TEM_{00} Gaussian profile: the excitation beam. The generated thermal distribution induces a proportional change in the refraction index value. Such a temperature gradient in the sample gives different optical path lengths, in each point, for the light of a probe laser beam, which passes through the heated space in the material.[10] During this time, the spot will increase in size, called thermal blooming. It is well known that TL technique is very sensitive that has been applied to study the samples with small absorption and to measure photo-thermal parameters because of the use of low power, stabilized laser source as the probe.

Essentially, the focal length F of the TL formed in a liquid when a CW laser beam irradiation at $t = 0$ is given by[11]

$$\frac{1}{F} = \frac{P_{abs}(\partial n/\partial t)}{\pi k w^2 (1+t_c/2t)} \quad (1.1)$$

where P_{abs} is the power absorbed by the sample and k is the heat conductivity of the sample, $\partial n/\partial t$ is the temperature coefficient of refractive index of the medium and w is beam size at the sample plane $t_c = w^2\rho c/4k$, where ρ is the density and c is the specific heat. The time variation of the TLS at the detector photodiode, $I(t)$, is now a well-known function. It is a function of time and depends on thermal end optical parameters

$$I(t) = I_0 \left[1 - \theta \left(1+\frac{t_c}{2t}\right)^{-1} + \frac{1}{2}\theta^2 \left(1+\frac{t_c}{2t}\right)^{-2} \right]^{-1} \quad (1.2)$$

where the parameter θ is related to the thermal power radiated as heat, P_{th}, and other thermo-optic parameters of the material as heat, P_{th}, and other thermo-optic parameters of the material as

$$\theta = P_{th} \frac{\partial n}{\partial T} \lambda_L k$$

We use a MATLAB-based curve fitting of the experimental data to eq (1.2) with t_c and θ as the free-fit parameters. With the help of obtained t_c, we can calculate the thermal diffusivity D of the sample as $t_c = w^2/4D$.

The experiment was performed for different concentrations of dye (Rh6g) in DEG and various relative volume fractions of the dye and ZnO nano/microsamples. Figure 1.3 shows that the variation of diffusivity with dye concentration revealed an enhancement in the nonradiative process with increase in concentration. We also attempted to study the diffusion measurements of various fractions of ZnO microrods in 10^{-3} and 10^{-5} mol L^{-1} concentrated Rh6g solution and the concentration dependence of their TL signal showed the same intensity results which depend strongly on the fluorescence of the same. Fluorescence of a molecule will be affected mainly by structural geometry, their environment such as electromagnetic interaction with the solvent and interaction among the optical fields. Rh6g

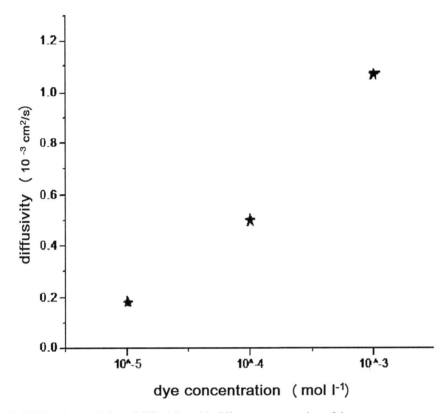

FIGURE 1.3 Variation of diffusivity with different concentration of dye.

is a highly soluble dye in DEG that may dissociate into cations and anions; therefore at lower concentration, intermolecular interaction between dyes molecules is negligible because of the large average distance between them, hence increases the fluorescence emission. On the other hand at higher concentration, the dye–dye interaction gains importance and, due to the phenomenon of reabsorption and re-emission and the formation of higher aggregation, decreases the fluorescence emission and enhances the nonradiative emission, and consequently, there is a peak shift in the fluorescence spectra (Fig. 1.4). It can be assumed that the presence of ZnO micromolecules enhances the optical field for the excited fluorophore and induces the formation of charge transfer complexes and the reduction of absorbed light due to the scattering created by the aggregation of the molecules also results in the reduced fluorescence.

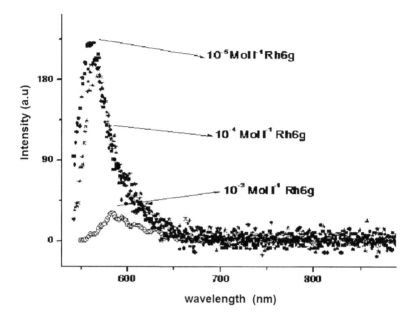

FIGURE 1.4 Fluorescence behavior of different concentration of dye with same volume fraction of microrod.

In general, there is a decrease in heat diffusion when we add ZnO samples to the dye solution. Various modes of energy transfer that are involved in sample solution are collision between DEG base-fluid molecules, thermal diffusion in the ZnO samples suspended in fluids, collision between the samples due to the Brownian motion, which is a very slow process, thermal interactions of dynamic or dancing samples with base-fluid molecules, and light-induced aggregation of samples.[12] If we compare the diffusivity D, we can find that further the addition of ZnO samples enhances the thermal signal and the thermal signal is size dependent. There were many reports which show size-dependent thermal conductivity.[13–16] Dye plays the role of radiation absorption element, more contributes to the ZnO structures only; the evidence for this is that at a particular concentration, the diffusivity is of the order of pure Rh6g, we mean that the heat released by the ZnO sample overcome to that of the dye molecules. Thus, thermal diffusion in the cluster of ZnO sample itself is responsible for the enhancement of the TL signal. The probe beam trace recorded on the oscilloscope as illustrated in Figure 1.5 showing the variation of ZnO low-dimensional structures with two different concentrations of dye revealed the same idea of diffusion. The diffusion graphs of

Thermal Lens Technique: An Investigation

two different concentration of dye are more or less same. Their threshold behavior has very good dependences on the aspect ratio of ZnO samples. The aspect ratio is a factor for the contribution to increase diffusivity, the volume of microrod with large aspect ratio is larger than that of the particle with same diameter, but with the same volume fraction, the number of microrods is smaller than that of nanoparticles. The SEM image of microsphere reveals that there are nanopetals over the microsphere; therefore, these nanopetals play an important role for the heat transport, so nanoparticles and microspheres show almost same behavior. The aspect ratio of the microsphere is less than that of nanoparticle. So, the nanoparticles show greater threshold around at 0.6 mol L^{-1}. From Figure 1.5, it is clear that the diffusivity for the ZnO DB microrod shows greater heat transfer than that of microrod. Higher concentration of the ZnO samples shows greater thermal conductivity; this depends on the thermal diffusivity as shown in Figure 1.6. Thermal diffusivities of the samples are different from each other and the thermal diffusivity as an important physical parameter is related to the presence of defects and amorphization.[13]

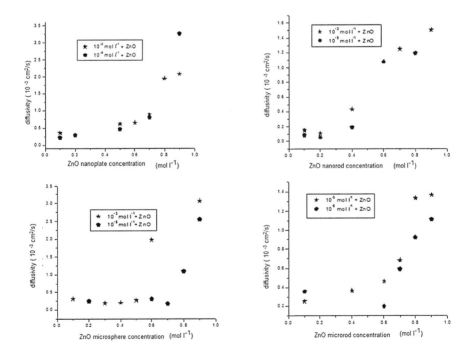

FIGURE 1.5 Variation of thermal diffusivity with different ZnO low-dimensional structures.

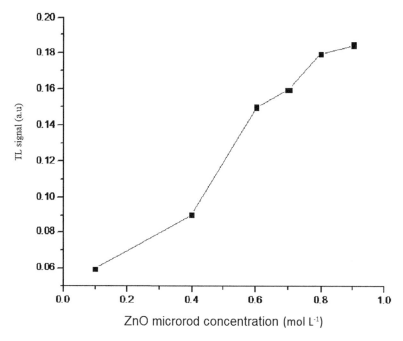

FIGURE 1.6 Variation of TL signal amplitude of the mixture of ZnO microrod and 10^{-3} mol L^{-1} concentrated Rh6g solution.

1.4 CONCLUSION

The TL technique has been successfully employed for the investigation of thermal diffusivity of the ZnO low-dimensional structures with Rh6g dye at various concentrations. For this, we have developed a simple, low-temperature, efficient, environmentally benign, wet chemical method to realize the large scale ZnO low-dimensional structures. At higher concentration, as the result of aggregate formation, re-absorption, and re-emission, the heat transfer becomes enhanced, the dye molecules acting as the heat source. We have also observed that the diffusivity is size dependent. The aspect ratio plays an important role for contributes to the enhancement of thermal diffusivity.

ACKNOWLEDGMENT

The author A. T. acknowledges the financial support from UGC-DSKPDF.

KEYWORDS

- **thermal lens**
- **ZnO**
- **low-dimensional structures**
- **SEM**
- **dye**

REFERENCES

1. Han, X.-G.; He, H.-Z.; Kuang, Q.; Zhou, X.; Zhang, X.-H.; Xu, T.; Xie, Z.-X.; Zheng, L.-S. *J. Phys. Chem. C* **2009**, *113*, 584–589.
2. Guo, J. H.; Vaysseiers, L.; Persson, C.; Ahuja, R.; Johansson, B.; Nordgren, J. *J. Phys. Condens. Matter* **2002**, *14*, 6969–6974.
3. Suehiro, J.; Nakagawa, N.; Hidaka, S.-I.; Ueda, M.; Imasaka, K.; Higashihata, M.; Okada, T.; Hara, M. *J. Nanotechnol.* **2006**, *17*, 2567–2573.
4. Konenkamp, R.; Word, R. C.; Godinez, M. *J. Nanolett.* **2005**, *5* (10), 2005–2008.
5. Sadek, A. Z.; Wlodarski, W.; Li, Y. X.; Yu, W.; Li, X.; Yu, X.; Kalantar-zadeh, K. *J. Thin Solid Films* **2007**, *515*, 8705–8708.
6. Baxter, J. B.; Walker, A. M.; van Ommering, K.; Aydil, E. S. *J. Nanotechnol.* **2006**, *17*, S304–S312.
7. Baxter, J. B.; Aydil, E. S. *J. Solar Energy Mater. Solar Cells* **2006**, *90*, 607–622.
8. Tian, J.-H.; Hu, J.; Li, S.-S.; Zhang, F.; Liu, J.; Shi, J.; Li, X.; Tian, Z.-Q.; Chen, Y. *J. Nanotechnol.* **2011**, *22*, 245601 (p 9).
9. Leonid, S. *Rec. Pat. Eng.* **2009**, *3*, 129–145.
10. Bernal-ALvarado, J.; Pereira, R. D.; Mansanares, A. M.; Da Silva, E. C. *Jpn. Soc. Anal. Chem. Anal. Sci.* **2001**, *17*, 19.
11. Joseph, S. A.; Mathew, S.; Sharma, G.; Hari, M.; Kurian, A.; Radhakrishnan, P.; Nampoori, V. P. N. *Plasmonics* **2010**, *5*, 63–68.
12. Joseph, S. A.; et al. *Optics Commun.* **2010**, *283*, 313–317.
13. Liang, L. H.; Wei, Y. G.; Li, B. *J. Phys.: Condens. Matter* 2008, *20*, 365201 (p 4).
14. Warrier, P.; Teja, A. *Nanoscale Res. Lett.* **2011**, *6*, 247.
15. Ju, H.-Y.; Zhang, S.-Y.; Li, Z.; Shui, X.-J.; Kuo, P.-K. *Appl. Phys. Lett.* **2009**, *94*, 031902.
16. Xue, Q.-Z. *Phys. Lett. A* **2003**, *307*, 313–317.

CHAPTER 2

THERMAL PROPERTIES OF POLYPROPYLENE HYBRID COMPOSITES

D. PURNIMA[*] and SAGARIKA TALLA

Department of Chemical Engineering, BITS-Pilani, Hyderabad, Telangana, India

[*]Corresponding author. E-mail: dpurnima@hyderabad.bits-pilani.ac.in

ABSTRACT

Polypropylene (PP) is a widely used polymer because of its properties such as low density and good chemical inertness which makes it particularly suitable for a wide range of applications. It has many potential applications in automobiles, appliances, and other commercial products in which creep resistance, stiffness, and some toughness are demanded in addition to weight and cost savings. However, use of polypropylene is limited by the lack of good impact resistance. Various elastomers such as ethylene propylene rubber, ethylene propylene diene rubber, styrene ethylene butylenes styrene copolymer (SEBS), ethylene vinyl acetate (EVA), polybutadiene, and natural rubber have been used to improve the impact strength of PP. However recently, coir, a natural fiber, has found to result in improvement in impact strength when added to PP; however, as in case of elastomers, the impact strength, increased with decrease in tensile properties. To get a material with balance of tensile and impact properties, inorganic filler talc was added to the composite.

Hybrid composites of PP were prepared by keeping % of coir constant at 25% and varying the talc content (5%, 10%, 20%, 30%, and 40%) from using compression molding technique. The crystallization characteristics of the composites were studied using differential scanning calorimety. Five parameters—T_{onset}, the onset temperature, T_p, the peak temperature, S_i, the initial slope, Δw, the width of the exotherm measured at half height, and A/m, the ratio of peak area (A) to the mass (m) of the crystallizable component, were found. The TGA results of the samples were analyzed and fitted into model equations which yielded activation energy (E) and order of reaction (n) and a best model was chosen based on correlation coefficient ($r^2 = 1$). It was found that E decreases at higher heating rates. The results are analyzed and discussed in the chapter.

2.1 PREFACE

The thermal properties of polypropylene (PP) hybrid composites containing coir and talc were characterized by means of thermogravimetric and differential scanning calorimetric analysis. Differential scanning calorimetry (DSC) helped in identifying the change in the pattern of crystallization of PP on the addition of talc and coir. Furthermore, thermoanalytic investigations on PP were carried out to evaluate the thermal stability and the respective activation energy of the material. The results prove that the thermal stability is necessary for the consolidation process of composite materials.

2.2 INTRODUCTION

PP is a widely used polymer because of its properties such as low density and good chemical inertness which make it particularly suitable for a wide range of applications like in automobiles, appliances, and other commercial products in which creep resistance, stiffness, and some toughness are demanded in addition to weight and cost savings. However, use of PP is limited by the lack of good impact resistance. Various elastomers, such as styrene ethylene butylene styrene copolymer, polybutadiene, and natural rubber, have been used to improve the impact strength of PP. Coir, an agriculture residue, is found to improve the impact strength in PP.[1] However, tensile strength decreases on addition of coir to PP. Inorganic fillers, such as mica, talc, etc., were added to improve tensile properties of PP. Crystallization properties of hybrid composites of PP containing coir and talc were studied using DSC which were analyzed using the five-parameter model.[1] Also, thermal properties were studied using the thermogravimetric analysis results of the samples where activation energy (E) and order of reaction (n) were determined.[2]

2.3 EXPERIMENT

2.3.1 MATERIALS

Isotactic PP, produced by Reliance Industries Ltd., India (grade H030SG) having MFI, 3 g/10 min was used. Talc was supplied by Hychem Laboratories, Hyderabad, India, and coir was obtained from a local place in Rajahmundry, Andhra Pradesh, India.

2.3.2 SAMPLE PREPARATION

Hybrid composites of PP were prepared by compression molding on a compression molding machine at a temperature of 200°C and pressure of 10 MPa at CIPET, Hyderabad, India. They were prepared by keeping % of coir constant at 25% and varying the talc content (0%, 5%, 10%, 20%, 30%, and 40%).

2.3.3 DIFFERENTIAL SCANNING CALORIMETRY

DSC measurements were done on a "TA 60 Thermal Analyzer" using DSC 60, Shimadzu. In a nitrogen atmosphere of 100 mL/min, the samples (5 mg) were first heated from 30°C to 200°C at a rate of 10°C/min and held at 200°C for 5 min to allow through melting and remove any thermal history. They were then cooled from 200°C to crystallization temperatures (T_c), 120°C at 10°C/min, and allowed to crystallize. The exotherm was recorded under identical instrumental settings and experimental conditions for all samples for reliable comparison and analysis of the exotherm parameters.

2.3.4 THERMOGRAVIMETRIC ANALYSIS

Sample (10 mg) was heated using DTG 60 Shimadzu at 10°C/min from 40 to 900°C, with nitrogen, carrier gas passing through the furnace at 50 mL/

min. Sample weight was recorded against time and temperature to produce a thermogram at four selected heating rates, 5, 10, 20, and 30°C/min. The results of the samples were analyzed and fitted into model equations which yielded activation energy (E) and order of reaction (n), and a best model was chosen based on correlation coefficient ($r^2 = 1$).[2]

2.4 RESULTS AND DISCUSSIONS

2.4.1 DIFFERENTIAL SCANNING CALORIMETRY

Crystallization exotherms of PP obtained in the DSC thermograms recorded during cooling cycle show that exotherms are analyzed using the five-parameter model namely T_{onset}, the onset temperature, T_p, the peak temperature, S_i, the initial slope, Δw, the width of the exotherm measured at half height, and A/m, the ratio of peak area (A) to the mass (m) of the crystallizable component.[1]

From Table 2.1, it is seen that T_{onset} increases with talc content as it speeds up the initial process, that is, the nucleation. T_p increases with talc content accounted for by variation of peak width. S_i decreases in PP 3, 4, and 5 as there is a slowing down of rate of nucleation causing nuclei to be created at different instants of time, growing to different sizes of the crystallites resulting in a wider crystal size distribution. Δw increases with talc content and A/m decreases as there is a decrease in crystallinity due to chain mobility restriction which hinders the formation of crystals.

TABLE 2.1 Variation of Crystallization Parameters with Coir and Variable Talc Content.

Sample	Coir (%)	Talc (%)	T_{onset} (°C)	T_p (°C)	$S_i = \tan \alpha$	Δw (°C)	A/m (mJ/g)
PP	0	0	122	118.9	2.145	0.81	105,000
PP-coir	25	0	122.3	119.1	2.605	1.95	77,723.3
PP-3	25	5	126.1	120.1	2.748	5.2	60,294.1
PP-4	25	10	125.1	119.6	2.605	3.28	58,979.3
PP-5	25	20	127.9	119.8	1.88	6.24	56,735.3
PP-6	25	30	129.3	123.4	2.145	7.46	67,912.7
PP-7	25	40	131.4	124.5	2.475	9.42	47,408.6

2.4.2 THERMOGRAVIMETRIC ANALYSIS

For many kinetic processes, rate of reaction, r, is given by

$$r = -\frac{dw}{dt} = Aw^n e^{-E/RT}$$

$$\ln r = \ln\left(-\frac{dw}{dt}\right) = \ln A + n \ln w - \frac{E}{RT} \quad \text{(on applying log on both sides)}$$

where w is the weight fraction remaining in a TGA run, T is the absolute temperature, E is the activation energy of the kinetic process, A is the pre-exponential factor, and R is the universal gas constant.[2] The model equations used to analyze the data were as follows:

$$\left(\frac{\Delta \ln r}{\Delta(1/T)}\right) = n\left(\frac{\Delta \ln w}{\Delta(1/T)}\right) - \frac{E}{R} \qquad (2.1)$$

Thus, a plot of $[\Delta \ln r/\Delta(1/T)]$ versus $[\Delta \ln w/\Delta(1/T)]$ should be a straight line with slope equal to the order of reaction, n, and an intercept of $-E/R$.[2]

$$\left(\frac{\Delta \ln r}{\Delta \ln w}\right) = n + E\left(-\frac{\Delta(1/T)}{R\Delta \ln w}\right) \qquad (2.2)$$

A plot of $[\Delta \ln r/\Delta \ln w]$ versus $[-\Delta(1/T)/(R\Delta/(w))]$ with slope, E, and an intercept, n.[2]

$$\ln\left(\frac{r}{w}\right) = E\left(-\frac{1}{RT}\right) + \ln A \qquad (2.3)$$

Slope of plot of $\ln(r/w)$ versus $[-1/(RT)]$ should be equal to activation energy, E.[2]

From Table 2.2, it is observed that E decreases at higher heating rates due to lower activation barrier for crystallization because of weak links present in the polymer. E increases, as talc content increases because it decreases chain mobility which hinders crystal formation requiring higher energy to crystallize.[2] The third model is the best suited where $n = 1$, and also based on r^2 value and correlation coefficient, it shows reasonable values where $r^2 > 0.01$. It fits the experimental data well compared to the other two models.

TABLE 2.2 Kinetic Parameters Obtained Using above Equations Where E_1, E_2, E_3 are Activation Energies Obtained from Equations 1–3, respectively, and n_1, n_2, n_3 are Orders of Reactions Obtained from Equations 1–3, Respectively.

Samples	Heating rate (deg/min)	E_1 (J/mol)	n_1	E_2 (J/mol)	n_2	E_3 (J/mol)	n_3
PP	5	52,295.06	−0.0077	47,852	0.0434	3670	1
	10	50,373.69	9.00E − 05	370,333	−0.15	2156.8	1
	20	4960.38	0.0041	15,058	−0.315	2358.1	1
	30	9619.29	0.5085	80,887	−1.37	1000	1
PP-coir	5	6.65E − 09	2.00E − 17	0	0	3178.2	1
	10	191,488	0.003	0.001	5.1647	4045.3	1
	20	9978.46	−0.0033	40,825	0.0585	3248.2	1
	30	274.5	0.0024	29,219	0.0121	6138.5	1
PP-3	5	0	0	0	0	513.62	1
	10	24,171.29	7.5761	188,936	−0.014	4273.9	1
	20	1200.458	0.0477	79,085	0.034	4282.3	1
	30	724.98	−0.0154	6320	−0.024	4010.6	1
PP-4	5	0	0	0	0	311.81	1
	10	28,016.51	5.9275	276,420	−0.051	4231.4	1
	20	252.24	0.2816	2.00E − 05	0.1456	5322.1	1
	30	150.69	−0.2374	5.00E − 05	−0.399	6557.9	1

TABLE 2.2 (Continued)

Samples	Heating rate (deg/min)	E_1 (J/mol)	n_1	E_2 (J/mol)	n_2	E_3 (J/mol)	n_3
PP-5	5	504.11	0.0003	977.9	0.0004	2782.2	1
	10	8965.81	0.1191	87,697	0.0726	5164.9	1
	20	5042.1	7.00E − 05	28,753	− 0.093	12,438	1
	30	7.007	−0.0012	24,990	0.0007	11,291	1
PP-6	5	13,383	6.0414	0.0002	0.0189	−6407.4	1
	10	220,894.66	−0.0086	0	0	998.33	1
	20	7653.28	0.002	3.77E + 05	0.1242	2981.5	1
	30	142,925.9	0.3134	5.20E + 04	0.0567	6727.8	1
PP-7	5	6552.08	6.4187	44,391	1.094	4061.1	1
	10	22,203.36	1.3469	241,399	0.4017	1400.5	1
	20	73,027.68	4.37E − 01	35,266	0.0746	2144.8	1
	30	52,830.48	0.0049	5682.8	0.2005	8378.4	1

2.5 CONCLUSIONS

The crystallization properties of PP composites containing talc and coir were studied using DSC where the five-parameter model was used to analyze the models. It was found that the rate of nucleation decreases with increasing talc content, and there is an overall decrease in crystallinity due to chain-mobility restriction. Also, the temperature dependence of the nucleation process is noted. Thermogravimetric data were obtained for the composites at heating rates of 5, 10, 20, 30°C/min. An attempt made to fit the model for degradation behavior of the composites indicated in the third model given in this chapter in which an assumption that the degradation mechanism is first order fits better for given data.

KEYWORDS

- **polypropylene hybrid composites**
- **coir**
- **talc**
- **differential scanning calorimetry**
- **thermal stability**

REFERENCES

1. Purnima, D.; Maiti, S. N.; Gupta, A. K. *Interfacial Adhesion through Maleic Anhydride Grafting of EPDM in PP/EPDM Blend* [Online]. 2005; pp 5528–5532. http://www.interscience.wiley.com (accessed August 23, 2014).
2. Chan, J. H.; Balke, S. T. The Thermal Degradation Kinetics of Polypropylene: Part III. Thermo Gravimetric Analyses. *Polym. Degrad. Stab.* **1997**, *57*, 135–149. DOI: SO141-3910(96)00160.

CHAPTER 3

INVERSE RELAXATION IN POLYMERIC MATERIALS: SPECIAL REFERENCE TO TEXTILES

P. K. MANDHYAN[*], N. SHANMUGAM, P. G. PATIL, R. P. NACHANE, and S. K. DEY

ICAR-Central Institute for Research on Cotton Technology, Mumbai, Maharashtra, India

[*]Corresponding author. E-mail: pkmandhyan@gmail.com

ABSTRACT

Mechanical properties of elastic solids like metals can be explained by hook's law for infinitesimal strains. Similarly, the properties of viscous liquids can be explained by Newton's law for infinitesimal rates of strain. But for finite strains or finite rates of strains, these laws are incapable of explaining the mechanical behavior of elastic solids or viscous liquids, respectively. All the textile materials are made up of natural and synthetic polymers which are viscoelastic in nature, therefore exhibit several time-dependent properties such as stress relaxation, creep, creep recovery, inverse relaxation, etc. The stress develops in the fiber when extended due to increasing strain. If this straining is stopped before rupture of fiber and the fiber is constrained at that particular strain, the stress developed in the fiber starts decaying with time. This phenomenon is known as stress relaxation. If a fiber of any polymeric material is strained up to a predetermined level and then allowed to retract partially, that is, the strain is reduced but not removed completely amounting to some stress being still present in the fiber, then it is observed that the stress in the fiber increases with time. The rate of increase in stress is initially high but tends to decrease with time. The phenomenon of stress build up is called "inverse relaxation." The inverse relaxation in yarns and fabrics is explained through a theoretical model. A nonlinear spring placed along with Maxwell elements in parallel is the basis of the model. The transition of inverse relaxation from yarns to fabric has also been explained by an experiment wherein same yarns have been woven into the fabric. The role of different constants has been explained and it is shown that the dashpot component of the model, which experts the force in reverse direction, gives rise to inverse relaxation whenever the system is retracted from an extended position and then allowed to rest at a point.

3.1 INTRODUCTION

One of distinct features of textiles, leather, plastics, and other polymer materials is a well-marked viscoelasticity, manifesting in a high time-dependency of their mechanical behavior. Study of material's viscoelastic behavior is a subject of great importance from viewpoint of the viscoelasticity originated from the material structure as well as from viewpoint of the material processing or its usage according to the specific purpose. The viscoelastic behavior of a polymeric material various experimental techniques has been used, among which stress relaxation, crccp, and viscoelastic recovery (VR) after preceding sustaining at constant strain or at constant load are continually in common use. In manufacturing or processing, the VR of textile products is often blocked to take the product a shape or size coinciding with the specified mold (tube, drum, beam, etc.). Mechanical properties of elastic solids like metals can be explained by hook's law for infinitesimal strains. Similarly, the properties of viscous liquids can be explained by Newton's law for infinitesimal rates of strain. But for finite strains or finite rates of strains, these laws are incapable of explaining the mechanical behavior of elastic solids or viscous liquids, respectively.

Some materials exhibit behavior which is a combination of elastic solid-like and viscous liquid-like behavior, even when strain and rate of strain are infinitesimal. Such a material does not maintain constant deformation under constant stress. Instead, it creeps, that is, it goes on deforming more and more with time, the rate of deformation decreasing with time. Also, if it is constrained at a constant deformation, the stress required to keep it in the deformed state decreases with time, the rate of change of stress decreasing with time. The first phenomenon is known as creep while the second one is known as stress relaxation. The mechanical behavior of polymeric substance is dominated by viscoelastic phenomena.

3.2 STRESS RELAXATION

Stress relaxation, also referred to as stress (or force) decay, is the reduction in stress or load over time under a constant extension. The stress may drop to a limiting value over time or it may drop to zero depending on the material. The phenomenon of stress relaxation is very similar to creep, which is the deformation or increase in extension of a material under a constant load over time. With stress relaxation, as the material is held at constant dimensions, internal stresses in the structure of the material are slowly relieved through the continued spontaneous breakage of cross-links in the molecular assembly, leading to lower tension. When a crosslink breaks, it can no longer contribute to the stress. In a stress relaxation curve, where time or the log of time is plotted versus stress or load, the first 10 s or so shows a rapid initial decay of stress which quickly levels off to a rate of decay that is almost zero. No stress relaxation occurs for a purely elastic material because the strain is completely reversible, whereas for a purely viscous material, the stress can decay to zero. Generally, for a polymeric material, a combination of both elastic and viscous behavior is exhibited which affects the limiting value of the stress. In a polymer where viscous flow is more apparent, the stress can decay to zero after a sufficiently long time, but a polymer where no or little viscous flow is evident, the stress decays to a finite value. Figure 3.1 illustrates the stress relaxation behavior of purely elastic and purely viscous materials. No stress relaxation occurs for a purely elastic material because the strain is completely reversible, whereas for a purely viscous material, the stress can decay to zero. Generally, for a polymeric material, a combination

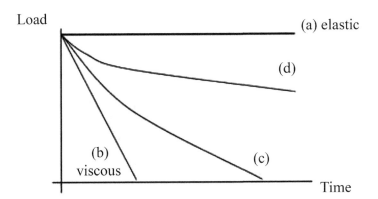

FIGURE 3.1 Stress relaxation behavior of (a) elastic, (b) viscous, (c) polymeric material with viscous flow, and (d) polymeric material with little viscous flow.

of both elastic and viscous behavior is exhibited which affects the limiting value of the stress. In a polymer where viscous flow is more apparent, the stress can decay to zero after a sufficiently long time, but a polymer where no or little viscous flow is evident, the stress decays to a finite value.[1]

3.3 INVERSE RELAXATION

Time-dependent mechanical properties such as stress-relaxation, creep, creep recovery, etc., in textile materials have been subjects of detailed study in the past. But a closely associated property, namely, inverse relaxation (IR) seems to have received little attention. The term IR refers to the building up of tension in the material which has been allowed to recover a part of the extension initially given to it and then constrained to remain at that level of extension. The occurrence of IR is often evident during cyclic loading tests as well as during the process of weaving. When a fiber specimen is initially loaded to a certain level and then the load is partly reduced, there would be an instantaneous recovery corresponding to the reduction in load. But the specimen may continue to contract over a period of time exhibiting what may be called "inverse creep." The inverse creep is different from the creep recovery because the later happens when the load is removed altogether. Similarly, in the process of weaving, the yarns are subjected to periodically varying tensions, though after being enmeshed in the fabric network, they are at relatively lower tension. If all the yarn segments have identical values of tension at any stage during the process and if their inverse creep or relaxation behavior is comparable, they would all contract by the same level. On the other hand, if the inverse creep or relaxation tendencies are dissimilar, a defective fabric could result. Strength and deformability of textile materials are important because these are the only mechanical properties determining the behavior of yarns in woven or knitted fabrics making up as well as the behavior of a fabric in end use. The behavior of textile materials as of any polymeric bodies is mostly viscoelastic. The response of a material to the specific mechanical action depends not only on the action itself but also on the former actions undergone, that is, it depends on mechanical prehistory of material. This implies that time-dependence of the response of any textile material opposing the applied forces should be taken into account. Viscoelastic properties can serve as an index for various purposes, for example, for comparative evaluation of materials or as a criterion at the control of the specific process. Independent of the stress level or amount of deformation involved, the origin of the deformation of polymer materials lies in their ability to adjust their chain conformation on a molecular level by rotation around single covalent bonds in the main chain. This freedom of rotation is, however, controlled by intra-molecular (chain stiffness) and intermolecular (interchain) interactions. Together, these interactions give rise to an energy barrier that restricts conformational change(s) of the main chain. The rate of conformational changes, that is, the molecular mobility, is determined totally by the thermal energy available in the system. Increasing the thermal energy increases the rate of change which, on a fixed time scale, allows for larger molecular rearrangements and, thus, accommodation of larger deformations. Since thermal energy is determined by temperature, there will be a relatively strong relation between temperature and mobility, and thus also with macroscopic deformation (in fact, polymers are known for their pronounced temperature dependence). In addition to this, there is also a strong influence of stress on molecular mobility since polymers allow for "mechanical" mobility when secondary bonds are broken by applying stress (rather than by increasing the thermal mobility).

In elastic solids, each atom is in some mean position with respect to the surrounding atoms. Any deformation applied to the solid (or inflicted on the solid) results in changes in relative mean positions of the adjacent atoms. The forces involved in such a change are quite local in character. Similarly in a liquid, the viscous behavior is reflection of intermolecular forces of attraction. Here also, the forces of importance are only between the nearest neighbors and hence local in character.

But in the case of a polymeric substance, the forces between neighboring atoms are not alone important. For example, as shown in Figure 3.2, a polymeric substance consists of polymer molecules, which in turn is made up of monomeric units linked to each other through covalent bonding. Any monomer molecule of a polymeric chain which is in vicinity of another molecule of another chain attracts the other molecule through a weak interaction called van der Waals force. Thus, the force field around monomer unit is highly asymmetric in nature, having maximum values in the directions of the covalent bonding. Therefore, when molecule monomer number one gets affected by deformation inflicted on the polymeric substance, its position relative to molecule monomer number two gets affected. This in turn affects position of molecule monomer number two with respect to monomer number three, etc. Thus, the deformation affects the entire chain length of the polymer molecule and is not at all localized.

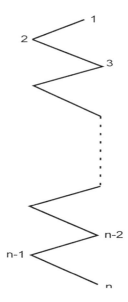

FIGURE 3.2 A linear polymer molecule. Each point corresponds to a monomer unit.

3.3.1 MODELS REPRESENTING VISCOELASTIC MATERIALS

Generally, the linear viscoelastic models, for the static tests, are developed from two elements: a spring and a hydraulic dashpot. It is considered that the first one works according to Hooke's law, and the second element obeys the Newton's law. A spring and a dashpot in series (Fig. 3.3) constitute the model called Maxwell element and in parallel (Fig. 3.4) the Kelvin or Voigt[2] element. These are the simplest models based on the theory of linear viscoelasticity to explain fiber behavior.

It was found that the phenomenon of IR is dependent on the material of the fiber, the extension level up to which it is stretched and the retraction level up to which it is retracted. This is a time-dependent phenomenon much like stress relaxation and creep. At any particular level of extension and retraction, the rate of increase in stress is fast initially but decreases with time. A polymeric fiber's behavior can be best represented by number of Maxwellian elements in parallel to each other and also a spring S_3 as shown in Figure 3.5.

As per the theory,[3] a specimen is represented by a spring in parallel with two Maxwellian elements. Element $S_1 D_1$ has a relaxation period much shorter than the element S_2D_2. Force in the specimen at any instant of time is $F = F_1 + F_2 + F_3$, where F_1, F_2, and F_3 are tensions in the S_1, S_2, and S_3, respectively. In the process of continuous deformation of the system, the extensions in the S_1 and S_2 do not occur at the same rate as that in S_3 because the plungers in the dashpots are also in motion. The movement of dashpots gives a much lower rate of extension for S_1 as compared to S_2, since the relaxation period of S_1D_1 is small as compared to that of S_2D_2. When the extension is stopped and specimen is kept in this extended state, movements of plungers of dashpots tend to make F_1 and F_2 zero. This corresponds to the stress relaxation.

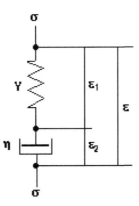

FIGURE 3.3 Maxwell model.

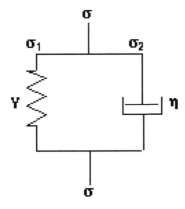

FIGURE 3.4 Kelvin (Voigt) model.

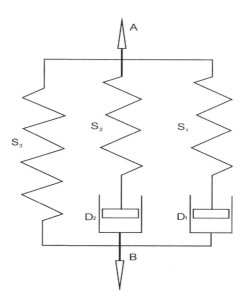

FIGURE 3.5 Model explaining the phenomena of inverse relaxation.

Now, if the specimen is retracted, the reverse movement of point B is superimposed on the upward movements of pistons in the dashpot D_1 and D_2. If the retraction is stopped at a level, where S_1 and S_2 are still under tension, stress relaxation would result. But if the retraction is allowed up to a level where S_1 has already crossed its equilibrium position of zero tension and is instead compressed while S_2 is still in the extended condition, the load in S_1 would oppose that in S_2 and S_3. Thus, the total load now would become $F' = F_3' + F_2' - F_1'$. But F_1' on account of differences in relaxation times, giving rise to IR in the beginning, followed by stress-relaxation, the latter being due to the fact that tension F_2' in S_2 reduces to zero as time passes. As per the theory cited above, it is the motion of the plunger of the dashpot which gives rise to IR in the system. The plunger is supposed to move in the opposite direction in which whole system is moving so that the tension is developed in the system and thereafter it relaxes resulting in the stress relaxation.

The force during IR is governed by the equation reproduced below. To calculate the force at time t during IR, it is required that the values of constants involved should be known.

$$F = K_0\alpha(2t_1 - t_2) + \sum \cap_i \alpha\left[\left(2e^{(-(t_2-t_1)/\tau_i)}\right) - e^{(-t_2/\tau_i)} - 1\right]e^{-(t_2-t_1)/\tau_i}$$

The constants involved in the equation are as follows:

a) K_0 is the spring constant of linear spring in the Maxwellian element.
b) η_i is the dashpot (viscosity) constant of ith Maxwellian element.
c) τ_i is the relaxation time of ith Maxwellian element.

The other quantities such as rate of traverse, time of extension, and time of retraction are known from the experiment.

3.3.2 EXPERIMENTAL REALIZATION OF INVERSE RELAXATION

If a fiber of any polymeric material is strained up to some extension level and is then allowed to retract partially so that there still exists some strain in the fiber, also amounting to some stress present in the fiber, it is observed that the stress in the fiber goes on increasing with time. The rate of increase in stress is high in the beginning but it decreases progressively, so that the stress almost levels off after a long time interval. This particular phenomenon called IR has received very little attention in the field of polymer research. IR was also observed in the Central Institute for Research on Cotton Technology, independently and has already been reported by Nachane et al.[4-6] The work carried out till now has been mainly of an exploratory nature. ICAR-CIRCOT first reported this phenomenon in cotton fibers and cotton yarns.[4] Subsequently, the study was extended to IR in spun yarns.[5] The study included ring spun yarns of cotton, polyester, viscose, and jute. A further investigation was carried out in the fibers of cotton, polyester, viscose, wool, and ramie.[6] All these polymeric fibers which included natural as well as synthetic fibers exhibited the phenomenon of IR. It was observed in these preliminary studies that any polymeric substance which shows delayed recovery mechanism active in it exhibits this phenomenon. To understand the change in stress with time in an IR experiment, let us consider a typical curve observed in such an experiment. The rate of strain r is assumed to be constant. As can be seen from Figure 3.6, the portion of stress–time curve OA corresponds to continuous straining up to αt_1 over a period of time zero to t_1. The stress developed at A is S_a. The portion AB corresponds to retraction from time t_1 to t_2 at the same rate of straining. Also, $t_2 < 2t_1$, that is, the retraction is not complete. At point B, the stress in the fiber is S_b.

Also, the strain in the fiber at this time t_2 is $\alpha(2t_1 - t_2)$. The fiber is now constrained to remain at this extension of $\alpha(2t_1 - t_2)$. From this point, as the time increases, the stress also goes increasing. If the stress is observed at a sufficiently long interval of time (of the order of a few thousand seconds), say, point C as shown, then we would have almost reached a steady state point. Let this stress be S_c. Then, we have $S_c > S_b$.

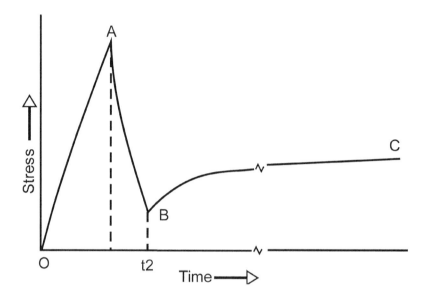

FIGURE 3.6 A typical stress versus time curve in an inverse relaxation experiment.

Due to variability of fiber specimen in different types of fibers studied, in the actual experiments carried out, instead finding out stresses S_a, S_b, S_c developed the loads W_a, W_b, and W_c developed in individual fiber specimen were measured and a ratio $(W_c - W_b)/W_a$ is defined as an IR index. Index was experimentally determined for all the fibers and yarns stated above, at different levels of extension and retraction. It was found that the phenomenon of IR is dependent on the material of the fiber, the extension level up to which it is stretched and the retraction level up to which it is retracted. To understand the behavior of any polymeric fiber at different retraction levels but the same extension level, let us consider the curve in Figure 3.7. Extension and retraction are at the same constant rate α. The section OA represents increasing stress with time as the strain increases at the rate α.

The total strain developed at point A in the fiber is αt_1. The curved portion ABP–FGHI represents stress during retraction. As retraction goes on increasing in time and the total strain remaining in the fiber goes

Inverse Relaxation in Polymeric Materials

on decreasing, the stress changes along the curve as shown in Figure 3.7. To reach the point I where the retraction is complete, that is, when the extension in the fiber is reduced to zero, the total time taken from start at 0 is $2t_1$.

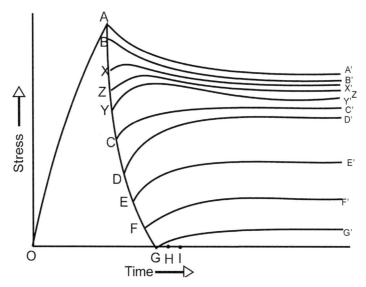

FIGURE 3.7 Stress versus time curve in inverse relaxation experiment at different retraction levels corresponding to a given extension level.

If the specimen is held in the extended condition at A, we get continuous decrease in load. The rate of decrease goes on falling. In effect, it is pure stress relaxation represented by the section AA'. If the fiber is retracted by a small amount $a(t_B - t_1)$ up to point B and is constrained to remain at that extension, that is, at $a(2t_1 - t_B)$ extension, then also the stress in the fiber continuously decreases. Of course, the fall in stress for the corresponding time periods for the stress relaxation observed at point B is smaller as compared to that observed at point A.

Instead of constraining the fiber at point B, if it is retracted by the amount $a(t_x - t_1)$ up to point X as shown in the figure, we find that there is an initial increase in stress (followed by almost constant stress) for some time and then decrease in stress. The initial increase corresponds to IR and the final decrease corresponds to stress relaxation. Here, the stress S_x'; observed at X' after a considerable amount of time is less than the stress S_x observed at X. This means that we have an overall stress relaxation when retraction is restricted to $\alpha(t_X - t_1)$. Similarly, when retraction is allowed up to point Y, that is, by an amount $\alpha(t_y - t_1)$, then also we get IR in the initial phase just after constraining the fiber at extension $\alpha(2t_1 - t_y)$, followed by stress relaxation, as is observed for the retraction up to point X. But then stress S_Y'; at point Y' is greater than the stress S_Y at point Y, corresponding to an overall IR.

Now, let us consider a case where the retraction is carried out for a still longer time $(t_C - t_1)$ where $t_C > t_y$. Here, we get an increase in stress which almost levels off after a long time. It can be said that the stress S_C goes on continuously increasing with time up to S_C'. The rate of increase also goes on decreasing with time. Here, we have pure IR. As we go down the retraction curve say up to points D, E, F, etc., we find that the extent of IR which can be gauged by $S_D' - S_D$, $S_E' - S_E$, $S_F' - S_F$, etc. increases reaches a maximum and then decreases. At point G, even when the stress has just become zero, we find IR taking place. In fact, it is observed up to point H, that is, for retraction of $\alpha(t_H - t_1)$. The retraction levels which fall between the points H and I represent the permanent deformation zone. Therefore, no relaxation is observed for retractions between points H and I.

From the retraction curve, it can therefore be concluded that for retraction from point A to point P, we observe only stress relaxation. From point P to point Q along the retraction curve, we observe both stress relaxation and IR. In the initial part, that is, near P, there is overall stress relaxation, while in the later part as we approach Q there is overall IR. A point Z exists in between PQ such that the stress $S_Z' = S_Z$. S_Z' and S_Z are stresses at point Z on the traction curve and point Z' after a considerable amount of time allowed for relaxation. The region PQ of retraction is therefore called transition zone. For retractions between point Q and point H, there is only IR. Here, we have increase in IR with increasing retraction, followed by a decrease in IR, after reaching maximum IR value. For retractions beyond point H, there is no relaxation observed.

The above discussion in brief gives behavior of a polymeric fiber in general as we go on changing the retraction level for any particular extension level.

3.4 FLOW OF INVERSE RELAXATION INDEX FROM YARN TO FABRIC

An experiment was designed to study the flow of IR index from yarn to fabric. The fabric samples were woven in such a way that warp and weft yarns of the fabrics were derived from the same yarn. Two 100% cotton yarns, one coarse, that is, 24s Ne Count and other fine, that is, 60s Ne count yarn were selected for the study. For carrying out IR test three, extension levels with the interval of 1.0 mm were selected for both the samples after considering extension at break for both the samples. In each case, relaxation of the yarn was studied first then retraction was carried out first by 0.3 mm and thereafter at the interval of 0.5 mm in each case. For example, yarn was first extended to 8.3 mm and then retracted to 8.0 mm, resulting into a retraction of 0.3 mm, etc. Instron Universal Tester was utilized to carry out the tests. Following important conclusions emerged out of the experiment were carried out.

Following are some major outcomes of the study.

- The value of dashpot constant ($\cap$) is negative in all the cases wherever the IR index is positive. It increases as the IR index increases and decreases as the IR index decreases. Therefore, it confirms the theory that it is the dashpot which exerts the force in reverse direction giving rise to IR whenever the system is retracted from an extended position and then allowed to rest at a point.
- The relaxation time is a measure of the time required for the energy stored in the spring to shift to the dashpot and dissipate. The relaxation time constant (τ) is positive in all the cases. It is higher in the vicinity of the position where IR is higher.
- The linear spring constant K_0 is also positive in all the cases. For a given extension, it increases as the difference between extension and retraction decreases. This is because the force retained in the system is higher when retraction is low, whereas the force retained in the system is lower when the retraction is high.
- The values of dashpot constant ($\cap$) is much higher (negative) in case of fabric compared to yarn from which it is woven. Similar is the case with K_0 the linear spring constant. This may be due to higher force required to extend the fabric to the same level as that of the yarn.
- The IR index is higher in case of yarn compared to corresponding fabric samples for a given extension and retraction level.
- There is not much difference in inverse relaxation index of warp and weft way of the fabric for a given extension and retraction level. It may due to the same material used in both ways of the fabric.

KEYWORDS

- **textiles**
- **leather**
- **plastics**
- **polymer materials**
- **viscoelasticity**

REFERENCES

1. Mishra, S. P. Developments in MMF Productions. Part XXX: Stress Relaxation Behaviour. *Asian Text. J* **1997**, *1*, 59–61.
2. Ferry, J. D. *Viscoelastic Properties of Polymers*. John Wiley & Sons: New York, 1980.
3. Nachane, R. P.; Sundaram, V. *J. Textile Inst.* **1995**, *86* (1), 10–19.
4. Nachane, R. P.; Hussain, G. F. S.; Krishna Iyer, K. R. *Textile Res. J.* **1982**, *52*, 483.
5. Nachane, R. P.; Hussain, G. F. S.; Patel, G. S.; Krishna Iyer, K. R. *J. Appl., Polym. Sci.* **1986**, *31*, 1101.
6. Nachane, R. P.; Hussain, G. F. S.; Patel, G. S.; Krishna Iyer, K. R. *J. Appl., Polym. Sci.* **1989**, *38*, 21.

CHAPTER 4

SYNTHESIS AND CHARACTERIZATION OF NANOCRYSTALLINE ZnO THIN FILM PREPARED BY ATOMIC LAYER DEPOSITION

T. V. LIDIYA[*] and K. RAJEEV KUMAR

Department of Instrumentation, Cochin University of Science and Technology, Cochin, Kerala, India

[*]*Corresponding author. E-mail: tvlidiya@gmail.com*

ABSTRACT

Nanocrystalline zinc oxide thin film was prepared on glass substrate by homemade atomic layer deposition system using diethyl zinc and deionized water as precursors and high pure Argon gas as purging gas. The deposited thin film was investigated for structural and optical and morphological studies using glancing angle X-ray diffraction, stylus profilometer, UV–vis spectrophotometer, photoluminescence spectrometer, scanning electron microscope, etc.

4.1 INTRODUCTION

Zinc oxide (ZnO), a II–VI semiconductor with a wide and direct band gap of 3.37 eV at room temperature, enables the absorption of UV radiation and thus is very suitable for short wavelength applications.[1–3] It has a large exciton-binding energy of 60 meV.[4] The stable exciton state even at room temperature assures more efficient exciton emission at higher temperatures. The crystalline structure of ZnO is wurzite with lattice constants a = 3.24 Å and c = 5.19 Å.[5] In particular, the ZnO is an environmental-friendly material and is largely available. ZnO exhibits different forms like powder, thin films, nanorods, nanoflowers, nanopillers, quantum dots, etc.,[6] which enables it suitable for numerous applications such as solar cells, light-emitting diodes, transparent electrodes, gas sensors, etc.[7,8] ZnO thin films can be prepared easily by different techniques such as sol–gel, RF–magnetron sputtering, chemical vapor deposition, pulsed laser deposition, molecular beam epitaxy, and atomic layer deposition (ALD).[9,10] Each growth technique has its own impact on the properties of the film, among which ALD is a growth technique that utilizes sequential, self-limiting surface reactions. ALD offers many advantages, including accurate thickness control, excellent conformality, high uniformity over a large area, good reproducibility, dense and pinhole-free structures, low deposition temperatures, etc.

The as-grown ZnO thin film was characterized for the degree of crystallinity and crystal orientation by X-ray diffraction (XRD: PANalytical X'Pert PRO with Cu $K\alpha$ radiation, λ = 1.5418 Å) and for the thickness by a stylus profilometer (Hitachi U-2000), and scanning electron microscope (SEM-JEOL Model JSM-6390LV) measurements were performed to study the surface morphology of the selected sample. Elemental compositions of thin-film sample are investigated with energy dispersive X-ray analysis (EDAX) measurements (JEOL Model JED-2300). Optical properties, such as band gap, transparency, etc., were studied with UV–vis spectrometer.

Due to the coexistence of transparency and conductivity, ZnO films are used in various applications including LEDs, solar cells, lasers, etc. Wide varieties of properties are shown by ZnO films, which made them important in different applications. This chapter deals with synthesis of ZnO thin films by homemade ALD system and is characterized for knowing the optical and electrical parameters.

4.2 EXPERIMENTAL

ZnO thin film was deposited on glass substrates (microscope slides) using homemade ALD system. For depositing ZnO, diethyl zinc (DEZn–[$Zn(C_2H_5)_2$]) is used as zinc precursor and deionized water (DI water) as oxygen precursor. Prior to deposition, glass substrates were first cleaned by DI water two times and kept it in the soap solution for few minutes. Then, it was rinsed thoroughly using DI water and dichromate solution was prepared by adding sulfuric acid to potassium dichromate solution until solution become dark brown. The substrates were dipped in the solution for 30 min and rinsed six or seven times in DI water. After cleaning the samples, they pretreated with heat to makes the surface oxygen free. Also, etching with Ar plasma inside the chamber for 5 min made the precursor ultracleaned. Before placing the substrate in the coating unit, ensure vacuum (low pressure) is inside the coating unit about 0.05 mbar.

Each precursor pulse is separated by purging with an inert gas. Here, we used high pure Argon gas for purging. Both gaseous precursors were introduced alternatively into the ALD reactor as pulses during the growth process so that substrate surface would be saturated by the precursors sequentially as shown in the following equations[11]:

$$ZnOH^* + Zn(CH_2CH_3)_2 \rightarrow ZnOZn(CH_2CH_3)^* + C_2H_6$$

$$Zn(CH_2CH_3)^* + H_2O \rightarrow ZnOH^* + C_2H_6$$

where * indicates surface species.

Ar gas purges out the excess precursors and reaction byproducts from the reaction chamber. A delay time of 5 s is given after purging for avoiding the inter mixing of precursors. Pulsing time of each precursor is listed in Table 4.1.

TABLE 4.1 Pulsing Time of Each Precursor Gas in ALD.

DEZn	0.5 s
Ar purge	5 s
Delay	5 s
DI water	0.5 s
Ar purge	5 s
Delay	5 s

Deposition was carried out at low deposition temperature 100°C and cycle number 250. Deposition parameters were listed in Table 4.2.

TABLE 4.2 Deposition Parameters.

Process parameters	Conditions
Zinc source	Diethyl zinc
Oxygen source	Deionized water
Substrate	Glass
Deposition temperature	100°C
Chamber pressure	0.048 mbar
ALD cycle sequence	DEZn/Ar/Delay/H$_2$O/Ar/Delay
ALD cycle timing	0.5 s/5 s/5 s/0.5 s/5 s/5 s
No. of ALD cycles	250

4.3 RESULTS AND DISCUSSION

To measure the thickness of the ZnO sample in stylus profilometer, a sharp step should be made at the sample surface. For this purpose, some portion of the sample was etched with chromic acid solution. Thickness of the sample is found to be 45 nm. The growth rate is obtained as 0.18 nm. This is in agreement with the reported values.

4.4 STRUCTURAL CHARACTERIZATION

XRD spectra of the ZnO thin film grown at 100°C show crystalline nature. Binary oxides often have a tendency to crystallize, even at low process temperatures. Diffraction peaks are related to the hexagonal wurtzite ZnO structure. The structures of the films were characterized by XRD with Cu $K\alpha$ radiation, $\lambda = 0.15406$ nm. From the peaks position and the integral width at half maximum of the respective peak, the grain sizes of prepared ZnO films were calculated; the average grain size in the c-axis orientation is estimated using the Debye–Scherrer relation[12]:

$$D_{XRD} = \frac{0.9\lambda}{B\cos\theta}$$

where D_{XRD} is the mean particle size, θ is the Bragg diffraction angle, and B is the full width at half maximum (FWHM) of the diffraction peak. Distance (d) calculated for all set of films using the Bragg equation[11]:

$$d = \frac{n\lambda}{2\sin\theta}$$

Figure 4.1 shows XRD pattern of ZnO thin film. It shows polycrystalline nature with major peak at (0 0 2) direction and other peaks at (1 0 0), (1 0 1), (1 1 0), (1 0 3), and (1 1 2) directions. Using Scherrer's equation, crystallite size was calculated as 201 nm.

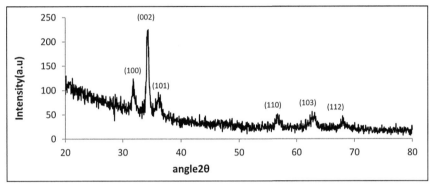

FIGURE 4.1 XRD of ZnO thin film.

Comparison of peak positions with JCPDS data is listed in Table 4.3. It clearly supports the formation of ZnO thin film on glass substrate.

TABLE 4.3 Comparison of XRD Pattern of ZnO Thin Film with Standard Hexagonal ZnO (Joint Committee on Powder Diffraction Standard (JCPDS) Card No. (36-1451)).

	JCPDS (36-1451)		As grown film	
hkl	2θ	I	2θ	I
1 0 0	31.76	57	31.82	123
0 0 2	34.42	44	34.42	224
1 0 1	36.25	100	36.46	91
1 1 0	56.60	32	56.66	51
1 0 3	62.86	29	62.7	56
1 1 2	67.96	23	67.98	47

SEM image of the sample A is shown in Figure 4.2. It indicates the formation of a uniform crystalline film on the glass substrate. The film surface is granular in nature with grain size of the order of 70 nm.

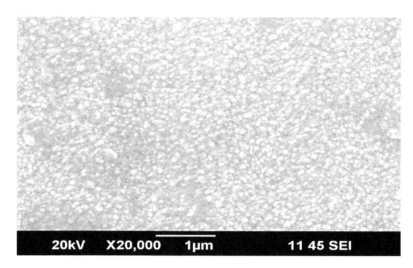

FIGURE 4.2 SEM image of ZnO thin film.

Compositional analyses of the samples were done with energy dispersive X-ray spectrometer. The Zn/O ratio was found to be nearly 1 but with some excess zinc atoms. Stoichiometry can be controlled by annealing the samples at oxygen ambient as shown in Figure 4.3.

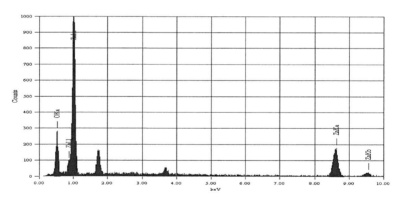

FIGURE 4.3 EDX spectrum of ZnO thin film.

TABLE 4.4 Compositional Analysis of the Sample.

Atom	Mass %	Atom %	Zn/O ratio
Zn	84.56	57.26	1.2
O	15.44	42.74	

4.5 OPTICAL CHARACTERIZATION

The optical properties of the samples were carried out. Figure 4.4 shows the optical transmittance spectrum of ZnO thin film in the wavelength range from 200 to 1500 nm. The transparency of the film is found to be low in the visible range of the electromagnetic spectrum. Only 34% is observed for the sample. This problem can be overcome by growing films by controlling the stoichiometry or by doping with Al.[13] Figure 4.5 shows the optical absorbance spectrum of ZnO thin-film sample.

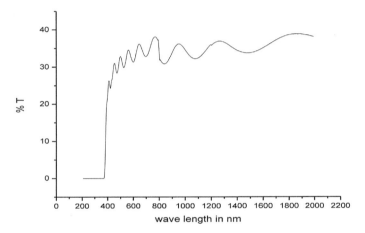

FIGURE 4.4 The optical transmission spectra of ZnO thin film.

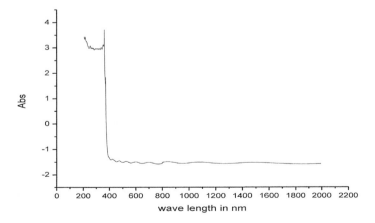

FIGURE 4.5 The optical absorbance spectra of ZnO thin film.

The optical energy gap (E_g) determined by using Tauc equation (Fig. 4.6)[14]:

$$(\alpha h\upsilon) = B(h\upsilon - E_g)^n$$

where B is Tauc constant and $h\upsilon$ is the photon energy, α is the absorption coefficient. $n = 1/2$ yields linear dependence, which describes the allowed direct transition. The optical energy gap of the film can be obtained by plotting $(\alpha h\upsilon)^2$ versus $h\upsilon$ and extrapolating the straight line portion of this plot to the energy axis. The energy gap E_g of the sample was estimated to be 3.34 eV. This is in agreement with the room temperature energy band gap of ZnO, that is, 3.37 eV.

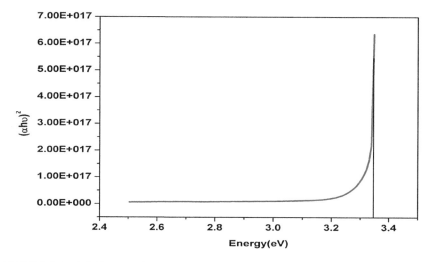

FIGURE 4.6 $(\alpha h\upsilon)^2$ versus $h\upsilon$ plot of ZnO thin film.

Photoluminescence spectra of ZnO are given below. For the ZnO sample, a sharp violet emission peak at 390 nm (3.178 eV) was observed. It is known that ZnO displays three major PL peaks: a UV near-band-edge emission peak around 380 nm, a green emission around 510 nm, and a red emission around 650 nm. It is generally accepted that the green and red emissions are associated with oxygen vacancies and interstitial Zn ions in the ZnO lattice as shown in Figure 4.7.

The absence of deep-level emission and small FWHM in our sample indicates that the concentrations of the defects responsible for the deep level

Synthesis and Characterization of Nanocrystalline

emissions are negligible. The sharp peaks at 3.178 eV presumably result from free excitons.

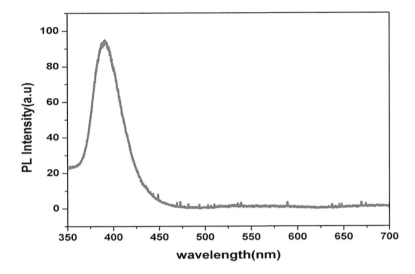

FIGURE 4.7 PL spectra of ZnO thin film.

4.6 ELECTRICAL CHARACTERIZATION

Electrical parameters of LT–ZnO thin films were obtained from the Hall effect measurements, which were carried out by the direct current four-probe method at room temperature in the magnetic field of 5.70E − 01 T. A linear current–voltage characteristic of these contacts was obtained.

TABLE 4.5 The Electrical Parameters of the ZnO Thin Film.

Carrier concentration (cm^{-3})	Resistivity (Ω cm)	Conductivity (Ω cm)$^{-1}$	Mobility (cm^2/V s)
2.48E + 18	4.47E − 02	2.24E + 01	5.64E + 01

We obtained n-type ZnO. This behavior may be attributed to the hydrogen concentration in our samples. Moreover, interstitial zinc and oxygen vacancies can also contribute to the observed n-type conductivity of the film. In the present sample, the concentration of electrons is in 2.48E + 18. For the transparent electrode applications, carrier concentration of the ZnO films should be of the order of 10^{20} cm^{-3} and resistivity of the order of 10^{-5} Ω cm. The resistivity of the as-grown sample is found to be higher. The sample shows higher mobility as 56. The electrical parameters of the sample are given in Table 4.5.

4.7 CONCLUSION

In this chapter, nanocrystalline ZnO thin film of 45-nm thickness was prepared by ALD system fabricated in the lab at relatively small deposition temperature (100°C). It is evident from XRD that the film is polycrystalline hexagonal wurtzite ZnO with *hkl* value (0 0 2) as preferred growth direction and average crystallite size is 142 nm. SEM image confirms the uniformity and granular nature of the film. EDAX analysis confirms that the sample is composed of Zn and O with excess zinc atoms. The metal-rich film shows less transparency in the visible region with 3.35 eV as band gap. Undoped ZnO thin film shows excellent electrical property with high mobility as 56.

KEYWORDS

- ZnO thin film
- atomic layer deposition
- GAXRD
- Hall measurement
- photoluminescence

REFERENCES

1. Cao, Y.; et al. *Appl. Phys. Lett.* **2006,** *88,* 251116.
2. Alias, M. F. A.; Aljarrah, R. M.; Al-Lamy, H. K. H.; Adem, K. A. W. *Int. J. Appl. Innov. Eng. Manage.* **2013,** *2* (7), 198–203.
3. Bouchenak Khelladi, N.; Chabane Sari, N. E. *Am. J. Optics Photonics* **2013,** *1* (1), 1–5.
4. Özgür, Ü.; Alivov, Y. I.; Liu, C.; et al. *J. Appl. Phys.* **2005,** *98,* 041301–041402.
5. Izyumskaya, N.; Avrutin, V.; Özgür, Ü.; Alivov, Y. I.; Morkoç, H. *Phys. Stat. Sol. B* **2007,** *244* (5), 1439–1450.
6. Wang, Z. L. *J. Phys.: Condens. Matter* **2004,** *16,* R829–R858.

7. Chu, M. C.; You, H. C.; Meena, J. S.; Shieh, S. H.; Shao, C. Y.; Chang, F. C.; Ko, F. H. *Int. J. Electrochem. Sci.* **2012**, *7*, 5977–5989.
8. Nithya, N.; Radhakrishnan, R. *Adv. Appl. Sci. Res.* **2012**, *3*, 4041–4047.
9. Rajesh Kumar, B.; Subba Rao, T. *Dig. J. Nanomater. Biostruct.* **2011**, *6* (3), 1281–1287.
10. Raghavan, R.; Bechelany, M.; Parlinska, M.; Frey, D.; Mook, W. M.; Beyer, A.; Michler, J. *Appl. Phys. Lett.* **2012**, *100*, 191102.
11. Krajewski, T. A.; Luka, G. *Acta Phys. Pol. A* **2011**, *120*, A-17–A-21.
12. Bindu, P.; Thomas, S. *J. Theor. Appl. Phys.* **2014**, *8*, 123–134.
13. Qian, X.; Cao, Y.; Guo, B.; Zhai, H.; Li, A. *Chem. Vap. Depos.* **2013**, *19*, 180–185.
14. Tauc, J. *Amorphous and Liquid Semiconductor* (Chapter 4); Plenums Press: New York and London, 1974; pp 159–214.

CHAPTER 5

VIBRATIONAL PROPERTIES OF A GLASS FABRIC/CASHEW NUT SHELL LIQUID RESIN COMPOSITE

SAI NAGA SRI HARSHA CH.[1*], K. PADMANABHAN[2*], and R. MURUGAN[3*]

[1]*Srujana-Innovation Center, L. V. Prasad Eye Institute, Hyderabad, India*

[2]*Centre for Excellence in Nano-Composites, School of Mechanical and Building Sciences, VIT University, Vellore, India*

[3]*Department of Mechanical Engineering, Sri Venkateswara College of Engineering, Sriperumbudur 602117, India*

*Corresponding authors. E-mail: sainagasriharsha@gmail.com, padmanabhan.k@vit.ac.in, muruga@svce.ac.in

ABSTRACT

The search for natural and biobased plastics is enforced due the decrease in the petroleum resources, and environmental and social concern. Cashew nut shell liquid (CNSL) is one of the most promising natural biopolymers which has wide applications in the use of composite materials. This chapter gives a detailed discussion of the dynamic testing of glass fabric/CNSL matrix composites using an impulse hammer technique. Vibrational properties like natural frequencies, Q factors, and damping percentages of the glass fabric/CNSL matrix composite were recorded and analyzed for the first time and reported. The CNSL resin is found to be a good damping material. The composite's dynamic characteristics were compared with those of glass/epoxy and carbon/epoxy specimens.

5.1 INTRODUCTION

One of the main requirements of fiber-reinforced composite materials to be successfully used in practice is their dynamic mechanical performance.[1,2] This chapter focuses on evaluation of vibration properties of a natural composite material with cashew nut shell liquid (CNSL) as the matrix[3] and glass fibers as the reinforcement. The findings reported form the first of their kind.

The mechanical properties of glass fabric used in the layup of the composite are elasticity modulus = 35 GPa, shear modulus = 14 GPa, density = 2.52 g/cm^3, and Poisson's ratio = 0.25. The mechanical properties of CNSL matrix used are elastic modulus = 1.5 GPa, density = 0.95–1.00 g/cm^3, and Poisson's ratio = 0.35.

5.2 EXPERIMENTAL DETAILS

Impact modal analysis is one experimental modal analysis technique that is widely used. The vibrational response of the structure to the impact excitation is analyzed and measured through the signal analyzer and transformed into frequency–response function using force frequency time (FFT) technique. The measurement of the frequency response function is the heart of modal analysis and the force response function (FRF) is used to extract the frequency modal parameters such as natural frequency and mode shape. The experimental setup is shown in Figure 5.1.

The vibration properties like natural frequencies and damping percentage are extracted from the experimental modal analysis. The prepared composite[4] is cut into pieces to make specimens for impact hammer modal analysis. The dimensions of the specimens are 250 mm along the length and 25 mm along the width. To carry out the modal analysis, five such specimens are prepared.

The experimental modal analysis of the specimen is carried out in cantilever condition which is fixed to a trestle. A three-dimensional accelerometer (KISTLER 8778A500) is glued to the specimen which sends the vibration response to the connected signal analyzer (DEWE 501). Impact hammer (DYTRAN 1051V) is used for the excitation of the specimen due to which the specimen vibrates. The vibration response sent by the accelerometer is amplified by the signal analyzer and forwarded to the computer for the post processing. Postprocessing of the vibration response is done by the software

Vibrational Properties of a Glass Fabric/Cashew

(RT Pro Photon) to plot FFT, FRF, etc. Through the plots (Figs. 5.2 and 5.3), the vibration properties such as natural frequency, damping percentage, etc. of the corresponding mode can be extracted.

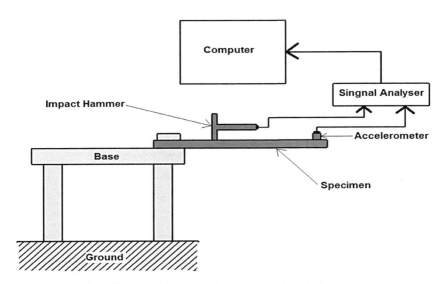

FIGURE 5.1 Block diagram of the impact hammer modal analysis.

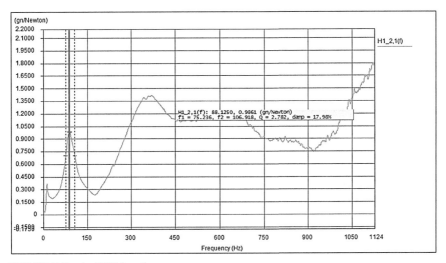

FIGURE 5.2 An FRF function.

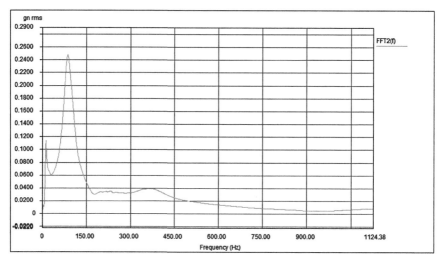

FIGURE 5.3 An FFT plot.

5.3 RESULTS AND DISCUSSION

The accelerometer is fixed at the edge of the specimen and excitation is given on the top surface of the specimen. The excitation is given on the specimen by increasing the distance from the accelerometer till the support to attain vibration responses for different modes. The results from tests are plotted.

From the vibration response plots, the vibration properties are tabulated through Tables 5.1–5.3. The first natural frequency, Q factor, and damping percentage indicated in Tables 5.1–5.3 establish a better damping than glass/epoxy specimens.

TABLE 5.1 Natural Frequencies.

Speci-men	Mode I			Mode II			Mode III		
	Trial 1	Trial 2	Trial 3	Trial 1	Trial 2	Trial 3	Trial 1	Trial 2	Trial 3
1	10	10	10	83.75	83.75	83.125	339.375	332.5	318.1250
2	10.625	10.625	10.625	86.25	85	85	337.5	315	315
3	10.625	10.625	10	89.375	89.375	88.125	416.25	572.5	630.625
4	10	10	9.375	83.75	81.25	80	309.375	304.375	288.75
5	10.9421	10.625	10.9421	88.125	88.125	86.875	361.875	369.375	367.5

TABLE 5.2 Q Factors.

Speci-men	Mode I Trial 1	Trial 2	Trial 3	Mode II Trial 1	Trial 2	Trial 3	Mode III Trial 1	Trial 2	Trial 3
1	1.637	1.566	1.528	3.959	9.934	3.887	1.747	0.663	0.669
2	2.137	1.603	1.838	3.88	3.41	3.374	2.041	2.334	2.277
3	1.691	1.955	2.296	3.349	3.027	2.326	0.658	1.364	1.61
4	2.079	1.858	1.638	3.155	2.923	3.002	0.604	0.529	1.871
5	1.804	1.929	1.924	2.746	2.782	2.469	0.805	0.863	0.855

TABLE 5.3 Damping Percentages.

Speci-men	Mode I Trial 1	Trial 2	Trial 3	Mode II Trial 1	Trial 2	Trial 3	Mode III Trial 1	Trial 2	Trial 3
1	30.54	31.93	32.72	12.63	12.71	12.86	28.62	75.38	74.70
2	23.40	31.18	27.21	12.86	14.66	14.82	24.49	21.42	21.96
3	29.57	25.57	21.77	14.93	16.52	21.5	75.95	36.67	31.05
4	24.05	26.91	30.53	15.85	17.10	16.66	82.72	94.60	26.72
5	27.72	25.92	25.99	18.21	17.98	20.25	62.10	57.97	58.50

5.4 CONCLUSIONS

The vibration properties such as natural frequencies, Q factors, and damping percentages of a CNSL-based glass fabric composite show only three modes. The CNSL composite is a high-damping material wherein the natural frequencies are low, the Q factor is lower than glass/epoxy for the same volume fraction, and the damping percentage is higher.

KEYWORDS

- **CNSL resin**
- **glass fabric**
- **biocomposite**
- **vibrations**
- **first natural frequency**
- **damping ratio**
- **loss factor**

REFERENCES

1. Murugan, R.; Ramesh, R.; Padmanabhan, K. *Proc. Eng.* **2014,** *97*, 459–468.
2. Murugan, R.; Ramesh, R.; Padmanabhan, K.; Jeyaram, R.; Krishna, S. *Appl. Mech. Mater.* **2014,** *529*, 96–101.
3. Gedam, P. H.; Sampathkumaran, P. S. *Progr. Organ. Coat.* **1986,** *14*, 115–157.
4. Sai Naga Sri Harsha, Ch.; Padmanabhan, K. In: *ICNP Proceedings*, Kottayam, Kerala, 2015.

CHAPTER 6

A COMPARATIVE STUDY OF EFFECT OF DYE STRUCTURE ON POLYELECTROLYTE-INDUCED METACHROMASY

R. NANDINI[1*] and B. VISHALAKSHI[2]

[1]*Department of Chemistry, MITE, Moodabidri 574226 (DK), Karnataka, India*

[2]*Department of Chemistry, Mangalore University, Mangalagangotri, Karnataka, India*

*Corresponding author. E-mail: hodche@mite.ac.in

ABSTRACT

The interaction of two cationic dyes, namely, toluidine blue (TB) and methylene blue (MB) with an anionic polyelectrolyte, namely, sodium heparinate (NaHep), has been investigated by spectrophotometric method. The polymer induced metachromasy in the dyes resulting in the shift of the absorption maxima of the dyes toward shorter wavelengths. The stability of the complexes formed between TB and NaHep was found to be lesser than that formed between TB and NaHep. This fact was further confirmed by reversal studies using alcohols, urea surfactants, and electrolytes. The interaction parameters revealed that binding between TB and NaHep was mainly due to electrostatic interaction, while that between MB and heparinate is found to involve both electrostatic and hydrophobic forces. The effect of the structure of the dye and its relation to metachromasy has been discussed.

6.1 INTRODUCTION

Metachromasy is a well-known phenomenon in the case of dye–polymer interactions and is generally found in the case of aggregation of cationic dyes on anionic polymers.[1–4] Metachromasy is related to the interaction of cationic dyes with polyanions where a single individual compound is formed by the interaction of the dye cation and the chromotrope polyanionic polymer. Several physiochemical parameters such as molecular weight of each repeating unit, stoichiometry of the dye–polymer complex, binding constant, and other related thermodynamic parameters like free energy, enthalpy, and entropy changes can be evaluated using polymer–dye interactions. Biological activity of macromolecules depends on its tertiary conformation. Conformation of the polyanions controls the induction of metachromasy of aqueous dye solution. Although there are several reports on metachromasy of various classes of acidic polysaccharides and different synthetic polyanions,[5,6] cyanine dyes, which are cationic in nature, have widely been used to probe biological systems such as helical structure of DNA, tertiary conformation of bacterial polysaccharides and other polymers.

As these dye have high light absorptivity, they can be used as optical probes in studying membranes, micelles, and other host systems.[7–10] Studies on polymer–surfactant interaction in aqueous solution have been attracting widespread attention due to multiple practical uses in biology.[11–14] Such studies are also assumed to be important as the mixed systems/aggregates can give rise to advanced functions that are unobtainable from single component.[15] Several physicochemical properties of macromolecule, surfactants, are quite relevant in this context. Formulation procedures based on suitable mixture may have been appealing applications.[16–18] It has been noted that oppositely charged surfactant binds to polymer surfaces through both electrostatic and hydrophobic interaction.[19] Different techniques for their isolation and stability determination of metachromatic complexes have been reported in literature.[20] The phenomena of reversal of metachromasy by addition of urea, alcohol neutral electrolytes,[21] and also by increasing the temperature of the system may be used to determine the stability of the metachromatic compounds. The nature of dye–polymer interaction in the metachromatic complex formation and also the conditions for the interaction between the cationic dye and the anionic site of the macromolecules has been studied by determining the thermodynamic parameters of the interaction.[22] The interaction between polyelectrolyte's and oppositely charged surfactants[23,24] has been investigated due to its importance in both fundamental and applied fields.[25] Comprehensive studies of a variety of cationic surfactants with anionic polyelectrolytes have been reported.[26–32] Cationic dyes that interacted with Heparin mostly fell in the thiazine group such as methylene blue (MB), toluidine blue (TB).[33–35] azure A,[36,37] and azure B.[38,39] Hence, the objective of the present study is to compare the extent of metachromasy induced by sodium heparinate (NaHep), in the cationic dyes TB and MB, and to evaluate the thermodynamic parameters of interaction and to study the extent of reversal by using alcohols, urea, surfactants, and electrolytes which is an indirect evidence for the stability of the metachromatic complex formed. The effect of dye structure on metachromasy has also been discussed.

6.2 MATERIALS AND METHODS

6.2.1 APPARATUS

The spectral measurements were carried out using Shimadzu UV-2500 Spectrophotometer using a 1-cm quartz cuvette.

6.2.2 REAGENTS

MB was obtained from HiMedia, Germany and used as received. TB was obtained from Acros Media, Germany. NaHep (HiMedia, India) was used without further purification; methanol (MeOH), ethanol (EtOH), and 2-propanol (PrOH) (Merck, India) were distilled before use. Urea, sodium chloride, and potassium chloride were obtained from Merck, India. Sodium lauryl sulfate and sodium dodecyl benzene sulfonate were obtained from Loba Chemie, India.

6.3 METHODS

6.3.1 DETERMINATION OF STOICHIOMETRY OF POLYMER–DYE COMPLEX

Increasing amounts of polymer solution (0.0–5 × 10^{-3} M) were added to a fixed volume of dye solution (0.5 mL, 1 × 10^{-3} M) in case of MB and

(0.5 mL, 1×10^{-3} M) of dye to increase the amount of polymer solution (0.0–2.5 × 10^{-3} M) in case of TB in different sets of experiments and the total volume was made up to 10 mL by adding distilled water in each case. The absorbances were measured at 628 and 528 nm in case of MB–Hep and at 606 and 516 nm in case of TB–Hep complex.

6.3.2 STUDY OF REVERSAL OF METACHROMASY USING ALCOHOLS AND UREA

For measurements of the reversal of metachromasy, solutions containing polymer and dye in the ratio 2:1 were made containing different amount of alcohol. The total volume was maintained at 10 mL in each case. The absorbances were measured at 628 and 528 nm in case of MB–NaHep and at 606 and 516 nm in case of TB–NaHep. A similar procedure is repeated with urea also.

6.3.3 STUDY OF REVERSAL OF METACHROMASY USING SURFACTANTS AND ELECTROLYTES

For measurements of the reversal of metachromasy, solutions containing polymer and dye in the ratio 2:1 were made containing different amount of surfactants or 1×10^{-7} M–1×10^{-2} M were made in case of MB–NaHep and (1×10^{-6} M-0.1 M) in case of TB–NaHep. The total volume was maintained at 10 mL in each case. The absorbances were measured at 628 and 528 nm in case of MB–NaHep at 606 nm and at 516 nm in case of TB–NaHep. Similarly, polymer–dye solutions containing different amounts of electrolytes (1×10^{-6} M–0.1 M) in case of TB–NaHep and 1×10^{-7} M–1×10^{-2} M in case MB–NaHep were made, and the absorbances were measured at the wavelengths as mentioned previously.

6.3.4 DETERMINATION OF THERMODYNAMIC PARAMETERS

The thermodynamic parameters were determined by measuring the absorbances of the pure dye solution at the respective monomer band and metachromatic band in the temperature range (36–54°C). The above experiments were repeated in presence of the polymers at various polymer–dye ratios (2, 5, 8, and 10).

6.4 RESULTS AND DISCUSSION

The absorption spectra of MB and TB show an absorption maximum at 628 nm in case of MB and 606 nm in case of TB indicating the presence of a monomer dye species in the concentration range studied. On adding increasing amounts of polymer solution, the absorption maxima shifts to 528 nm in case of MB–NaHep and at 516 nm in case of TB–NaHep complex. The blue shifted band is attributed to the stacking of the dye molecules on the polymer backbone and this reflects high degree of cooperativity in binding.[40] Appearance of multiple banded spectra proposed that the polymer might have a random coil structure in solution, whereas at higher concentration of the polymer, almost a single-banded spectrum was observed due to possible change from random coil to helical form.[41] The absorption spectra at various P/D ratios are shown in Figures 6.1 and 6.2, respectively.

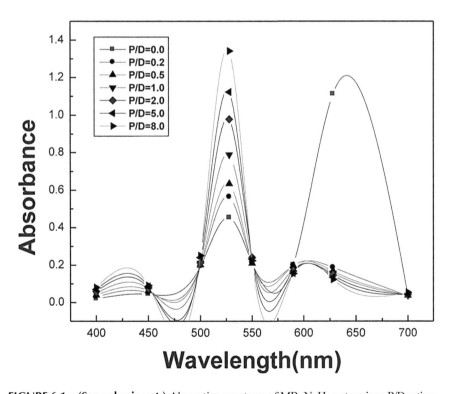

FIGURE 6.1 (**See color insert.**) Absorption spectrum of MB–NaHep at various P/D ratios.

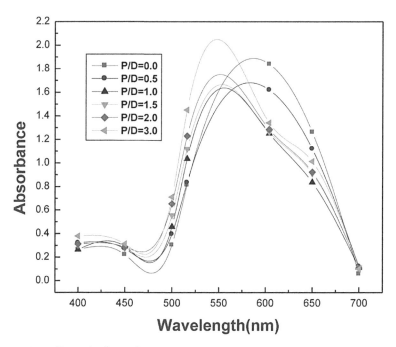

FIGURE 6.2 **(See color insert.)** Absorption spectrum of TB–NaHep at various P/D ratios.

6.4.1 DETERMINATION OF STOICHIOMETRY

To determine the stoichiometry of the polymer–dye complex, a plot of $\mathbf{A}_{528}/\mathbf{A}_{628}$ versus the polymer/dye ratio was made for MB–NaHep system. A similar procedure was repeated with TB–NaHep complex also. The stoichiometry of MB–NaHep complex was found 1:2 which indicates that the binding is at alternate anionic sites. This indicates that every potential anionic site of the polyanion was associated with the dye cation and aggregation of such dye molecules was expected to lead to the formation of a card pack stacking of the individual monomers on the surface of the polyanion so that the allowed transition produces a blue-shifted metachromasy.[42] The results were in good agreement with the reported values for interaction of similar dyes with polyanions.[43] While in case of Pcyn–NaCar complex, the stoichiometry is 1:3 and the binding is at alternate anionic sites. This indicates that there is lesser overcrowding and more aggregation of the bound dyes on the polymer chain in the latter case than in the former case. Similar results were reported in case of binding of pinacyanol chloride on poly(methacrylic acid) and poly(styrene sulfonate) systems.[44] The stoichiometry results are obtained by

A Comparative Study of Effect of Dye Structure

plotting A_{628}/A_{528} or A_{606}/A_{516} versus P/D ratio in each case. The results are shown in Figures 6.3 and 6.4, respectively.

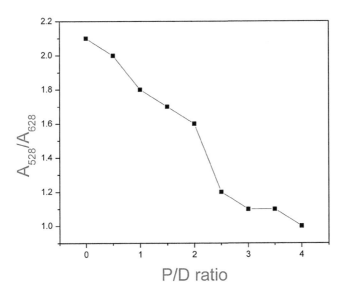

FIGURE 6.3 Stoichiometry of MB–NaHep complex.

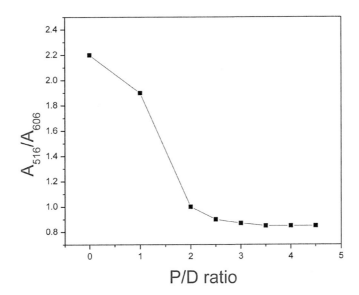

FIGURE 6.4 Stoichiometry of TB–NaHep complex.

6.4.2 REVERSAL OF METACHROMASY USING ALCOHOLS AND UREA

The metachromatic effect is presumably due to the association of the dye molecules on binding with the polyanion which may involve both electrostatic and hydrophobic interactions. The destruction of metachromatic effect may occur on addition of low-molecular weight electrolytes, alcohols, or urea. The destruction of metachromasy by alcohol and urea attributed to the involvement of hydrophobic bonding has already been established.[45–49] The efficiency of alcohols in disrupting metachromasy was found to be in the order methanol < ethanol < 2-propanol, indicating that reversal becomes quicker with increasing hydrophobic character of the alcohols. The above facts are further confirmed in the present system. On addition of increasing amount of alcohol to the polymer/dye system at P/D = 2.0, the original monomeric band of dye species is gradually restored. The efficiency of the alcohols, namely, methanol, ethanol, and 2-propanol, on destruction of metachromasy was studied. In case of AB–NaCar system, 45% methanol, 35% ethanol, 25% 2-propanol were sufficient to reverse metachromasy 60% methanol, 50% ethanol, 40% 2-propanol were required to reverse metachromasy in Pcyn–NaCar system. From the plot of A_{528}/A_{628} or A_{516}/A_{606} (Figs. 6.5 and 6.6) against the percentage of alcohols or molar concentration of urea, the percentage of alcohols or molar concentration of urea, needed for

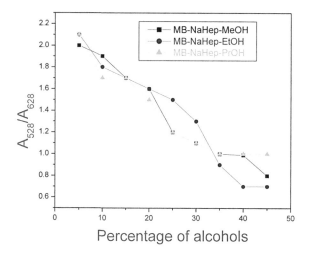

FIGURE 6.5 Reversal of metachromasy on addition of alcohols (MB–NaHep).

A Comparative Study of Effect of Dye Structure

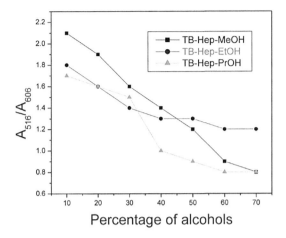

FIGURE 6.6 Reversal of metachromasy on addition of alcohols (TB–NaHep).

complete reversal has been determined. The concentration of urea to reverse metachromasy is found to be as high as 5 M in MB–Hep system and 4.5 M in case of TB–NaHep system (Fig. 6.7). This indicates that the metachromatic complex formed between MB and NaHep is more stable than that between TB and NaHep complexes. Similar reports are available in literature for reversal of metachromasy in anionic polyelectrolyte/cationic systems by addition of alcohols or urea.[50,51]

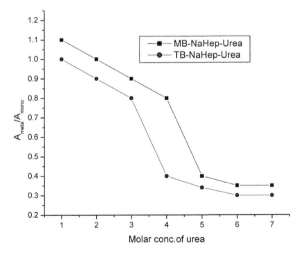

FIGURE 6.7 Reversal of metachromasy on addition of urea.

6.4.3 REVERSAL OF METACHROMASY USING SURFACTANTS

The strength and nature of interaction between water-soluble polyelectrolytes and oppositely charged surfactants depend on the characteristic features of both the polyelectrolytes and the surfactants. The charge density, flexibility of the polyelectrolyte, and the hydrophobicity of the nonpolar part and the bulkiness of the polar part also play a vital role in the case of polysaccharide–surfactant interaction.[52] On adding increasing amounts of sodium lauryl sulfate and sodium dodecyl benzene sulfonate to TB–NaHep complex, the molar concentrations of sodium lauryl sulfate and sodium dodecyl benzene sulfonate needed to cause reversal were found to be 1×10^{-4} M and 1×10^{-3} M in case of TB–NaHep and 1×10^{-3} M and 1×10^{-2} M in case of MB–NaHep system. These results agree with those reported earlier in literature.[53] Thus, the addition of surfactants causes the production of micelles and the surfactant molecules interact with the polymer by replacing the cationic dye. The release of dye molecules from the dye–polymer complex in presence of cationic surfactants revealed that surfactants interacted electrostatically[54] with the anionic site of the polymer and thus the dye becomes free. The ease of reversal of metachromasy can be correlated with its chain length. Therefore, the binding between oppositely charged polymer surfactant is primarily by electrostatic forces which are reinforced by hydrophobic forces. From the plot of absorbance at 628 nm in case of azure B or absorbance at 506 nm in case of MB against molar concentration of surfactants, the molar concentration of surfactants needed for complete reversal of metachromasy has been determined. The results are shown in Figures 6.8 and 6.9, respectively.

6.4.4 DETERMINATION OF INTERACTION PARAMETERS

The interaction constant K_C for the complex formation between TB and NaHep and MB–NaHep was determined by absorbance measurements at the metachromatic band at four different temperatures taking different sets of solutions containing varying amounts of polymer (C_S) in a fixed volume of the dye solution (C_D). The value of K_C was obtained from the slope and intercept of the plot of $C_D C_S/(A - A_O)$ against C_S shown in Figures 6.10 and 6.11.

A Comparative Study of Effect of Dye Structure

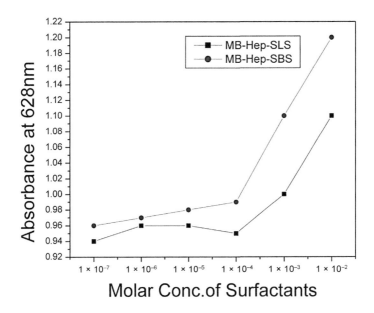

FIGURE 6.8 Reversal of metachromasy on addition of surfactants.

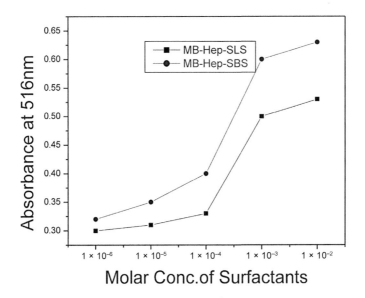

FIGURE 6.9 Reversal of metachromasy on addition of surfactants.

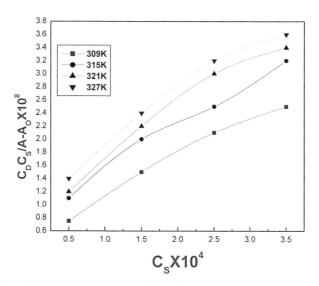

FIGURE 6.10 Effect of temperature on MB–NaHep system.

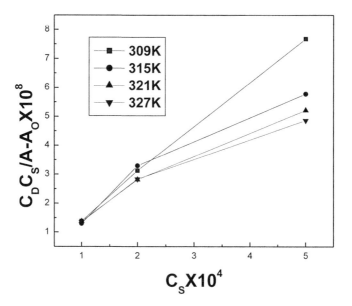

FIGURE 6.11 Effect of temperature on TB–NaHep system.

In each case, the thermodynamic parameters of interaction, namely, ΔH, ΔG, and ΔS, were also calculated. The results are given in Table 6.1.

A Comparative Study of Effect of Dye Structure

TABLE 6.1 Thermodynamic Parameters of Interaction of MB–NaHep and TB–NaHep Systems.

System Temp. (K)	K_c (dm³ mol⁻¹) I	K_c (dm³ mol⁻¹) II	ΔG (kJ mol⁻¹) I	ΔG (kJ mol⁻¹) II	ΔH (kJ mol⁻¹) I	ΔH (kJ mol⁻¹) II	ΔS (J mol⁻¹ K⁻¹) I	ΔS (J mol⁻¹ K⁻¹) II
309	5400	7000	−4.156	−5.255				
315	3500	2933	−4.577	−5.455	−2.4576	−8.945	−0.046	−0.023
321	2644	1600	−4.714	−5.815				
327	1811	1062	−4.856	−6.095				

I, TB–NaHep system; II, MB–NaHep system.

a) Calculated from Figures 6.11 and 6.12 according to Rose–Drago equation
b) Calculated from the thermodynamic equation, $\Delta G = -RT \ln K_c$
c) Calculated graphically by plotting $\ln K_c$ against $1/T$ according to van't Hoff equation
d) $\ln K_c = -\Delta H/RT + C$
e) Calculated from the thermodynamic expression, $\Delta G = \Delta H - T\Delta S$

6.4.5 EFFECT OF STRUCTURE OF DYE

The structures of and TB and MB are given in Figures 6.12 and 6.13, respectively. It is evident that MB, being a rigid planar cationic dye and hence a small distance between the adjacent anionic sites on the polyanion, will be more favorable for binding resulting in stacking arrangement; on the other hand, TB being larger in size is more hydrophobic and hence induces greater aggregation. Thus in this case, the distance between the two adjacent dye molecules will be greater and the dye molecules are oriented like a stair case which agrees with the reported literature.[55] Furthermore, the dye molecules are arranged in a parallel and stacked manner within the aggregates resulting in a hypsochromic shift.

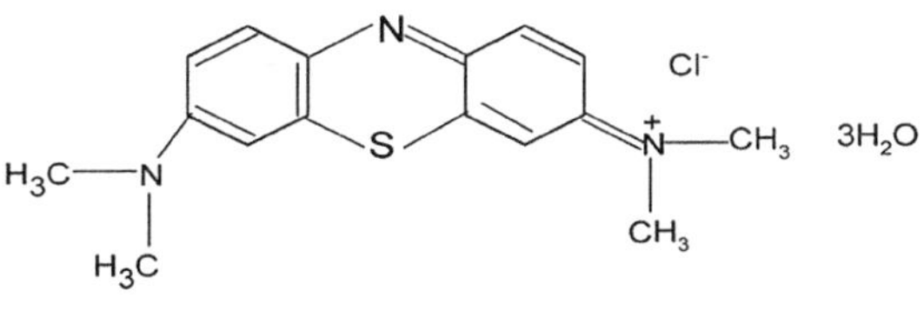

FIGURE 6.12 Structure of Methylene blue.

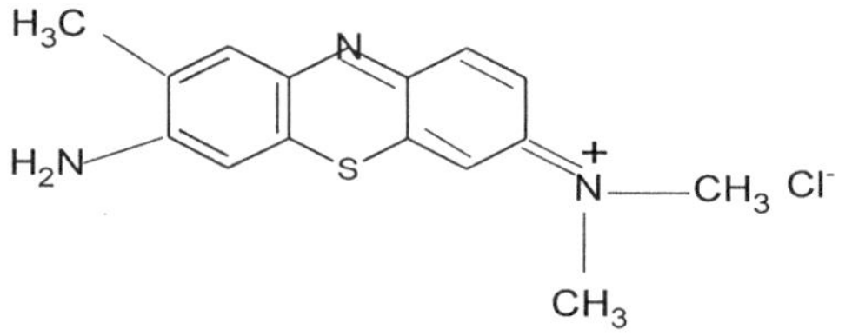

FIGURE 6.13 Structure of Toluidine blue.

6.5 CONCLUSIONS

The polymers NaHep induced metachromasy in the dye TB and MB. The monomeric band occurs at 628 and 606 nm, while the metachromatic band occurred at 528 nm in the case of MB–NaHep and at 586 nm in the case of TB–NaHep. The spectral shifts are higher in the case of MB–NaHep (100 nm) than in case of TB–NaHep (90 nm). These results are further confirmed by the reversal studies using alcohols, urea, electrolytes, and surfactants. The thermodynamic parameters further confirm the above fact. It is thus evident from the above studies that both electrostatic and nonionic forces contribute toward the binding process. Based on the results, it can be concluded that MB is more effective in inducing metachromasy in NaHep than cooperativity in binding is observed to occur due to neighbor interactions among the bound dye molecules at lower P/D ratios leading to stacking. The stacking tendency is enhanced by the easy availability and close proximity of the charged sites.

KEYWORDS

- cationic dyes
- metachromasy
- sodium heparinate
- dye structure
- aggregation

REFERENCES

1. Mitra, A.; Nath, R. K.; Biswas, S.; Chakraborty, A. K.; Panda, A. K.; Mitra, A. *J. Photochem. Photobiol. A: Chem.* **1997**, *111*, 157.
2. Mitra, A.; Chakraborty, A. K. *J. Photochem. Photobiol. A: Chem.* **2006**, *178*, 98.
3. Horbin, R. W. *Biochemie. Biochem.* **2002**, *77*, 3.
4. Bergeron, J. A.; Singer, M. *J. Biophys. Biochem. Cytol.* **1958**, *4*, 433.
5. Pal, M. K.; Ghosh, B. K.; *Makromol. Chem.* **1980**, *181*, 1459.
6. Norden, B.; Kubista, M. In: *Polarized Spectroscopy of Ordered Systems* Samori, B., Thulstrup, E. W., Eds.; Kulwer Academic Publisher: Dordrecht, Holland, 1988; vol 242.
7. Sabate, R.; Esterlich, J. *J. Phys. Chem B* **2003**, *107*, 4137.
8. von Berlepsch, H.; Kirstein, S.; Bottcher, C. *Langmuir* **2002**, *18*, 769.
9. Sabate, R.; Gallardo, M.; De la Maza, A.; Esterlich, J. *Langmuir* **2001**, *17*, 6433.
10. Berret, J. F.; Cristobal, G.; Hervel, P.; Oberdisse, J.; Grillo, I. *Eur. Phys. J.* **2002**, *E9*, 301.
11. Meszaos, R.; Varga, I.; Gilanyi, T. *J. Phys. Chem. B* **2005**, *109*, 13538.
12. Monteux, C.; Williams, C. E.; Meunier, J.; Anthony, O.; Bergeron, V. *Langmuir* **2004**, *20*, 57.
13. Honda, C.; Kamizono, H.; Matsumoto, K.; Endo, K. *J. Colloid Interface Sci.* **2004**, *278*, 310.
14. Moulik, S. P.; Gupta, S.; Das, A. R. *Makromol. Chem.* **1980**, *181*, 1459.
15. Lee, J.; Moroi, Y. *J. Colloid Interface Sci.* **2004**, *273*, 645.
16. Mesa, C. L. *J. Colloid Interface Sci.* **2005**, *286*, 148.
17. Konradi, R.; Ruhe, J. *Macromolecules* **2005**, *38*, 6140.
18. Villeti, M.; Borsali, A.; Crespo, R.; Soldi, V.; Fukada, K. *Macromol. Chem. Phys.* **2004**, *205*, 907.
19. Fundin, J.; Hansson, P.; Brown, W.; Lidegran, I. *Macromolecules* **1997**, *30*, 1118.
20. Mitra, A.; Nath, A. R.; Chakraborty, A. K. *Colloid Polym. Sci.* **1993**, *271*, 1042.
21. Panda, A. K.; Chakraborty, A. K. *J. Colloid Interface Sci.* **1998**, *203*, 260.
22. Chakraborty, A. K.; Nath, R. K. *Spectrochim. Acta* **1989**, *45A*, 981.
23. Jain, N.; Trabelsi, S.; Guillot, S.; Meloughlin, D.; Langevin, D.; Leteiller, P.; Turmine, M. *Langmuir* **2004**, *20*, 8496.
24. Bakshi, M. S.; Varga, I.; Gilanyi, T. *J. Phys. Chem. B* **2005**, *109*, 13538.
25. Balomenou, L.; Bokias, G. *Langmuir* **2005**, *21*, 9038.
26. Bakshi, M. S.; Sachar, S. *Colloid Polym. Sci.* **2005**, *283*, 671.
27. Romani, A. P.; Gehlen, M. H.; Itri, R. *Langmuir* **2005**, *21*, 127.
28. Wang, C.; Tam, K. C. *Langmuir* **2002**, *18*, 6484.
29. Sjogren, H.; Ericsson, C. A.; Evenas, J.; Ulvenlund, S. *Biophys J.* **2005**, *89*, 4219.
30. Zhu, D. M.; Evans, R. K. *Langmuir* **2006**, *22*, 3735.
31. Chatterjee, A.; Moulik, S. P.; Majhi, P. R.; Sanyal, S. K. *Biophys Chem.* **2002**, *98*, 313.
32. Mata, J.; Patel, J.; Jain, N.; Ghosh, G.; Bahadu, P. *J. Colloid Interface Sci.* **2006**, *297*, 797.
33. Jiao, Q. C.; Liu, Q.; Sun, C.; He, H. *Talanta* **1999**, *1095*, 48.
34. Liu, Q.; Jiao, Q. C. *Spectrosc. Lett.* **1998**, *31*, 913.
35. Jiao, Q. C.; Liu, Q. *Spectrochim. Acta* **1999**, *55A*, 1667.
36. Jiao, Q. C.; Liu, Q. *Anal. Lett.* **1998**, *31*, 1311.
37. Jiao, Q. C.; Liu, Q. *Spectrosc. Lett.* **1998**, *31*, 1353.
38. Hugglin, D.; Seiffert, A.; Zimmerman, W. *Histochemistry* **1986**, *86*, 71.
39. Mitra, A.; Nath, R. K.; Chakraborty, A. K. *Colloid Polym. Sci.* **1993**, *271*, 1042.

40. Basu, S.; Gupta, A. K.; Rohatgi-Mukherjee, K. K. *J. Indian Chem. Soc.* **1982,** *59,* 578.
41. Pal, M. K.; Schubert, M. *J. Histochem. Cytochem.* **1961,** *9,* 673.
42. Pal, M. K.; Ghosh, B. K. *Macromol. Chem.* **1980,** *181,* 1459.
43. Pal, M. K.; Ghosh, B. K. *Macromol. Chem.* **1979,** *180,* 959.
44. Pal, M. K.; Schubert, M. *J. Phys. Chem.* **1963,** *67,* 182.
45. Frank, H. S.; Evans, M. W. *J. Chem. Phys.* **1945,** *13,* 507.
46. Kauzmann, W. *Adv. Protein. Chem.* **1959,** *14,* 1.
47. Bruning, W.; Holtzer, A. *J. Am. Chem Soc.* **1961,** *83,* 4865.
48. Whitney, P. L.; Tanford, C. *J. Biol. Chem* **1962,** *237,* 1735.
49. Mukerjee, P.; Ray, A. *J. Phys. Chem.* **1963,** *67,* 190.
50. Frank, H. S.; Quist, A. S. *J. Chem. Phys.* **1961,** *34,* 604.
51. Rabinowitch, E.; Epstein, L. F. *J. Am. Chem. Soc.* **1941,** *63,* 69.
52. Romani, A. P.; Gehlen, M. H.; Itri, R. *Langmuir* **2005,** *21,* 127.
53. Levine, A.; Schubert, M. *J. Am. Chem. Soc.* **1958,** *74,* 5702.
54. Nandini, R.; Vishalakshi, B. E-J. Chem. **2012,** *9* (1), 1–14; Rose, N. J.; Drago, R. S. *J. Am. Chem. Soc.* **1959,** *81,* 6138.
55. Nandini, R.; Vishalakshi, B. *e-J. Chem.* **2011,** *8,* S253.

CHAPTER 7

NANOSTRUCTURED EPOXY/ BLOCK COPOLYMER BLENDS: CHARACTERIZATION OF MICRO- AND NANOSTRUCTURE BY ATOMIC FORCE MICROSCOPY, SCANNING ELECTRON MICROSCOPY, AND TRANSMISSION ELECTRON MICROSCOPY

RAGHVENDRAKUMAR MISHRA[1*], REMYA V. R.[1], JAYESH CHERUSSERI[3], NANDAKUMAR KALARIKKAL[1], and SABU THOMAS[1,2]

[1]International and Inter University Centre for Nanoscience and Nanotechnology, Mahatma Gandhi University, Kottayam 686560, Kerala, India

[2]School of Chemical Science, Mahatma Gandhi University, Kottayam 686560, Kerala, India

[3]Materials Science Programme, Indian Institute of Technology Kanpur, Kanpur 208016, Uttar Pradesh, India

*Corresponding author. E-mail: raghvendramishra4489@gmail.com

ABSTRACT

This chapter mainly focuses on the toughening of epoxy. These areas are typically concentrated by the incorporation of different block copolymers as toughening agents employed in epoxy. The general objective of this chapter is to investigate and identify the toughening mechanisms of the modified epoxies and also to understand the relationship between the micro- and nanostructures and the resulting properties of the modified epoxies. The specific problems and objectives involved in the toughening of epoxies by various toughening agents are discussed briefly.

7.1 INTRODUCTION

The term "epoxy resin" is used to describe a class of thermosetting materials used in industries because of their high modulus, low creep, high strength, and good thermal and dimensional stabilities.[1] The epoxy is found useful for adhesive bonding, coating, electrical encapsulation (dielectric properties), and composite matrix. The desired properties of epoxy-based components are achieved by proper selection of the epoxy resin and hardener as well as the optimized curing procedures.[2–5] As shown in Figure 7.1 and Table 7.1, the diglycidyl ether of bisphenol S (DGEBS)/tetrahydrophthalic anhydride (THPA)/*N,N*-dimethylbenzylamine (DMBA) system offers a relatively higher glass transition temperature and storage modulus than those of the DGEBS/phthalic anhydride (PA)/DMBA system. In this study, the authors used PA and THPA as a curing agent for DGEBS epoxy resin; DMBA was used as an accelerator. The network construction of the DGEBS/THPA/DMBA system was established at comparatively higher temperature than that of the DGEBS/PA/DMBA system and the DGEBS/THPA/DMBA system offered better thermal properties, including glass transition temperature, initial decomposition temperature, and activation energy for decomposition than those of the DGEBS/PA/DMBA system, which could be referred to the higher cross-linking density of the DGEBS/THPA/DMBA system and separate activities of the anhydride (Fig. 7.2, Table 7.2).

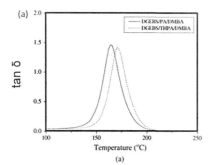
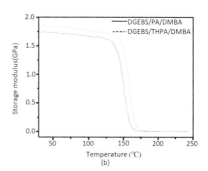

FIGURE 7.1 (a) The tan δ curve and (b) storage modulus of the cured diglycidyl ether of bisphenol S (DGEBS)/anhydride/dimethylbenzylamine (DMBA) systems as a function of temperature (adapted from Ref. [5] with permission).

TABLE 7.1 Dynamic Mechanical Analysis of the Cured DGEBS/Anhydride/DMBA Systems (adapted from Ref. [5] with Permission).

System	T_g (°C)	Storage modulus (GPa) Glassy region (storage modulus at 30°C)	Storage modulus (GPa) Rubbery region (storage modulus at T_g + 30°C)	Cross-linking (ρ) (10^{-3} mol/cm³)
DGEBS/PA/DMBA	164	1.74	0.006	0.32
DGEBS/THPA/DMBA	171	1.85	0.008	0.42

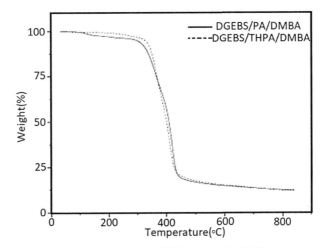

FIGURE 7.2 TGA thermograms of the DGEBS/anhydride/DMBA systems (adapted from Ref. [5] with permission).

TABLE 7.2 Thermal Properties of the Cured DGEBS/Anhydride/DMBA Systems (adapted from Ref. [5] with Permission).

System	IDT (°C)	T_{max} (°C)	E_t (kJ/mol)	Char (%) at 700°C
DGEBS/PA/DMBA	294	415	96	13.4
DGEBS/THPA/DMBA	324	413	130	13.5

The highly cross-linked microstructure makes the epoxy very brittle and hence it suffers from poor resistance to crack initiation and propagation. To improve the performance and feasibility of pristine epoxies, they are modified by using toughening agents. The addition of toughening agents may lead to improve the performance and become suitable for a variety of applications. Generally, materials such as thermoplastics, microphase of a dispersed rubber, block copolymer, nanoparticles, and rubber core–shell nanoparticles are used for the toughening of epoxy resin.[6–11] In one of the studies by the authors, an epoxy resin, cured using an anhydride hardener, has been adapted by the inclusion of preformed core–shell rubber (CSR) particles which were comparatively 100 or 300 nm in diameter.[6] The glass transition temperature, T_g, of the cured epoxy polymer was 145°C. Microscopy indicated that the CSR particles were well spread through the epoxy matrix. Young's modulus and tensile strength were decreased, and the glass transition temperature of the epoxy was unaltered by the addition of the CSR particles. The fracture energy improved from 77 J/m² for the unmodified epoxy to 840 J/m² for the epoxy with 15 wt% of 100-nm diameter CSR particles. A review of these authors' studies is shown in below.[6] Authors mentioned that the atomic force microscopy recognized the soft cores, which are detectable as the dark, circular areas. The mean diameter of the cores was determined to be 58 ± 19 nm for the MX 125 and 58 ± 13 nm for the MX 156 particles as shown in Figure 7.3.

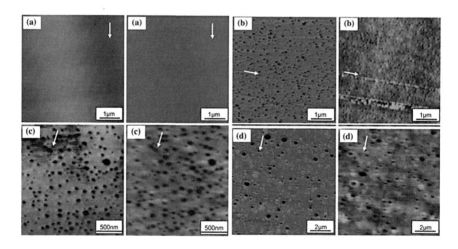

FIGURE 7.3 (See color insert.) AFM phase (left) and height (right) images of the (a) unmodified epoxy and epoxy modified with (b) 9 wt% MX 125 [100 nm styrene–butadiene core and poly(methyl methacrylate) (PMMA) shell 85–115 nm], (c) 9 wt% MX 156 (100 nm polybutadiene core and PMMA shell 85–115 nm), and (d) 9 wt% MX 960 (300 nm siloxane and PMMA shell 250–350 nm) (the arrows indicate the cutting direction) (adapted from Ref. [6] with permission).

Authors used differential scanning calorimetry (DSC) to find out the glass transition temperature, T_g, of the cured epoxy polymers, as reviewed in Table 7.3.

Nanostructured Epoxy/Block Copolymer Blends

TABLE 7.3 Glass Transition Temperature, T_g, Young's Modulus, E, and Fracture Energy, GC, for the CSR- and CTBN-Modified Formulations (adapted from Ref. [6] with Permission).

	Wt% CSR	Wt% CTBN	T_g (°C)	E (GPa) Mean	SD	GC (J/m²) Mean	SD	KC (MPa m^{1/2}) Mean	SD
Unmodified	0	0	145	2.76	0.08	77	15	0.51	0.02
MX 125 (100 nm styrene–butadiene core and PMMA shell 85–115 nm)	9	0	146	2.32	0.00	596	38	1.16	0.15
MX 156 (100 nm polybutadiene core and PMMA shell 85–115 nm)	9	0	143	2.33	0.04	485	41	1.31	0.19
MX 960 (300 nm siloxane and PMMA shell 250–350 nm)	9	0	146	2.21	0.02	500	33	1.25	0.16

Authors indicated that Young's modulus decreased linearly with including CSR particles, as reviewed Figure 7.4, and the fracture energy was obtained to increase steadily with the CSR concentration for the three types of core–shell, as mentioned in Figure 7.5.

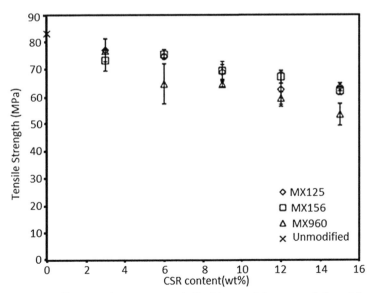

FIGURE 7.4 Tensile strength versus core–shell rubber particle content (adapted from Ref. [6] with permission).

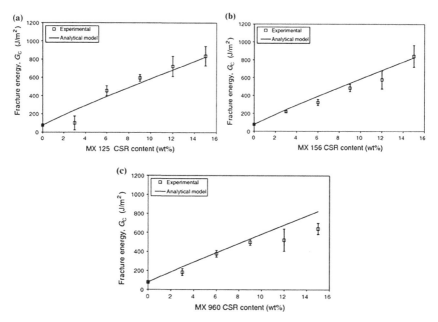

FIGURE 7.5 Fracture energy versus CSR content showing experimental data (points) compared to the predictive model (solid line) for (a) the MX 125, (b) MX 156, and (c) MX 960 CSR-modified epoxies (adapted from Ref. [6] with permission).

Nowadays, CSR particles, PMMA-b-poly(butyl acrylate)-b-PMMA (MAM) tri-block copolymers, amphiphilic block copolymers, glass particles, glass microparticles, hybrid materials, fiber (it can restrict crack propagation), silica nanoparticles, homopolymer-grafted hairy NPs and various type of hierarchical self-organized structure are also employed in the toughening of epoxy resins, some case, these materials are very favorable to improve the cyclic-fatigue of epoxies. The toughening phenomenon can increase the fracture energy and toughness of the epoxy system over a wide range of temperatures and is found to depend on the cross-linking density as well as the morphologies of the modified epoxies, however, some of the toughening agents have the ability to alter the viscosity of epoxies, that is the main drawback with toughening agent at higher concentration (Fig. 7.6).[12–14]

The way by which toughening (energy absorption or improvement in fracture energy) the epoxy resin (single-phase) is to incorporate a second tough/soft phase in the epoxy matrix to synthesize a multi-phase morphology system (by induced phase separation). In convention way, the reactive liquid rubbers (contain –OH, –OOH, and –C=O group in the isoprene chain) or thermoplastics and various types above-mentioned structured materials have been used as the second phase to toughen the epoxy resins. These toughening agents are incorporated in the epoxy resin before its curing and the resin is toughened after the dispersed particles form a separate phase during the curing process. Mostly, three main mechanisms take part in epoxies toughening including particle debonding, plastic yielding of particles, and localized shear banding, because these mechanism are responsible for energy dissipation and improvement in fracture toughness.[15–17]

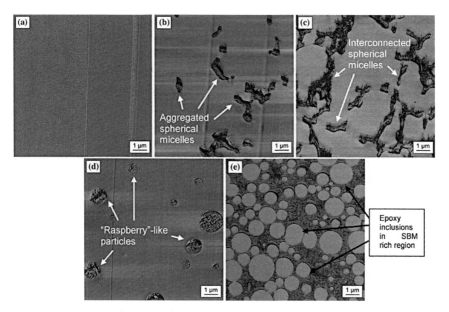

FIGURE 7.6 (See color insert.) AFM phase images of (a) unmodified epoxy (appears flat and homogeneous thermoset), (b) 5 wt% E21 grade (SBM) modified epoxies [E21 SBM phase separated as a network of aggregated spherical micelles], (c) 10 wt% E21 grade (SBM)-modified epoxies (spherical micelles become increasingly interconnected at higher E21 content due to reduction of secondary phase separating into individual particles, which is result of increase in viscosity), (d) 5 wt% E41 grade (SBM)-modified epoxies (SBM particles phase separated with a "raspberry"-like microstructure "sphere-on-sphere"), and (e) 10 wt% E41 grade (SBM)-modified epoxies (particle diameter increased and partially phase-inverted microstructure) (adapted from Ref. [12] with permission). [SBM—(tri-block copolymer of poly(styrene)-*b*-1,4-poly(butadiene)-*b*-poly(methyl methacrylate modifier, E21 SBM has low polarity higher butadiene content and molecular weight than E41).]

Carboxyl-terminated butadiene-acrylonitrile (CTBN)[4–9] and amine-terminated butadiene-acrylonitrile[18–20] are commonly used as reactive liquid rubber toughening agents, considering PMMA,[21,22] poly(ether sulfone),[23,24] and poly(ether imides),[25–28] etc. More than 50 thermoplastic toughening agents are employed in the toughening of the epoxy resin, but the main problem with thermoplastic spheres is much higher modulus. High modulus does not allow stress concentration cavitations; therefore, crack pinning or crack deflection is a less effective mechanism in the case of thermoplastics toughening. Alternatively, preformed elastomeric particles,[6,28–34] block copolymers,[35–43] nanoparticles (like silica),[44–47] homopolymer-grafted hairy NPs, and several other nanoparticles have also been employed as toughening agents.[44,48–53]

Preformed elastomeric particles are normally core–shell particles, which consist of a soft rubbery core and a thin glassy shell and are found to improve the fracture toughness of epoxies. These preformed particles are produced by emulsion polymerization and are dispersed in the epoxy resin. Polysiloxane,[6] polybutadiene,[34] and poly(butyl acrylate)[53] are used to make soft core and PMMA and polyvinyl alcohol are commonly employed to prepare the shell of the core–shell particles. Dissolution of highly cross-linked rubber cores in epoxy resins is the main problem.

The shell has to react with the epoxy matrix to improve its compatibility. The CSR particles are capable to increase the toughness of both bulk thermosetting polymers and composites significantly along with a slight compromise in the stiffness.[29,31,32,53–55] Degradation in the thermomechanical properties after curing is a major drawback of the rubber toughened epoxies. To eliminate this drawback, different types of block copolymers have been used as toughening agents.

Recently, hybrid toughening strategies such as a combination of nanoparticles and block copolymers are found to enhance the toughness of epoxies. The block copolymers are capable to generate distinct nanostructure phases in a moderate-to-high concentration of the epoxy precursor and are potential candidates to produce the product with novel and improved properties.

7.1.1 BLOCK COPOLYMER

Block copolymers are composed of long two or more chemically distinct polymer chains (blocks) and these blocks are connected by covalent bonds to form a single macromolecule. The block copolymers can be synthesized by methods, such as anionic polymerization, controlled radical polymerization, etc. The blocks can be arranged in various sequences, such as di-block, ABC tri-blocks, ABA tri-block, $(AB)_n$ multiblocks, $(AB)_n$ star blocks, $(ABC)_n$ star blocks. Amphiphilic block copolymers consisting epoxy-philic and epoxy-phobic blocks have been used as modifiers for toughening the epoxies, it

Nanostructured Epoxy/Block Copolymer Blends

participates in generating self-assembled nanostructures in an undiluted melt or in solution. Figure 7.7 depicts the monolayers, cylindrical, and spherical morphologies of the block copolymers.

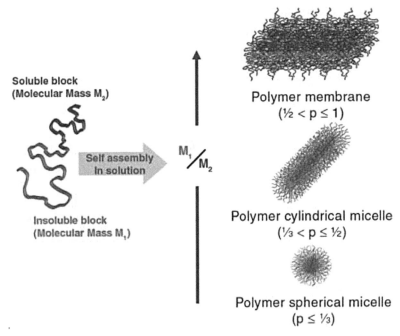

FIGURE 7.7 (See color insert.) Various geometries produced by block copolymers in a particular solvent (membranes include two monolayers of soluble block–insoluble block–soluble block copolymers aligned to produce a sandwich-like membrane, and both cylindrical and spherical micelles consist of a nonsoluble core surrounded by a soluble corona) (adapted from Ref. [59] with permission).

A block copolymer (consist two or more distinct polymer chains connected together) having at least one segment that is miscible in the epoxy resin and doesn't undergo phase separation and another one is immiscible in the epoxy resin during the network formation is reported recently. At a certain concentration, block copolymers start to self-assemble so as to separate the insoluble blocks from the epoxy.[58,59] As a result, the self-assembled constituent blocks can be attained as an identical set of ordered nanostructured morphologies (i.e., spheres, cylinders,[60–62] vesicles,[63–64] etc.), which are observed in homologous tri-block (ABA) and di-block (AB).[60–66] The geometry and degree of order of these morphologies depend on various factors, such as concentration, time, the volume ratio between immiscible and miscible blocks, and short-range repulsion between adjacent same natures of blocks. A critical aggregation concentration is defined as block copolymers initiate to self-assemble so as to separate the insoluble blocks from the host matrix or solvent. The formations of nanostructures in block copolymer toughened epoxies have shown a significant increase in the fracture toughness of materials.[67–70] A change in the block copolymer micelle morphology is found to be triggered by a physical stimulus such as heating,[71,72] extrusion,[73] shearing,[74–77] and photo-irradiation.[78,92] Figures 7.8–7.10 illustrate the morphologies of the block copolymer and self-assembled nanostructure.

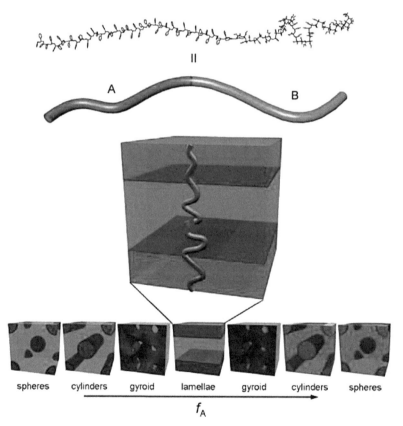

FIGURE 7.8 (See color insert.) Schematics representation of thermodynamically stable di-block copolymer phases. The A–B di-block copolymer, such as the PS-*b*-PMMA molecule indicated at the top, is represented as a simple two-color chain. The chains self-organize such that contact between the immiscible blocks is minimized, with the structure determined primarily by the relative lengths of the two polymer blocks (f_A) (adapted from Ref. [91] with permission).

Nanostructured Epoxy/Block Copolymer Blends

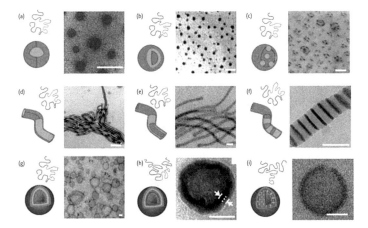

FIGURE 7.9 **(See color insert.)** Assemblies formed in a particular solvent conditions by multiblock copolymers: (a) Janus sphere,[79] (b) core–shell spheres,[80] (c) raspberry-like spheres,[69] (d) Janus cylinders,[81] (e) core–shell cylinder,[82] (f) segmented cylinders,[83] (g) asymmetric (Janus) membrane vesicles,[84] (h) double-layer membrane vesicles, and (i) vesicles with hexagonally packed cylinder.[85] Scale bar 50 nm (adapted from Ref. [66] with permission).

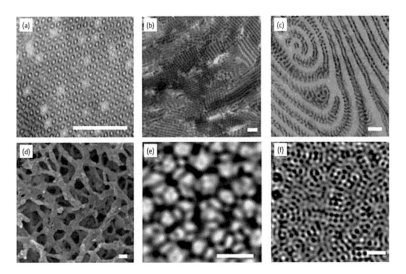

FIGURE 7.10 (a) Cubic micellar phase formed by poly(ethylene oxide)-poly(ethyl ethylene) in an epoxy network.[86] (b) Hexagonally packed cylinders formed by poly(ethylene oxide)-poly(butadiene) in water.[87] (c) Disordered lamellar phase formed by poly(styrene)-block-poly(butadiene)-block-poly(methyl methacrylate) in an epoxy network.[69] (d) Disordered network formed by poly(ethylene oxide)-poly(butadiene) in water.[88] (e) Hexagonally packed vesicles formed by poly(ethylene oxide)-poly(butylene oxide) in water.[89] (f) Im3m bi-continuous phase formed by poly(ethylene oxide)-poly(butylene oxide) in water[90] (adapted from Ref. [59] with permission).

7.1.2 POLYMER BLENDS

In the past decade, polymer blend technology has been considered as a major topic of polymer research and development. The main motives of polymer blending are cost reduction and achievement of properties more than the constituent polymers. The miscibility of blend's polymer partners allowed to useful properties over the composition range. In the early 1940s, it was studied that miscible blends of different polymers help to improve the properties when compared to that of their unblended counterpart. It is important to understand the nature of polymer blends to obtain the required properties. It is studied that the properties of final blends depend on the miscibility and compatibility of constituent polymer.[93–97] Accordingly, the blends are classified into three major categories:

- miscible,
- partially miscible, and
- immiscible blends.

Generally, immiscible blends are said to be technologically incompatible and phase separated, and phase separation morphology can be controlled by using reactions, reaction-induced phase separation, and reactive blending which are explained (Fig. 7.11).[98,99]

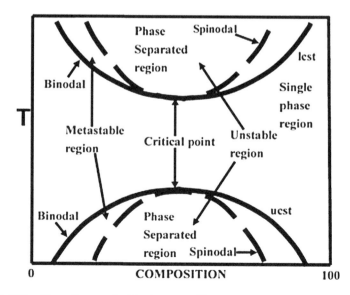

FIGURE 7.11 Phase diagram for blends system.[128]

The basic thermodynamic free energy, enthalpy, and entropy relationship for the mixture is given by the equation,

$$\Delta G_m = \Delta H_m - T\Delta S_m \qquad (7.1)$$

where ΔG_m is the free energy of mixing, ΔH_m is the enthalpy of mixing, ΔS_m is the entropy of mixing, and T is the temperature.

For miscibility of two phases, $G_m < 0$. This is a necessary criterion, while the following term must also be acquired:

$$\left(\frac{\partial^2 \Delta G_m}{\partial \varnothing_i^2}\right)_{T,P} > 0$$

From phase diagram, as well $G_m < 0$, but negative values of $\left(\partial^2 \Delta G_m / \partial \varnothing_i^2\right)_{T,P}$ can give an area of the phase diagram, where the blend will separate into a phase rich in component 1 and a phase rich in component 2. For low molecular weight materials, rising temperature commonly serves to increase in miscibility as the TS_m expression rises; thus, free energy G_m indicates further negative values. For higher molecular weight constituents, the TS_m expression is negligible and additional considerations (such as noncombinatorial entropy contributions and temperature dependent H_m conditions) can influence and direct to the reverse behavior, especially losing miscibility with rising temperature. Consequently, liquid–liquid and polymer–solvent mixtures, which are borderline in miscibility and consistently reveal upper critical solution temperatures, and polymer–polymer mixtures generally show lower critical solution temperatures.

The spinodal curve is described to the point where

$$\left(\frac{\partial^2 \Delta G_m}{\partial \varnothing_i^2}\right)_{T,P} = 0$$

The binodal curve is displayed to the equilibrium phase boundary between the single phase and the phase-separated region. The values of the binodal curve can be measured from the double tangent to the $\Delta G_m/RT$ versus concentration ($\varnothing$) curve. The point, where the binodal and spinodal intersect is called critical, measured from the following equation:

$$\left(\frac{\partial^3 \Delta G_m}{\partial \varnothing_i^3}\right)_{T,P} = 0$$

Tie-line is used to calculate phase compositions; therefore, we can calculate the volume fraction of component 1 rich phase and component 2 rich phase. Highly miscible polymers show single-phase nature at entire temperature–volume fraction; for highly immiscible polymer blends, the phase curve is basically all in the two-phase zone with the binodal curves.

The basic theory of polymer blends is proposed by Flory and Huggins and is known as Flory–Huggins theory.[129–132]

$$\Delta G_m = kTV\left[\frac{\varnothing_1}{V_1}\ln\varnothing_1 + \frac{\varnothing_2}{V_2}\right] + \varnothing_1\varnothing_2\frac{\chi_1\chi_2 kTV}{v_r} \quad (7.2)$$

where V is the total volume, V_1 is the molecular volume of component 1, V_2 is the molecular volume of component 2, φ_i is the volume fraction of component i, k is the Boltzmann's constant, χ_1 and χ_2 are the Flory–Huggins interaction parameter, and v_r is the interacting segmental volume. Initially, the equation was applied for solvent–polymer mixtures. But it is also valid for higher molecular weight polymer blends. According to this equation, a decrease in miscibility of solvent–polymer mixtures when compared with solvent–solvent mixtures and a decrease in the combinatorial entropy of mixing is obtained. With polymer-polymer mixtures, the combinatorial entropy of mixing approaches insignificant values in the limit of high molecular weight polymer mixtures. To achieve miscibility, a negative heat of mixing is required (i.e., $\chi_{12} < 0$). It is important to match the solubility parameter to achieve miscibility. With polymer–polymer mixtures, the solubility parameters need to be virtually identical to achieve miscibility in the absence of strong polar or hydrogen-bonding interactions. Hydrogen bonding is an important specific interaction often noted in the miscible polymer blends. There are many methods are available to make the system compatible. Many of these techniques involve modification of the interface between the immiscible and incompatible components. If the interfacial energy is reduced, the ability to transfer stress across the interface can be dramatically improved. This can be achieved by adding a ternary polymer to the blend offering good adhesion to both phases and the ability to concentrate at the interface. In the case of a copolymer composed of monomers where the homopolymers are highly immiscible, there exists an intramolecular repulsion. An example involves the styrene-acrylonitrile copolymer where the homopolymers (polystyrene and polyacrylonitrile) are highly immiscible. In blends with a homopolymer (like PMMA) where the MMA monomer unit has more favorable interactions with styrene and acrylonitrile, a window of miscibility can be achieved.

7.1.3 MICRO- AND NANOSTRUCTURES OF EPOXY AND BLOCK COPOLYMER BLENDS

The ability of block copolymers to generate nanostructures in epoxy thermosetting systems is helpful to obtain new functional properties of block copolymers/epoxy blends. Both type, amphiphilic and chemically modified di- or tri-block copolymers, have been employed for generating nanostructure in epoxy. In epoxy/block copolymer blends, one of the blocks is miscible or can react with the epoxy system prior and after curing, if both blocks are being an initially miscible help to develop self-assembled morphology and the other one is immiscible in epoxy prior and after curing, to develop reaction-induced phase separation morphology, which generates ordered and/or disordered nanostructures in thermosets. In some cases, development of nanostructures with different morphologies may occur by a combination of both mechanisms; also the improper mixing of the miscible block also may produce various morphologies. Other parameters such as the type of hardener, cure cycle, and the ratio of copolymer blocks are also found to influence on the morphology (such as spherical or worm-like micelles, hexagonally packed cylinders, bilayer micelles, or mixtures) and properties of modified epoxies. Last few decades, various block copolymers have been demonstrated to generate stable nanostructures in epoxy matrix.[88–98]

The selection of the miscible block and immiscible blocks (to produced generate stable nanostructures) is strongly dependent on the epoxies and hardener selected to perform the curing process,[116,117] in the case of immiscible, poly(ethyl ethylene), poly(ethylene-altpropylene), polyisoprene, poly(styrene-*b*-butadiene), poly(propylene oxide), poly(2-ethylhexylmethacrylate), and polyethylene are used to cure and produce the stable nanostructures in diglycidyl ether of bisphenol A (DGEBA)-based epoxies system.[69,80,89,118,119] Both inert and reactive miscible blocks have been used for specific interactions in bisphenol A based on epoxies. The search of a miscible block for a specific DGEBA-hardener combination is not a trivial task due to the variety of mechanisms of network formation involving different types of hardeners and the fact that commercial formulations frequently contain other epoxy monomers apart from DGEBA (e.g., brominated DGEBA for flame retardation). PMMA may be a suitable choice as a miscible block since it is soluble with DGEBA in all proportions.[21,120,121]

However, for most hardeners, it becomes phase separated during the polymerization well before gelation.[22,122–125] On the other hand, poly(*N,N*-dimethyl acrylamide) (PDMA) is miscible both in nonpolar solvents (such as in cyclohexane) and in highly polar solvents (such as in water, methanol, etc.).[126] Therefore, the family of random copolymers poly(MMA-co-DMA) with different proportions of both monomers must be a useful choice as universal miscible block for the synthesis of nanostructured epoxy. The polymerization induced phase separation of blends of the random copolymer in a reactive solvent based on DGEBA and 4,4'-diaminodiphenylsulfone as hardener are mentioned.[127]

Self-organization of amphiphilic block copolymers in the precursors of thermosets won't happen in every case. In many cases, all the subchains of block copolymers are actually miscible with the precursors of thermosets. The miscibility is ascribed to the non-negligible entropic contribution (ΔS_m) to the free energy of mixing (ΔG_m) in the mixtures of block copolymers, if the precursors (or monomers) of thermosets are generally compounds of low-molecular weights. In addition, the presence of self-organized microphases formed at lower temperatures doesn't assure the survival of these structures at elevated temperatures and elevated temperatures are required for curing of some thermosets; intermolecular hydrogen-bonding interactions are favorable for the miscible and are reduced for some thermoses at elevated temperature, which can reduce the miscibility. Increasing molecular weight at curing temperature, the phase separation is explained by the decrease of the entropic contribution to free energy of mixing (ΔG_m).[133–134]

Furthermore, a new concept of block copolymers has been established dealing with an idea of chemical compatibilization. This concept includes active groups into one block to support covalent bonding with the growing epoxy network without loss of ordering in the proceeding blends. This concept tends to promote the mechanical properties and durability of nanostructured materials. Ionic and established free radical approaches can presently be applied to produce block copolymers with appropriately specified proportions, molecular weights, and arrangements of very extensive structures. However, micellization of block copolymers in a particular solvent of one of the blocks is a typical feature of their colloidal characteristics.[135,136]

In state of things, a block copolymer is suspended in a liquid, which is a thermodynamical suitable solvent for one block and a precipitant for the other block, the copolymer chains would combine reversibly to produce micellar structure which are similar to most of their characters to those concerned with traditional low molecular weight surfactants. The micelles exhibit a richer or lesser swollen core of the insoluble blocks closed in by a flexible tape of soluble blocks. These micelles are usually spherical with definite size distribution but may be converted in various shape and size under specific circumstances (Figs. 7.12–7.14; Table 7.4).

Nanostructured Epoxy/Block Copolymer Blends 87

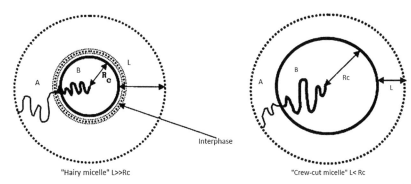

FIGURE 7.12 Schematic description of AB di-block copolymer micelles in a particular solvent of the A block. Rc: core radius; L: shell (corona) thickness (adapted from Ref. [136] with permission).

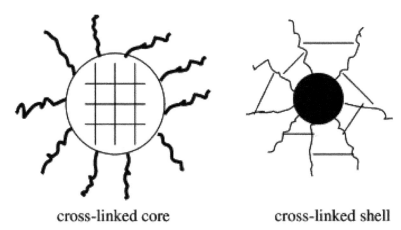

FIGURE 7.13 Schematic representation of photo cross-linking nanostructured blocks copolymer micelles (in the case of hydrophobic–hydrophilic or hydrophilic–hydrophilic copolymers) (adapted from Ref. [136] with permission).

Different nanostructures can be formed on the basis of the restriction of topological structures of block copolymers and subchain sequence during the formation of nanophases. The thermosets show a strong relationship between morphology and properties.[158] It has been observed that the establishment of the nanostructures in the multicomponent thermosets can greatly enhance the interactions between thermosetting matrix and modifiers and therefore the mechanical properties of materials were further increased, which has been described as a term nanostructured toughening of the thermosets (Figs. 7.14–7.18).[159]

TABLE 7.4 Various Complex Structure of Poly A/Poly B Blocks Copolymers (adapted from Ref. [136] with permission).

Structure	Type	A	B	Polymerization technique	References
	Tapered, overlap	PS	PB	Anionic	[137]
	Ring di-block	PEO	PBO	Anionic	[138]
	Coil-cycle-coil	Phenyl-ethynyl	PS	Coupling	[139]
	Catenated di-block	PS	P2VP	Anionic	[140]
	AB_2 star (miktoarm star)	$PS\,(PS)_2$	PI_2 PMMA, POE, PCL	Anionic Anionic	[141] [142]
	Heteroarm Star $AnBn$	$PIB\,(A_2)$ $PS\,(A_2)$	$PMeVE\,(B_2)$ $PB, PI\,(B_2)$	Cationic Anionic	[143] [144]
	H-shaped B_2AB_2	PS	PI	Anionic	[145]

TABLE 7.4 (Continued)

Structure	Type	A	B	Polymerization technique	References
	"Palm-tree" ABn	PB	PEO	Anionic	[146]
	Dumb-bell "pom-pom"	PDMS	PS	Anionic	[147]
	Star-block (A–B)$_n$	PnBMA PMMA	PDMAEMA PAA	ATRP ($n = 3$) ATRP ($n = 6$)	[153] [148]
	Star-block A$_2$(BA)$_2$	PS	PB	Anionic	[149]
	Dendrimer-linear	Styrene and (meth) acrylic polyamido-amine dendrimer	styrene and (meth) acrylic poly(2-methyl-2-oxazoline)	Nitroxide ATRP	[150] [151]

TABLE 7.4 (Continued)

Structure	Type	A	B	Polymerization technique	References
	Linear dendrimer	PEG	Poly(chloromethylstyrene)	ATRP	[152]
		PEO	Poly(benzyl ether)	Coupling	[153]
	Arborescent	PIB	PS	Cationic inimers, initiator-monomer	[154]
		PS	PLA methacrylic macromonomer	CRP (TEMPO)	[155]
		PMAA	POE methacrylic macromonomer	GTP	[156]
		PEO	PPO methacrylic macromonomer	ATRP	[157]

Nanostructured Epoxy/Block Copolymer Blends

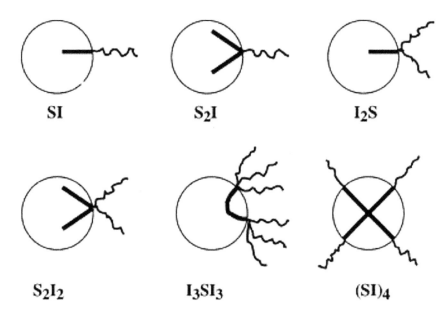

FIGURE 7.14 Schematic representation of star-block copolymers micellar structures for polystyrene (S)-polyisoprene (I) copolymers (in the case of AB2, AB3, and A2B2, as well water soluble as or nanosoluble copolymers, star architectures show higher CMC values and hydrodynamic radius R_h and the aggregation number Z increased in the $I_2S < S_2I < SI$ manner) (adapted from Ref. [136] with permission).

- Z (the aggregation or association number): It indicates the average number of polymer chains in a micelle, deduced from M_m and the molecular weight M_u of the unimer with $Z = M_m/M_u$.
- R_h: the total hydrodynamic radius of the micelle;
- R_g: the radius of gyration of the micelle; and
- R_g/R_h: it offers the information about shape.

In biological materials such as bones and shells, the properties of biological polymers are augmented through the inclusion of mineral components like calcium carbonate or silica.[162] These inorganic materials are integrated at the molecular level with proteins and peptides directing the assembly of nanometer-sized inorganic particles into complex, hierarchical structures.[162] As the resulting organic/inorganic composites have outstanding material properties,[161] they have considerable interest in mimicking the biological self-assembly processes. A significant step in this direction is taken by "Mobile Oil Corporation" where researchers used microphase separation in a surfactant solution to synthesize well-ordered mesoporous silicates.[162–164] Because the accessible pore sizes (2–10 nm) are much larger than the molecular scale (<1.3 nm) pores in zeolites, surfactant-templated silicates are found to use large pore molecular sieves.[165] In this pioneering work, the structure-directing properties of a range of surfactants, block copolymers, peptides, and other organic molecules have been extensively investigated. These self-assembled organic/inorganic materials are of utmost importance in various applications including chemical sensors, catalysis, low di-electric insulators, solid-state electrolytes, and optical materials.[166–168] Amphiphilic block copolymers can be thought of as giant surfactants (~50–1000 times the molecular volume of simple surfactants) and their use to structure silica-type materials permits access to larger (10–100 nm) mesoporous structures.[168–170] Nowadays, researchers are mainly focused to toughen the epoxy by incorporating amphiphilic block copolymers. Hence in this chapter, we are focusing on some microscopic techniques which are used to study the morphology of epoxy/block copolymer blends.

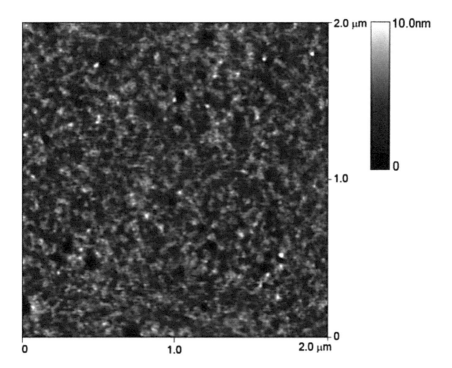

FIGURE 7.15 (See color insert.) AFM image of PS (polystyrene)-*alt*-PEO (poly(ethylene oxide)) alternating multiblock copolymer, the microphase separation at the nanometer scale took place and the co-continuous nanophases (dark and light regions) can be attributed to PS and PEO microdomains, separately (adapted from Ref. [159] with permission).

Nanostructured Epoxy/Block Copolymer Blends 93

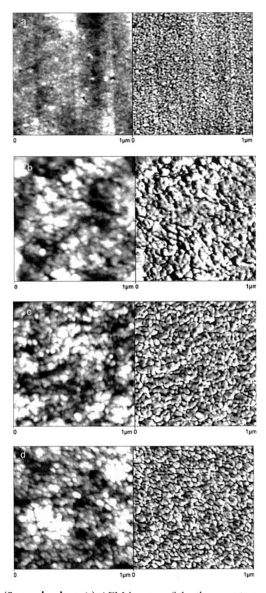

FIGURE 7.16 **(See color insert.)** AFM images of the thermosets containing (a) 10 wt% (small and uniform dispersed PS nanodomains), at higher concentration of PS-*alt*-PEO alternating multiblock copolymer, coagulated and the size of the nanodomains is increased, as shown in (b) 20, (c) 30, and (d) 40 wt% of PS-*alt*-PEO alternating multiblock copolymer, PS-*alt*-PEO multiblock copolymer displayed disordered nanostructures and the formation of the nanostructures depends on the block topology. Left image: topography. Right image: phase contrast images (adapted from Ref. [159] with permission).

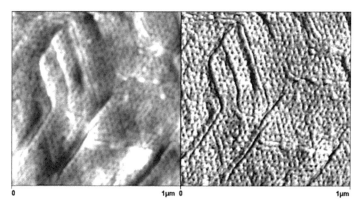

FIGURE 7.17 (See color insert.) AFM image of epoxy thermoset containing 40 wt% PS-*b*-PEO di-block copolymer, the formation of the long-ranged ordered nanostructures attributed to the specific topology of the block copolymer (i.e., di-block). The di-block topology offers an ordered arrangement of PS nanodomains in the epoxy matrix during the curing reaction (adapted from Ref. [159] with permission).

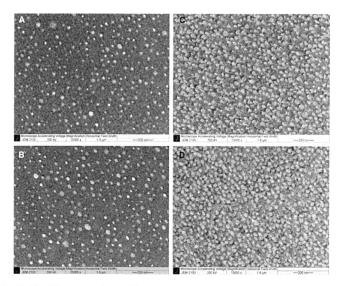

FIGURE 7.18 Thermosetting blends show microphase-separated structure (i.e., nanostructures). TEM micrograph of PBa thermoset containing (a) 10 wt% PVPy (poly(*N*-vinyl pyrrolidone))-*b*-PS (polystyrene) di-block copolymer (spherical PS nanophases with 20–30 nm in diameter), (b) 20 wt% PVPy-*b*-PS di-block copolymer, number of PS nanodomains increased (spherical PS nanophases with 20–30 nm in diameter), (c) 30 (PS nanodomains became increasingly interconnected and some cylindered nanophases are also examined exhibited), and (d) 40 wt% PVPy-*b*-PS di-block copolymer. (More PS nanodomains became interconnected and some cylindered nanophases are also examined.) The sections of samples were stained with RuO_4 (adapted from Ref. [160] with permission).

7.2 EPOXY/SBS BLOCK COPOLYMER SYSTEM

Block copolymers have been used as the chemical compatibilizers to epoxy resins.[65] Polystyrene-block-polybutadiene-block-polystyrene (SBS) tri-block copolymer is an amphiphilic block copolymer, which has great importance in the toughening of epoxy resin.[140] Miscibility is one of the major factors for obtaining the nanostructured morphology. To achieve this, researchers do some chemical modification as in the case of styrene–butadiene–styrene (SBS) tri-block copolymer. For this chemical modification, one of the convenient methods is the epoxidation of SBS block copolymer. The percentage of epoxidation degree could affect the nanostructured morphology and properties of epoxidized SBS (e.g., cohesive energy, solubility parameter, the glass transition temperature, and a molecular weight between cross-linking points) depends on the total oxygen weight content (Figs. 7.19–7.23; Tables 7.5 and 7.6).[141]

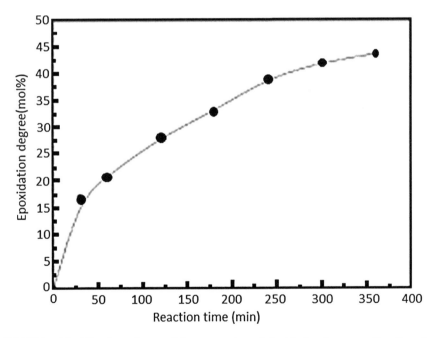

FIGURE 7.19 The dependence of the conversion of double bonds on reaction time for the epoxidation of C500 tri-block copolymer (conversion of double bonds increased with increasing reaction time, but the reaction rate decreased with increasing time. This can be attributed to the decrease in concentration of double bonds and H_2O_2), and it is used to produce copolymers with different degrees of functionality (adapted from Ref. [172] with permission).

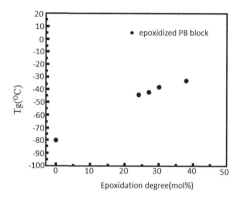

FIGURE 7.20 The glass transition temperature of the PB block (The glass transition temperature of the epoxidized PB increased with the degree of epoxidation for parent C500 due to an increase of the chain rigidity by the epoxies rings.) (adapted from Ref. [172] with\ permission).

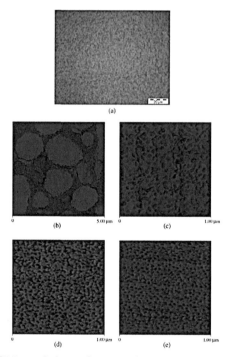

FIGURE 7.21 (a) OM morphology of epoxy mixture containing 30 wt% C500ep24; and TM-AFM phase images for epoxy mixtures containing 30 wt% of (b) C500ep24, (c) C500ep27, (d) C500ep30, and (e) C500ep38. (Increase of epoxidation produced a decrease in domains size and nanostructures from around 35 nm domain size for the mixture modified with C500ep27 tri-block copolymer to around 20 nm for the mixture modified with C500ep30 and C500ep38 tri-block copolymers, decrease in domains size due to a smaller amount of phase-separated nonepoxidized PB units around PS domains.) (adapted from Ref. [172] with permission).

Nanostructured Epoxy/Block Copolymer Blends

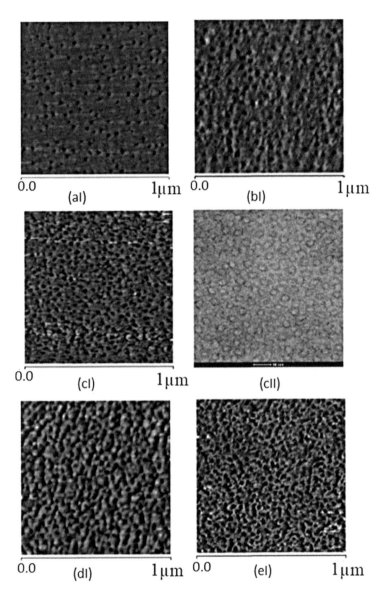

FIGURE 7.22 TM-AFM phase (I) and TEM (II) images of epoxy mixtures containing different contents of C500ep38: (a) 10 (micelles of PS in the epoxy matrix without long-range order), (b) 20 (micelles of PS in the epoxy matrix without long-range order), (c) 30 (micelles of PS in the epoxy matrix without long-range order), (d) 40 (micelles of PS in the epoxy matrix without long-range order) and (e) 50 wt%(worm-like morphology). Scale bar: 50 nm for TEM image (microphase-separated morphologies and nanodomains of PS with diameters less than 30 nm) (adapted from Ref. [172] with permission).

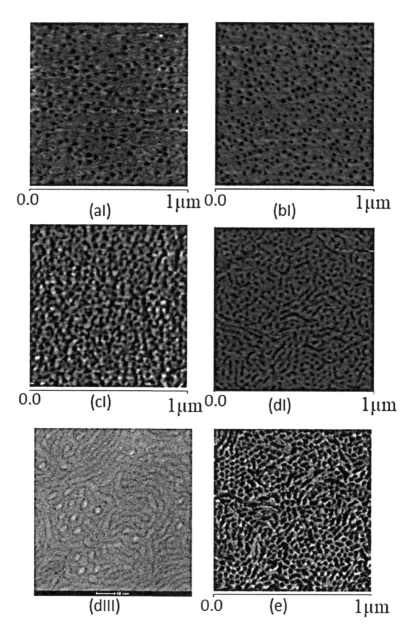

FIGURE 7.23 TM-AFM phase (I) and TEM (II) images for epoxy mixtures containing different contents of C540ep44: (a) 10 (micellar morphology), (b) 20 (micellar morphology), (c) 30 (micelles and some vesicles of PS dispersed), (d) 40 (worm-like morphology) and (e) 50 wt% (hexagonally packed cylinder morphology). Scale bar: 50 nm for TEM image (adapted from Ref. [172] with permission).

TABLE 7.5 Properties of the Two Polystyrene-Block-Polybutadiene-Block-Polystyrene Linear Tri-Block Copolymers, C500 and C540, with 70 and 60 wt% PB (adapted from Ref. [172] with Permission).

Block copolymer	[a]Epoxidation degree	[b]PS (wt%)	[b]PB (wt%)	[b]Epoxidized PB (wt%)	T_g [c]PB (°C)
C500	0	30	70	0	−80
C500ep24	24	30	50	20	−44
C500ep27	27	30	47	23	−42
C500ep30	30	30	45	25	−38
C500ep38	38	30	39	31	−33
C540	0	40	60	0	−80
C540ep44	44	40	30	30	−36

[a]Defined as mol% of epoxidized PB units, as determined by ^{1}H NMR analysis.
[b]As determined by ^{1}H NMR analysis.
[c]Glass transition temperatures for PB/epPB blocks, as determined by DMA measurements (heating rate of 5°C min^{-1}). That corresponding to PS block appears around 70°C for the two copolymers.

TABLE 7.6 T_g Values before Curing of the Mixtures Containing 30 wt% of Neat and Epoxidized C500 Block Copolymers after Mixing with DGEBA/MCDEA at 140°C (adapted from Ref. [172] with Permission).

Block copolymer (30 wt%)	Epoxidized PB in the overall mixture (wt%)	T_g of PB block before mixing (°C)	T_g[a] after mixing (°C)	State of the mixture (before curing)
C500	0	−80	−11	Opaque, inhomogeneous
C500ep24	4.2	−44	−25	Transparent
C500ep27	4.8	−42	−21	Transparent
C500ep38	6.5	−33	−15	Transparent

[a]T_g of neat epoxy = −12°C; glass transition temperatures for the mixtures determined by DSC measurements (heating rate of 10°C/min).

The microhardness properties and nanostructured morphology of epoxidized copolymers and the composites can also be increased and altered respectively by the incorporation of the nanoparticles.[173] The transparency and stiffness of neat epoxy can be enhanced by the nanostructured epoxy systems consisting SBS epoxidized tri-block copolymer and Al_2O_3 nanoparticles (Figs. 7.24 and 7.25).[174]

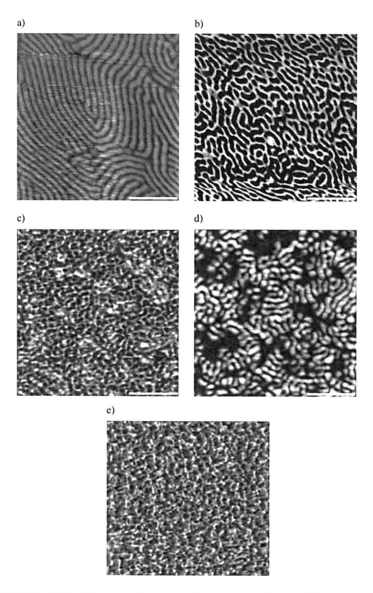

FIGURE 7.24 TM-AFM phase images for improvement in miscibility of epoxidized styrene–butadiene (SB) copolymers with epoxy monomers (epoxidation process improves the miscibility of the copolymers with the epoxy resin) and formation of self-assembled nanostructures in the uncured state: (a) parent SB (original SB star block copolymer self-assembles into a lamellar nanostructure), (b) 5 mol%, (c) 29 mol%, (d) 40 mol%, and (e) 75 mol% epoxidized copolymers, annealed at 80°C in vacuum for 3 h. Scale bar = 300 nm (adapted from Ref. [175] with permission).

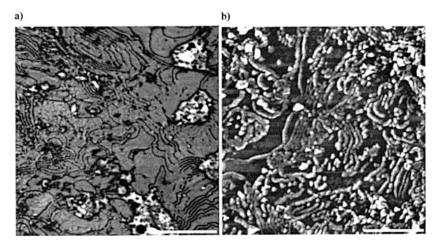

FIGURE 7.25 TM-AFM phase images for uncured blends containing 10 wt% DGEBA (bisphenol A diglycidyl ether) in (a) parent SB (small lamellar regions in a continuous disordered matrix); (b) 50 mol% epoxidized copolymer (copolymer can self-assemble into a poorly ordered hexagonal morphology, in which the domains shows higher thickness than those corresponding to the epoxidized copolymers, it is due to partial miscibility between the copolymer and the DGEBA), annealed at 80°C in vacuum for 3 h. Scale bar = 1 μm (adapted from Ref. [175] with permission).

7.3 EPOXY/POLY(ETHYLENE OXIDE) BLOCK BASED COPOLYMER SYSTEM

Mesostructured organic materials can be prepared by polymerizing an organic monomer in either hydrophobic or aqueous phase of a surfactant/water/organic ternary mixture. The advantage of this process is that it can avoid any possible ionic interactions between monomer and amphiphile and the polymerization takes place without altering the polymerization kinetics in the presence of other components.[176–179]

Generally, interactions between the organic reactive component and the amphiphile are altered by the increasing molecular weight of the polymerizing material, which tend to decrease its miscibility with other compounds. Rubber-modified epoxy systems are critical examples of this mechanism, which are produced by blending a reactive epoxy system with a miscible rubber. After curing of epoxies, cured epoxies associate the increase in molecular weight forces, which tend to macroscopic phase separation of the rubber from the thermosets, this phenomenon is known as "polymerization-induced phase separation."[180–183]

In this context, many studies have been reported based on creating the nanostructured thermosets using an amphiphilic block copolymer, poly(ethylene oxide)–poly(ethylene-*alt*-propylene) (PEO–PEP), and a reactive epoxy resin that selectively mixes with the PEO block. These are widely used in various applications due to their commercial availability, easy synthesis routes, including different ratios of each block and without participating any chemical synthesis or reaction with the epoxy system. One of the studies, transparent coatings were produced by the cross-linked unsaturated polyester and staggered platelets of polyethylene-*block*-poly(ethylene oxide).[90,184,185] The generating of nanostructure in cured blends of epoxy resin and di-block copolymer was first mentioned by Hillmyer et al. (Figs. 7.26 and 7.27; Table 7.7).[90]

TABLE 7.7 Morphology and Behavior of Poly(Ethylene Oxide) Block-Based Reactive and Nonreactive Block Copolymers in Thermosets (adapted from Ref. [120] with Permission).

Polymer	Abbreviation	Epoxy + hardener system	References
Poly(ethyleneoxide)-*b*-poly(propylene oxide)	PEO–PPO	BADGE + MDA	[186,187]
Poly(ethylene oxide)-*b*-poly(butylene oxide)	PEO–PBO	BADGE + PN	[188]
Poly(ethylene oxide)-*b*-poly(ethyl ethylene)	PEO–PEE	BADGE + PA	[90]
		BADGE + MDA	[189]
Poly(ethyleneoxide)-*b*-poly(ethylene-*alt*-propylene)	PEO–PEP	BADGE + PA	[90]
		BADGE + MDA	[189,190]
		BADGE + PN	[122]
Poly(ethyleneoxide)-*b*-poly(propylene oxide)-*b*-poly(ethylene oxide)	PEO–PPO–PEO	BADGE + MDA	[122,187–195]
Poly(ethylene oxide)-*b*-polyisoprene	PEO–PI	BADGE + MDA	[37]
Poly(ethylene oxide)-*b*-polybutadiene	PEO–PB	BADGE + MDA	[37]
Poly(ethylene oxide)-*b*-poly(ethylene)	PEO–PE	BADGE + MDA	[193]
Poly(ethylene oxide)-*b*-polystyrene	PEO–PS	BADGE + MOCA	[196]

Nanostructured Epoxy/Block Copolymer Blends

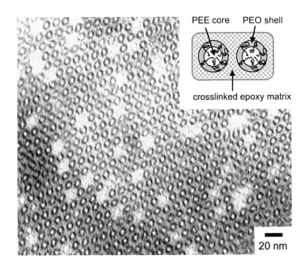

FIGURE 7.26 TEM image showing cylinders of PEO–PEE (25 wt.% PEO36PEE39) in an epoxy matrix. The core–shell morphology is clearly visible (cylinders with a core–shell morphology consisting of a nonpolar core surrounded by a corona of PEO)[90] (adapted from Ref. [120] with permission).

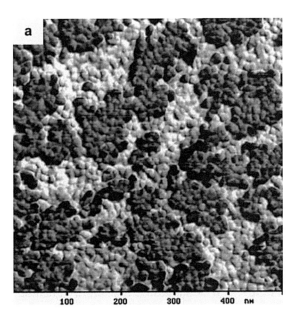

FIGURE 7.27 AFM phase image of 20 wt% PEO–PPO–PEO (80% PEO) cured in epoxy resin (spherical micelles of diameter ~10 nm). The lighter areas are harder (epoxy-rich) and the dark areas (copolymer-rich) are softer[193] (adapted from Ref. [120] with permission).

In the case of PEO–PEP di-block/epoxy system cured with aromatic amine demonstrates an increase in d-spacing as epoxy molecular weight is increased by the reaction.[162,163] These results are found to be consistent with swelling (or "wetting") of the PEO block by epoxy resin. The connectivity of the PEO–PEP blocks and the requirements to maintain constant density and minimize chain stretching lead to augmented interfacial curvature as the concentration of resin is increased. Examining the d-spacing against curing time indicates that swelling continues to occur long after the gel point. This is attributed to the fact that the gel point is a property of the bulk sample and doesn't necessarily coincide with the restriction of local mobility at the nanometer scale. This behavior may be due to initially epoxy miscible, PEO block is expelled from the resin as curing progresses. This expulsion leads to a "drying" or "de-swelling" of the PEO block and a subsequent reduction in the interfacial curvature. Consequently, as long as curing is not progressed to kinetically inhibit the phase transition, a morphology with reduced interfacial curvature may be obtained. Epoxy resins cured using aromatic cross-linking agents usually have high glass transition temperature (T_g), which is a desirable property for many applications. It also means that unmodified cured resins are extremely brittle. Traditionally, toughening (i.e., the resistance to propagation of a sharp crack) is achieved using rubbery modifiers, which are either thoroughly immiscible with the epoxy or undergo reaction induced macrophase separation.[31,197–200] However, poor miscibility between two materials may result in a microphase separation of one block of the block copolymer and epoxy blends, which minimizes the toughening effect of the block copolymer, although the blending of the block copolymer with epoxy may adversely affect Young's modulus and glass transition of the cured resin. Spherical micelles are found to give significantly greater improvements in toughness than vesicles. Even greater enhancement has been obtained in the case of worm-like micelles. The relative improvement in toughness is found to increase when the brominated epoxy content is increased. Increased bromination leads to a more brittle resin in the absence of a toughening agent. Enhanced brittleness means thinner samples are needed for the plane-strain condition. In the case of PEO–PBO di-block copolymers in nonbrominated BADGE þ phenol novolac, the worm-like micelles are found to prove the greatest improvement in toughening followed by spherical micelles and vesicles morphology.[197] It is suggested that the toughening observed with micelles may be due to cavitation processes. With worm-like micelles, SEM indicates that worms bridging the crack are "pulled-out." There is also evidence for nanometer scale crack deflection leading to the detachment of thin flakes of epoxy from the fracture surface. The T_g of a noncrystalline (amorphous) material can be defined as the critical temperature above which the material changes its behavior from "glassy" to "rubbery" state. When an amorphous material is cooled down, the segmental mobility decreases until T_g is reached, at that point the system drops out of thermodynamic equilibrium.[201] The resulting material, a glass, relaxes toward thermodynamic equilibrium by a structural relaxation process[202–204] also called "physical aging."[179,180] The nature of T_g and the structural relaxation process associated with it are currently considered as major challenges in condensed-matter physics.[205] The main requirement for these applications is that the material should be a rigid solid at a temperature of its use. This temperature must be below T_g (i.e., $T_{use} < T_g$). Thus, most of the work performed to date involving investigations of nanostructure formation in epoxy resins has been driven by the desire to modify the properties of the bulk material in a commercially useful manner. This includes increasing the T_g and improving the material fracture toughness simultaneously. The morphology adapted by block copolymer-modified epoxy systems has an effect not only on mechanical properties of resin but also on its T_g. Moreover, T_g is increased in systems containing spherical and worm-like micelles. Nevertheless, it is suggested that the presence of PEO may enhance the cross-linking in some manner, hence reducing polymer segment mobility and increase in T_g. As a result, it has been concluded that block copolymers are suitable candidates for use as a second phase in nanocomposites with the purpose of toughening by their nanostructure. The morphology and behavior of modified resins involving different matrices, curing agents and block copolymers have been well studied. The block copolymers can be divided into three categories. The first type can self-assemble in the uncured epoxy via a resinophilic part (epoxy-miscible block) and a resin phobic part (epoxy-immiscible block). The nanoscale structures are formed in the precure stage and fixed during cure. The second class is formed by the di-blocks where both blocks are miscible but one of them undergoes reaction induced microphase separation. The case where the epoxy–miscible block is reactive toward the resin or the curing agent conforms to the third type. The toughness attained with the incorporation of soft nanoinclusions depends on the morphology adopted by the block copolymers. It is reported that a vesicular morphology helps to improve in the fracture toughness significantly more than micelle morphology for nonreactive polymers. When comparing reactive with nonreactive inclusions, nonreactive vesicles provided poorer toughness than vesicles where one of the blocks is chemically bonded to the matrix. In general, for both reactive and nonreactive polymers, the morphology granting the best toughness is by the worm-like micelles. The glass transition in block copolymer-modified resins is found to rise in the systems containing spherical and worm-like micelles despite the intrinsically plasticizing nature of the resin ophilic polymer. This contradictory effect is not been explained fully. It has been reported that the thermosetting-mechanical behavior of nanoparticle-filled polymers should be similar to thin films. A better understanding is required on the physical mechanisms occurring at the soft nanoinclusion–matrix interface to interpret the data and tailoring of the polymer nanocomposites.

7.4 EPOXY/SIS BLOCK COPOLYMER SYSTEM

Block copolymers are commercially available at considerable low cost, which is an important aspect for potential applications. Styrene-butadiene and styrene–isoprene block copolymers, further identified as SBS and styrene–isoprene–styrene (SIS), two reported tri-block copolymers that include two hard polystyrene blocks at every end side of a molecular chain and a soft butadiene or isoprene organized in the middle. SBS and SIS are thermoplastic elastomers, blends that produce both the elasticity and resilience of butadiene rubber or isoprene rubber (natural rubber) and the sense of polystyrene to be processed as thermoplastics. It is necessary that the hard and soft blocks are immiscible; therefore, on a microscopic scale, the polystyrene blocks create disconnected regions in the rubber matrix, thereby offering practical cross-linkages to the rubber (Fig. 7.28).

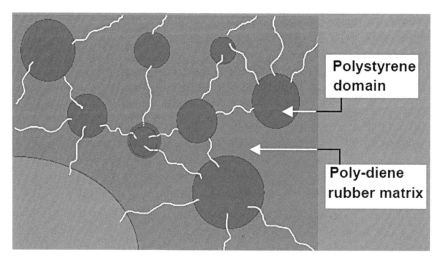

FIGURE 7.28 **(See color insert.)** Schematic representation of tri-block styrene–butadiene and styrene-isoprene block copolymers.

Nanostructured Epoxy/Block Copolymer Blends

Block copolymers mostly self-organize into nanoscale morphology, which depends on the interblock interaction parameter (χ), the degree of polymerization (N), molecular structure, and block formation and processing circumstances.

The chain-packing difficulty, the phase character in the block copolymers are considerably altered by the outside circumstances such as the existence of solvents and the rate of solvent evaporation, thermal process, and shear stress field at the nonequilibrium situation. The permanent nanodomain structures of common di-block or tri-block copolymers are essentially spherical domains, cylinders, bi-continuous morphology, or lamella grown up as an action of a rise in the concentration of the second component or an increase a separation between the blocks. Microphase-segregated block copolymers experience, during heating, an order-disorder transition (ODT) at a particular critical temperature, frequently pointed out to as the ODT temperature, which depends on molecular weight, block composition, and segment-segment interaction parameter of the constituent blocks as above mentioned and also on conformational inequality of the constituent blocks.[57,206–210] Spherical and cylindrical phases observed possess a volume of the spherical and cylindrical phases possess an interesting property that the volume of the internal phase (consisting of A segments) is identical to the volume of the external phase. Curved phases with such high internal volume, different interaction parameters and compositions could be of potential interest in separation and catalytic applications. Choice of parameters, it showed that significantly asymmetric lamellae can be achieved. In a similar manner, high volume cylindrical and spherical phases with the interior of the cylinder and spheres occupying of the order 65–75% of the total volume can also be obtained as shown in Figures 7.29–7.31.[211]

Epoxidation of polyisoprene block of SIS block copolymer same as polybutadiene block of poly(styrene-*b*-butadiene-*b*-styrene) block copolymer is turned up as an excellent approach to establish them miscible with the epoxy precursors. High epoxidation levels are an important point to provide miscibility of the epoxidized polyisoprene blocks with the resin and epoxidized polyisoprene (ePI) block as a reactive block of a poly(butadiene-*b*-isoprene) di-block copolymer. Epoxy/SIS system also exhibits the arrangement of nanostructures by self-assembly or by reaction-induced microphase separation. Epoxidation degree is observed to influence the obtained nanostructured pattern at different stages of the curing process (Fig. 7.32).[212]

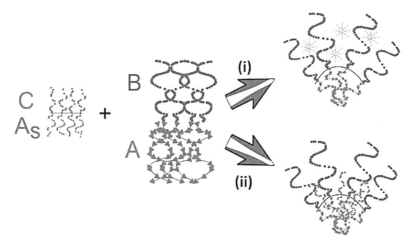

FIGURE 7.29 (See color insert.) Two possible mechanisms to explain the influence of polystyrene-block-poly(4-hydroxystyrene) copolymer (A_sC) chains on the morphologies of polystyrene-block-poly(2-vinyl pyridine) copolymer (AB) di-block copolymers and PS (A and A_s) blocks are immiscible with both poly(4-hydroxystyrene) and poly(2-vinyl pyridine) copolymer blocks, (i) the A_sC copolymers dissolve in the B phase and increase the effective volume fraction of the B component; (ii) the A_sC copolymers segregate to the AB interface and modify the curvature through hydrogen bonding interactions. The dissolution of A–C copolymers into the B phase is possible to establish particularly one of the contributions to the mechanisms controlling the phase transformation (adapted from Ref. [211] with permission American Chemical Society).

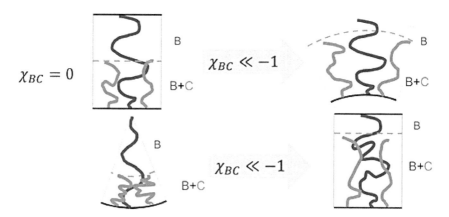

FIGURE 7.30 (See color insert.) Schematic of the proposed mechanism for transformation of lamellar morphologies to convex morphologies and concave morphologies to lamellar phases. A pure A–B block copolymer system that is either arranged in lamellar (symmetric) or curved (adapted from Ref. [211] with permission from American Chemical Society).

Nanostructured Epoxy/Block Copolymer Blends 109

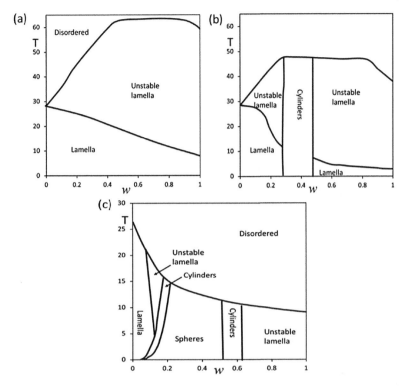

FIGURE 7.31 Phase diagram of a blend of AB + AC copolymers as a function of temperature and AC volume fraction w. The parameters correspond to $f = f_s = 0.5$, $\chi_{AB}N = 250$ T^{-1}, $\chi_{BC} = -2\chi_{AB}$. Key: (a) $\alpha = 0.48$; (b) $\alpha = 0.35$; (c) $\alpha = 0.07$ (adapted from Ref. [211] with permission from American Chemical Society).

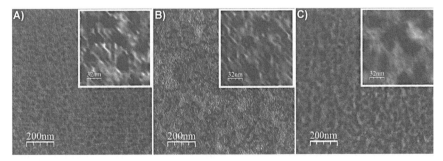

FIGURE 7.32 **(See color insert.)** Tapping mode-AFM phase image of nanostructured epoxy system/SIS85 (85:85% epoxidation degrees) (23 wt%) blends for (a) stage one (before curing, sphere-like nanostructures is observed), (b) stage two (after curing at 80°C for 100 min, less ordered sphere-like nanostructured is observed), and (c) stage three (after curing at 80°C for 180 min, bigger and less organized is observed) (adapted from Ref. [212] with permission from American Chemical Society).

This morphological transformation of the system greatly implies that phase separation further occurs as cross-linking continues, which is indicated the principles of the reaction-induced microphase separation mechanism. This builds bigger and less organized nanostructures. Here, ePI subchains are previously miscible with the epoxy precursors and later, immiscibility rises as the curing reaction progresses. This type of ePI block demixing points out to a broadening and suppression of homogeneity in the obtained nanodomains. This evolution is due to partial expulsion of ePI block.[212] A simplified representation of the nanostructure be the character of eSIS in the epoxy matrix is presented in Figure 7.31. Authors found that PS subchains self-assemble in sphere-like nanodomains owing to their high immiscibility of PS subchains with the epoxy precursors. When curing initiated, reaction-induced microphase separation of ePI subchains took place a gradual distortion of PS sphere-like nanostructures (Fig. 7.33).

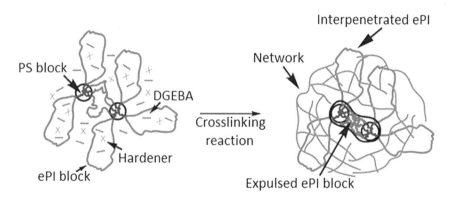

FIGURE 7.33 (See color insert.) Uncured (left, PS subchains self-assemble in sphere-like nanodomains (blue)) and cured (right, the interconnection of PS nanodomains) for SIS85/DGEBA/Hardener (adapted from Ref. [212] with permission from American Chemical Society).

The inclusion of carbon nanotubes (CNT) into a polymer matrix can have a significant influence on a wide variety of material features, such as mechanical strength, electric and thermal conductivities, thermal properties, etc. Addition of CNT in block copolymers can contribute to unique nanostructured composite materials with novel properties including higher elastic modulus and strain at break, possible electrical conductivities; it is due to excellent properties of CNT. At the same time, the block copolymer nanostructured matrix can produce various properties take into account of a homopolymer (Figs. 7.34–7.36).[213–218]

Nanostructured Epoxy/Block Copolymer Blends

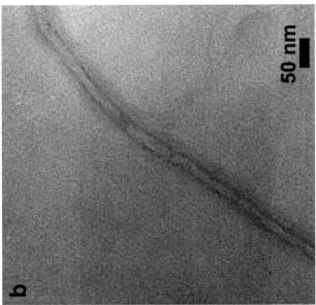

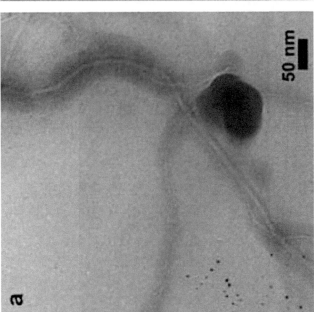

FIGURE 7.34 TEM microstructure: (a) PSg decorated MWCNT (discontinuous layer attributed to the surface grown PS and increments in the MWCNT thickness) and (b) purified MWCNT (adapted from Ref. [218] with permission).

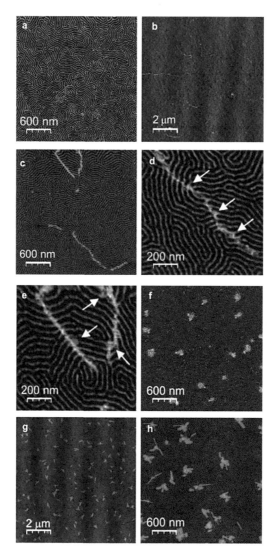

FIGURE 7.35 Tapping mode-AFM phase images for (a) neat SIS block copolymer (cylindrical PS domains parallel to the substrate), (b,c) DT/PSgMWCNT/SIS composite (similar self-assembled morphology as neat block copolymer), (d,e) expanding the PS cylinders of the SIS (arrows show the anchoring sites between PSgMWCNT and the PS domains of the SIS matrix), (f) PSgMWCNT/SIS composite (CNTs aggregate during the film preparation in absence of dodecanethiol (DT) surfactant) and (g,h) DT/purified MWCNT/SIS composite (good dispersion of purified MWCNT with PS in the SIS matrix, but aggregated MWCNTs). (PSgMWCNT aggregates appeared as brighter regions) (adapted from Ref. [218] with permission).

Nanostructured Epoxy/Block Copolymer Blends

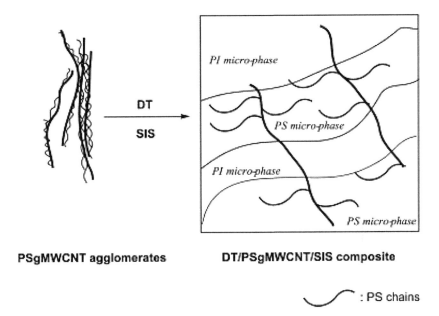

FIGURE 7.36 Mechanism of the dispersion of PSgMWCNT in the presence of DT (PS chains of PSgMWCNT are able to interlock with the PS of the SIS block copolymer, thus anchoring the nanotubes to the polymer matrix in the composite material after solvent evaporation, and DT assists the debundling of the PSgMWCNT) (adapted from Ref. [218] with permission).

7.5 CONCLUSION

The term "epoxy resin" is used to describe as a class of thermosetting materials used in industries due to their high modulus, low creep, high strength, and good thermal and dimensional stabilities. The epoxy is found useful for adhesive bonding, coating, electrical encapsulation (dielectric properties) and composite matrix. Block copolymers are composed of long two or more chemically distinct polymer chains (blocks) and these blocks are connected by covalent bonds to form a single macromolecule. The block copolymers can be synthesized by the methods such as anionic polymerization, controlled radical polymerization, etc. The blocks can be arranged in various sequences like di-block, ABC tri-blocks, ABA tri-block, $(AB)_n$ multi-blocks, $(AB)_n$ star blocks, $(ABC)_n$ star blocks. Amphiphilic block copolymers consisting of epoxy-philic and epoxy-phobic blocks have been used as modifiers for toughening the epoxies, it participates in generating self-assembled nanostructures in an undiluted melt or in solution. The ability of block copolymers to generate nanostructures in epoxy thermosetting systems is helpful to obtain new functional properties of block copolymers/epoxy blends. Both types, amphiphilic and chemically modified di- or tri-block copolymers, have been employed for generating nanostructure in epoxy. In epoxy/block copolymer blends, one of the blocks is miscible or can react with the epoxy system prior and after curing, if both blocks being miscible help to develop self-assembled morphology.

KEYWORDS

- **nanostructured**
- **block copolymer**
- **epoxy toughening**
- **characterization**
- **application**

REFERENCES

1. Petrie, E. *Epoxy Adhesive Formulations*; McGraw Hill Professional: New York, NY, 2005.
2. *Epoxy Adhesive Application Guide*. Retrieved 26 December 2016 from http://www.epotek.com/site/files/brochures/pdfs/adhesive_application_guide.pdf.
3. *What Is an Epoxy Adhesive?* Retrieved 26 December 2016 from http://www.masterbond.com/techtips/what-epoxy-adhesive.
4. *Dielectric Properties of Epoxies*. Retrieved 26 December 2016 from http://www.jpkummer.com/sites/default/files/Tech%20Tip%2025%20Dielectric%20Properties%20of%20Epoxies.pdf.
5. Park, S.-J.; Jin, F.-L. Thermal Stabilities and Dynamic Mechanical Properties of Sulfone-Containing Epoxy Resin Cured with Anhydride. *Polym. Degrad. Stab.* **2004,** *86* (3), 515–520.
6. Giannakopoulos, G.; K. Masania; A. C. Taylor. Toughening of Epoxy Using Core–Shell Particles. *J. Mater. Sci.* **2011,** 46 (2), 327–338.
7. Hsieh, T. H.; et al. The Toughness of Epoxy Polymers and Fibre Composites Modified with Rubber Microparticles and Silica Nanoparticles. *J. Mater. Sci.* **2010,** *45* (5), 1193–1210.
8. Thitsartarn, W.; et al. Simultaneous Enhancement of Strength and Toughness of Epoxy Using POSS-Rubber Core–Shell Nanoparticles. *Compos. Sci. Technol.* **2015,** *118,* 63–71.

9. Domun, N.; et al. Improving the Fracture Toughness and the Strength of Epoxy Using Nanomaterials—A Review of the Current Status. *Nanoscale* **2015**, *7* (23), 10294–10329.
10. Kargarzadeh, H.; Ahmad, I.; Abdullah, I. *Mechanical Properties of Epoxy–Rubber Blends*; 2015.
11. Chevalier, J.; et al. Micromechanics Based Pressure Dependent Failure Model for Highly Cross-Linked Epoxy Resins. *Eng. Fract. Mech.* **2016**, *158*, 1–12.
12. Chong, H. M.; Taylor, A. C. The Microstructure and Fracture Performance of Styrene–Butadiene–Methylmethacrylate Block Copolymer-Modified Epoxy Polymers. *J. Mater. Sci.* **2013**, *48* (19), 6762–6777.
13. Bray, D. J.; et al. The Modelling of the Toughening of Epoxy Polymers via Silica Nanoparticles: The Effects of Volume Fraction and Particle Size. *Polymer* **2013**, *54* (26), 7022–7032.
14. Tang, S.; et al. Self-Organization of Homopolymer Brush- and Mixed Homopolymer Brush-Grafted Silica Nanoparticles in Block Copolymers and Polymer Blends. *Polymer* **2016**, *90*, 9–17.
15. Bagheri, R.; Marouf, B. T.; Pearson, R. A. Rubber-toughened Epoxies: A Critical Review. *J. Macromol. Sci.; C: Polym. Rev.* **2009**, *49* (3), 201–225.
16. Yee, A. F.; Du, J.; Thouless, D. In *Polymer Blends: Performance*; Paul, D. R.; Bucknall, C. B.; Eds., 2000.
17. Seng, L. Y.; et al. Effects of Liquid Natural Rubber (LNR) on the Mechanical Properties of LNR Toughened Epoxy Composite. *Sains Malays.* **2011**, *40* (7), 679–683.
18. Chen, J. *Toughening Epoxy Polymers and Carbon Fibre Composites with Core–Shell Particles, Block Copolymers and Silica Nanoparticles*, 2013.
19. Kunz, S. C.; Sayre, J. A.; Assink, R. A. Morphology and Toughness Characterization of Epoxy Resins Modified with Amine and Carboxyl Terminated Rubbers. *Polymer* **1982**, *23* (13), 1897–1906.
20. Levita, G.; Marchetti, A.; Butta, E. Influence of the Temperature of Cure on the Mechanical Properties of ATBN/Epoxy Blends. *Polymer* **1985**, *26* (7), 1110–1116.
21. Ritzenthaler, S.; Girard-Reydet, E.; Pascault, J. P. Influence of Epoxy Hardener on Miscibility of Blends of Poly(Methyl Methacrylate) and Epoxy Networks. *Polymer* **2000**, *41* (16), 6375–6386.
22. Galante, M. J.; Oyanguren, P. A.; Andromaque, K.; Frontini, P. M.; Williams, R. J. J. Blends of Epoxy/Anhydride Thermosets with a High-Molar-Mass Poly(Methyl Methacrylate). *Polym. Int.* **1999**, *48* (8), 642–648.
23. Brooker, R. D.; Kinloch, A. J.; Taylor, A. C. The Morphology and Fracture Properties of Thermoplastic-Toughened Epoxy Polymers. *J. Adhes.* **2010**, *86* (7), 726–741.
24. Yoon, T. H.; Priddy, D. B.; Lyle, G. D.; McGrath, J. E. Mechanical and Morphological Investigations of Reactive Polysulfone Toughened Epoxy Networks. *Macromol. Symp.* **1995**, *98* (1), 673–686.
25. Hourston, D. J.; Lane, J. M. The Toughening of Epoxy Resins with Thermoplastics: 1. Trifunctional Epoxy Resin–Polyetherimide Blends. *Polymer* **1992**, *33* (7), 1379–1383.
26. Hourston, D. J.; Lane, J. M.; Macbeath, N. A. Toughening of Epoxy Resins with Thermoplastics. 2. Tetrafunctional Epoxy Resin–Polyetherimide Blends. *Polym. Int.* **1991**, *26* (1), 17–21.
27. Hourston, D. J.; Lane, J. M.; Zhang, H. X. Toughening of Epoxy Resins with Thermoplastics: 3. An Investigation into the Effects of Composition on the Properties of Epoxy Resin Blends. *Polym. Int.* **1997**, *42* (4), 349–355.

28. Becu, L.; Maazouz, A.; Sautereau, H.; Gerard, J. F. Fracture Behavior of Epoxy Polymers Modified with Core–Shell Rubber Particles. *J. Appl. Polym. Sci.* **1997**, *65* (12), 2419–2431.
29. Pearson, R. A.; Yee, A. F. Influence of Particle Size and Particle Size Distribution on Toughening Mechanisms in Rubber-Modified Epoxies. *J. Mater. Sci.* **1991**, *26* (14), 3828–3844.
30. Kim, D. S.; Cho, K.; Kim, J. K.; Park, C. E. Effects of Particle Size and Rubber Content on Fracture Toughness in Rubber-Modified Epoxies. *Polym. Eng. Sci.* **1996**, *36* (6), 755–768.
31. Lin, K. F.; Shieh, Y. D. Core–Shell Particles to Toughen Epoxy Resins. I. Preparation and Characterization of Core–Shell Particles. *J. Appl. Polym. Sci.* **1998**, *69* (10), 2069–2078.
32. Lin, K. F.; Shieh, Y. D. Core–Shell Particles Designed for Toughening the Epoxy Resins. II. Core–Shell-Particle-Toughened Epoxy Resins. *J. Appl. Polym. Sci.* **1998**, *70* (12), 2313–2322.
33. Lu, F.; Cantwell, W. J.; Kausch, H. H. The Role of Cavitation and Debonding in the Toughening of Core–Shell Rubber Modified Epoxy Systems. *J. Mater. Sci.* **1997**, *32* (11), 3055–3059.
34. Qian, J. Y.; Pearson, R. A.; Dimonie, V. L.; El-Aasser, M. S. Synthesis and Application of Core–Shell Particles as Toughening Agents for Epoxies. *J. Appl. Polym. Sci.* **1995**, *58* (2), 439–448.
35. Liu, J.; Thompson, Z. J.; Sue, H. J.; Bates, F. S.; Hillmyer, M. A.; Dettloff, M.; Pham, H. Toughening of Epoxies with Block Copolymer Micelles of Wormlike Morphology. *Macromolecules* **2010**, *43* (17), 7238–7243.
36. Liu, J.; Sue, H. J.; Thompson, Z. J.; Bates, F. S.; Dettloff, M.; Jacob, G.; Pham, H. Nanocavitation in Self-Assembled Amphiphilic Blocks Copolymer-Modified Epoxy. *Macromolecules* **2008**, *41* (20), 7616–7624.
37. Dean, J. M.; Grubbs, R. B.; Saad, W.; Cook, R. F.; Bates, F. S. Mechanical Properties of Block Copolymer Vesicle and Micelle Modified Epoxies. *J. Polym. Sci., B: Polym. Phys.* **2003**, *41* (20), 2444–2456.
38. Chen, J.; Taylor, A. C. Epoxy Modified with Triblock Copolymers: Morphology, Mechanical Properties and Fracture Mechanisms. *J. Mater. Sci.* **2012**, *47* (11), 4546–4560.
39. Thio, Y. S.; Wu, J.; Bates, F. S. The Role of Inclusion Size in Toughening of Epoxy Resins by Spherical Micelles. *J. Polym. Sci., B: Polym. Phys.* **2009**, *47* (11), 1125–1129.
40. Oldak, R. K.; Hydro, R. M.; Pearson, R. A. On the Use of Triblock Copolymers as Toughening Agents for Epoxies. In *Proceedings 30th Annual Meeting of the Adhesion Society, Inc.*; The Adhesion Society: Blacksburg, VA, 2007; pp 153–156.
41. Kishi, H.; Kunimitsu, Y.; Imade, J.; Oshita, S.; Morishita, Y.; Asada, M. Nano-Phase Structures and Mechanical Properties of Epoxy/Acryl Triblock Copolymer Alloys. *Polymer* **2011**, *52* (3), 760–768.
42. Barsotti, R.; Fine, T.; Inoubli, R.; Gerard, P.; Schmindt, S.; Navarro, C. Nanostrength Block Copolymers for Epoxy Toughening. In *Meeting of the Thermoset Resin Formulators Association.* Thermoset Resin Formulators Association: Chicago, IL, September 2008.
43. Bacigalupo, L. N. *Fracture Behavior of Nano-Scale Rubber-Modified Epoxies*, 2013.
44. Hsieh, T. H.; Kinloch, A. J.; Masania, K.; Taylor, A. C.; Sprenger, S. The Mechanisms and Mechanics of the Toughening of Epoxy Polymers Modified with Silica Nanoparticles. *Polymer* **2010**, *51* (26), 6284–6294.

45. Liang, Y. L.; Pearson, R. A. Toughening Mechanisms in Epoxy–Silica Nanocomposites (ESNs). *Polymer* **2009**, *50* (20), 4895–4905.
46. Johnsen, B. B.; Kinloch, A. J.; Mohammed, R. D.; Taylor, A. C.; Sprenger, S. Toughening Mechanisms of Nanoparticle-Modified Epoxy Polymers. *Polymer* **2007**, *48* (2), 530–541.
47. Ma, J.; Mo, M. S.; Du, X. S.; Rosso, P.; Friedrich, K.; Kuan, H. C. Effect of Inorganic Nanoparticles on Mechanical Property, Fracture Toughness and Toughening Mechanism of Two Epoxy Systems. *Polymer* **2008**, *49* (16), 3510–3523.
48. *Literature Review of Epoxy Toughening.* Retrieved 26 December 2016 from https://theses.lib.vt.edu/theses/available/etd-32398-61326/unrestricted/8-9.pdf.
49. Chaudhary, S.; et al. Toughening of Epoxy with Preformed Polyethylene Thermoplastic Filler. *Polym.–Plast. Technol. Eng.* **2015**, *54* (9), 907–915.
50. Hodgkin, J. H.; Simon, G. P.; Varley, R. J. Thermoplastic Toughening of Epoxy Resins: A Critical Review. *Polym. Adv. Technol.* **1998**, *9* (1), 3–10.
51. Han, J. G.; et al. Multiscale Analysis on the Toughening Behavior of Thermoplastic Modified Epoxy. In *21st International Conference on Composite Materials Xi'an*, 20–25th August 2017.
52. Remya, V. R.; et al. Biobased Materials for Polyurethane Dispersions. *Chem. Int.* **2016**, *2* (3), 158–167.
53. Day, R. J.; Lovell, P. A.; Wazzan, A. A. Toughened Carbon/Epoxy Composites Made by Using Core/Shell Particles. *Compos. Sci. Technol.* **2001**, *61* (1), 41–56.
54. Bécu-Longuet, L.; Bonnet, A.; Pichot, C.; Sautereau, H.; Maazouz, A. Epoxy Networks Toughened by Core–Shell Particles: Influence of the Particle Structure and Size on the Rheological and Mechanical Properties. *J. Appl. Polym. Sci.* **1999**, *72* (6), 849–858.
55. Day, R. J.; Lovell, P. A.; Pierre, D. Toughening of Epoxy Resins Using Particles Prepared by Emulsion Polymerization: Effects of Particle Surface Functionality, Size and Morphology on Impact Fracture Properties. *Polym. Int.* **1997**, *44* (3), 288–299.
56. Tiwari, A.; et al. Synthesis of Samarium (Sm) Doped Thin Film of $SrMnO_3$ by Pulse Laser Deposition and Its Structural and Magnetic Characterization. *Science* **2016**, *2* (1), 1–4.
57. Karande, R. D.; et al. Preparation of Polylactide from Synthesized Lactic Acid and Effect of Reaction Parameters on Conversion. *J. Mater. Sci. Eng. Adv. Technol.* **2015**, *12* (1–2), 1–37.
58. Ruzette, A.-V.; Leibler, L. Block Copolymers in Tomorrow's Plastics. *Nat. Mater.* **2005**, *4* (1), 19–31.
59. Smart, T.; et al. Block Copolymer Nanostructures. *Nano Today* **2008**, *3* (3), 38–46.
60. Hamley, I. W. *The Physics of Block Copolymers*; Oxford University Press: New York, 1998; Vol. 19.
61. Won, Y. Y.; Davis, H. T.; Bates, F. S. Giant Wormlike Rubber Micelles. *Science* **1999**, *283* (5404), 960–963.
62. Qian, J.; Zhang, M.; Manners, I.; Winnik, M. A. Nanofiber Micelles from the Self-Assembly of Block Copolymers. *Trends Biotechnol.* **2010**, *28* (2), 84–92.
63. Liu, G.; Yan, X.; Li, Z.; Zhou, J.; Duncan, S. End Coupling of Block Copolymer Nanotubes to Nanospheres. *J. Am. Chem. Soc.* **2003**, *125* (46), 14039–14045.
64. Discher, D. E.; Eisenberg, A. Polymer Vesicles. *Science* **2002**, *297* (5583), 967–973.
65. Grubbs, R. B.; Dean, J. M.; Broz, M. E.; Bates, F. S. Reactive Block Copolymers for Modification of Thermosetting Epoxy. *Macromolecules* **2000**, *33* (26), 9522–9534.

66. Lipic, P. M.; Bates, F. S.; Hillmyer, M. A. Nanostructured Thermosets from Self-Assembled Amphiphilic Block Copolymer/Epoxy Resin Mixtures. *J. Am. Chem. Soc.* **1998,** *120* (35), 8963–8970.
67. Maiez-Tribut, S.; Pascault, J.-P.; Soule, E. R.; Borrajo, J.; Williams, R. J. J. Nanostructured Epoxies Based on the Self-Assembly of Block Copolymers: A New Miscible Block That Can Be Tailored To Different Epoxy Formulations. *Macromolecules* **2007,** *40* (4), 1268–1273.
68. Thomas, S.; Mishra, R.; Karikal, N. *Micro and Nano Fibrillar Composites (MFCs and NFCs) from Polymer Blends*; Woodhead Publishing: Sawston, Cambridge, 2017.
69. Pearson, R. A.; Bacigalupo, L. N.; Liang, Y. L.; Marouf, B. T.; Oldak, R. K. Plastic Zone Size-Fracture Toughness Correlations in Rubber-Modified Epoxies. In *31st Annual Meeting of the Adhesion Society*: Austin, TX, 2008, February; pp 27–29.
70. Bacigalupo, L. N. *Toughening of Epoxies: Novel Self-Assembling Block Copolymers versus Traditional Telechelic Oligomers*, 2011.
71. Zhang, L.; Eisenberg, A. Multiple Morphologies of "Crew-Cut" Aggregates of Polystyrene-*b*-Poly(Acrylic Acid) Block Copolymers. *Science* **1995,** *268* (5218), 1728–1731.
72. LaRue, I.; Adam, M.; Pitsikalis, M.; Hadjichristidis, N.; Rubinstein, M.; Sheiko, S. S. Reversible Morphological Transitions of Polystyrene-*b*-Polyisoprene Micelles. *Macromolecules* **2006,** *39* (1), 309–314.
73. Bhargava, P.; Tu, Y.; Zheng, J. X.; Xiong, H.; Quirk, R. P.; Cheng, S. Z. Temperature-Induced Reversible Morphological Changes of Polystyrene-Block-Poly(Ethylene Oxide) Micelles in Solution. *J. Am. Chem. Soc.* **2007,** *129* (5), 1113–1121.
74. Chen, Q.; Zhao, H.; Ming, T.; Wang, J.; Wu, C. Nanopore Extrusion-Induced Transition from Spherical to Cylindrical Block Copolymer Micelles. *J. Am. Chem. Soc.* **2009,** *131* (46), 16650–16651.
75. Wang, C. W.; Sinton, D.; Moffitt, M. G. Morphological Control via Chemical and Shear Forces in Block Copolymer Self-Assembly in the Lab-On-Chip. *ACS Nano* **2013,** *7* (2), 1424–1436.
76. Yu, H.; Jiang, W. Effect of Shear Flow on the Formation of Ring-Shaped ABA Amphiphilic Triblock Copolymer Micelles. *Macromolecules* **2009,** *42* (9), 3399–3404.
77. Abitha, V. K.; et al. Influence of Hybrid Fillers on Morphological and Mechanical Properties of Carboxylated Nitrile Butadiene Rubber Composites. *J. Mater. Sci. Eng. Adv. Technol.* **2016,** *13* (1), 13–27.
78. Wang, C. W.; Sinton, D.; Moffitt, M. G. Flow-Directed Block Copolymer Micelle Morphologies via Microfluidic Self-Assembly. *J. Am. Chem. Soc.* **2011,** *133* (46), 18853–18864.
79. Erhardt, R.; et al. Amphiphilic Janus Micelles with Polystyrene and Poly(Methacrylic Acid) Hemispheres. *J. Am. Chem. Soc.* **2003,** *125* (11), 3260–3267.
80. Ritzenthaler, S.; et al. ABC Triblock Copolymers/Epoxy-Diamine Blends. 2. Parameters Controlling the Morphologies and Properties. *Macromolecules* **2003,** *36* (1), 118–126.
81. Liu, Y.; Abetz, V.; Müller, A. H. E. Janus Cylinders. *Macromolecules* **2003,** *36* (21), 7894–7898.
82. Stewart, S.; Liu, G. *Angew. Chem. Int. Ed.* **2000,** *39,* 340.
83. Cui, H.; et al. Block Copolymer Assembly via Kinetic Control. *Science* **2007,** *317* (5838), 647–650.
84. Stoenescu, R.; Graff, A.; Meier, W. Asymmetric ABC-Triblock Copolymer Membranes Induce a Directed Insertion of Membrane Proteins. *Macromol. Biosci.* **2004,** *4* (10), 930–935.

85. Li, Z.; Hillmyer, M. A.; Lodge, T. P. Laterally Nanostructured Vesicles, Polygonal Bilayer Sheets, and Segmented Wormlike Micelles. *Nano Lett.* **2006**, *6* (6), 1245–1249.
86. Hillmyer, M. A.; Lipic, P. M.; Hajduk, D. A.; Almdal, K.; Bates, F. S. Self-Assembly and Polymerization of Epoxy Resin-Amphiphilic Block Copolymer Nanocomposites. *J. Am. Chem. Soc.* **1997**, *119* (11), 2749–2750.
87. Förster, S.; et al. Lyotropic Phase Morphologies of Amphiphilic Block Copolymers. *Macromolecules* **2001**, *34* (13), 4610–4623.
88. Jain, S.; et al. Disordered Network State in Hydrated Block-Copolymer Surfactants. *Phys. Rev. Lett.* **2006**, *96* (13), 138304.
89. Blanazs, A.; Armes, S. P.; Ryan, A. J. Self-Assembled Block Copolymer Aggregates: From Micelles to Vesicles and their Biological Applications. *Macromol. Rapid Commun.* **2009**, *30* (4–5), 267–277.
90. Battaglia, G.; Ryan, A. J. Pathways of Polymeric Vesicle Formation. *J. Phys. Chem. B* **2006**, *110* (21), 10272–10279.
91. Darling, S. B. Directing the Self-Assembly of Block Copolymers. *Progr. Polym. Sci.* **2007**, *32* (10), 1152–1204.
92. Robin, Mathew, P.; et al. "Fluorescent Block Copolymer Micelles that Can Self-Report on their Assembly and Small Molecule Encapsulation. *Macromolecules* **2016**, *49* (2), 653–662.
93. Robeson, Lloyd. Historical Perspective of Advances in the Science and Technology of Polymer Blends. *Polymers* **2014**, *6* (5), 1251–1265.
94. Benten, Hiroaki, et al. Recent Research Progress of Polymer Donor/Polymer Acceptor Blend Solar Cells. *J. Mater. Chem. A* **2016**, *4* (15), 5340–5365.
95. Dutta, D.; et al. Polymer Blends Containing Liquid Crystals: A Review. *Polym. Eng. Sci.* **1990**, *30* (17), 1005–1018.
96. Mahalle, A. S.; Sangawar, V. S. Morphology and Conductivity Studies of PVC Based Micro-Porous Polymer Electrolyte. *Int. J. Eng. Res. Appl.* **2013**, *3* (3), 649–653.
97. Chandran, S.; et al. Confinement Enhances Dispersion in Nanoparticle–Polymer Blend Films. *Nat. Commun.* **2014**, *5*, 3697.
98. Utracki, L. A.; Wilkie, C. A.; Eds. Polymer Blends Handbook. Vol. 1. Kluwer Academic Publishers: Dordrecht, The Netherlands, 2002.
99. Frank, C. W.; Gashgari, M. A.; Semerak, S. N. Polymer Blend Thermodynamics: Flory Huggins Theory and Its Application to Excimer Fluorescence Studies. In *Photophysical and Photochemical Tools in Polymer Science*. Springer: Netherlands, 1986; pp 523–546.
100. Hildebrand, J. H. Solubility. *J. Am. Chem. Soc.* **1916**, *38* (8), 1452–1473.
101. Scatchard, G. Equilibria in Non-Electrolyte Solutions in Relation to the Vapor Pressures and Densities of the Components. *Chem. Rev.* **1931**, *8* (2), 321–333.
102. Hildebrand, J. H. *Solubility of Non-electrolytes*; 1936.
103. Hansen, C. M. The Three Dimensional Solubility Parameter-Key to Paint Component Affinities: I. Solvents, Plasticizers, Polymers, and Resins. *J. Paint Technol.* **1967**, *39* (505), 104–117.
104. Small, P. A. Some Factors Affecting the Solubility of Polymers. *J. Appl. Chem.* **1953**, *3* (2), 71–80.
105. Van Krevelen, D. W. Group Contribution Techniques for Correlating Polymer Properties and Chemical Structure. *Comput. Model. Polym.* **1992**, 55–123.
106. Hoy, K. L. New Values of the Solubility Parameters from Vapor Pressure Data. *J. Paint Technol.* **1970**, *42* (541), 76–118.

107. Coleman, M. M.; Graf, J. F.; Painter, P. C. *Specific Interactions and the Miscibility of Polymer Blends*. Technomic Publishing: Lancaster, PA, 1991.
108. Prigogine, I.; Bellemans, A.; Mathot, V. *The Molecular Theory of Solutions*; North-Holland: Amsterdam, 1957; Vol 4.
109. Flory, P. J.; Orwoll, R. A.; Vrij, A. Statistical Thermodynamics of Chain Molecule Liquids. I. An Equation of State for Normal Paraffin Hydrocarbons. *J. Am. Chem. Soc.* **1964,** *86* (17), 3507–3514.
110. Flory, P. J.; Orwoll, R. A.; Vrij, A. Statistical Thermodynamics of Chain Molecule Liquids. II. Liquid Mixtures of Normal Paraffin Hydrocarbons. *J. Am. Chem. Soc.* **1964,** *86* (17), 3515–3520.
111. Flory, P. D. Statistical Thermodynamics of Liquid Mixtures. *J. Am. Chem. Soc.* **1965,** *87* (9), 1833–1838.
112. McMaster, L. P. Aspects of Polymer–Polymer Thermodynamics. *Macromolecules* **1973,** *6* (5), 760–773.
113. Shaw, M. T. Studies of Polymer–Polymer Solubility Using a Two-Dimensional Solubility Parameter Approach. *J. Appl. Polym. Sci.* **1974,** *18* (2), 449–472.
114. Sanchez, I. C.; Lacombe, R. H. An Elementary Molecular Theory of Classical Fluids. Pure Fluids. *J. Phys. Chem.* **1976,** *80* (21), 2352–2362.
115. Lacombe, R. H.; Sanchez, I. C. Statistical Thermodynamics of Fluid Mixtures. *J. Phys. Chem.* **1976,** *80* (23), 2568–2580.
116. Ruiz-Pérez, L.; et al. Toughening by Nanostructure. *Polymer* **2008,** *49* (21), 4475–4488.
117. Garate, H.; et al. *Miscibility, Phase Separation, and Mechanism of Phase Separation of Epoxy/Block-Copolymer Blends*; Springer: Berlin-Heidelberg, 2016; pp 1–41.
118. Dean, J. M.; Verghese, N. E.; Pham, H. Q.; Bates, F. S. Nanostructure Toughened Epoxy Resins. *Macromolecules* **2003,** *36* (25), 9267–9270.
119. Meng, F.; Zheng, S.; Zhang, W.; Li, H.; Liang, Q. Nanostructured Thermosetting Blends of Epoxy Resin and Amphiphilic Poly(ε-Caprolactone)-block–Polybutadiene-block-Poly(ε-Caprolactone) Triblock Copolymer. *Macromolecules* **2006,** *39* (2), 711–719.
120. Gomez, C. M.; Bucknall, C. B. Blends of Poly(Methyl Methacrylate) with Epoxy Resin and an Aliphatic Amine Hardener. *Polymer* **1993,** *34* (10), 2111–2117.
121. Woo, E. M.; Wu, M. N. Blends of a Diglycidylether Epoxy with Bisphenol-A Polycarbonate or Poly(Methyl Methacrylate): Cases of Miscibility with or Without Specific Interactions. *Polymer* **1996,** *37* (12), 2485–2492.
122. García, F. G.; Soares, B. G.; Williams, R. J. J. Poly(ethylene-*co*-Vinyl Acetate)-Graft-Poly(Methyl Methacrylate) (EVA-Graft-PMMA) as a Modifier of Epoxy Resins. *Polym. Int.* **2002,** *51* (12), 1340–1347.
123. Mondragon, I.; Remiro, P. M.; Martin, M. D.; Valea, A.; Franco, M.; Bellenguer, V. Viscoelastic Behaviour of Epoxy Resins Modified with Poly(Methyl Methacrylate). *Polym. Int.* **1998,** *47* (2), 152–158.
124. Remiro, P. M.; Riccardi, C. C.; Corcuera, M. A.; Mondragon, I. Design of Morphology in PMMA-Modified Epoxy Resins by Control of Curing Conditions. I. Phase Behavior. *J. Appl. Polym. Sci.* **1999,** *74* (4), 772–780.
125. Galante, M. J.; Borrajo, J.; Williams, R. J. J.; Girard-Reydet, E.; Pascault, J. P. Double Phase Separation Induced by Polymerization in Ternary Blends of Epoxies with Polystyrene and Poly(Methyl Methacrylate). *Macromolecules* **2001,** *34* (8), 2686–2694.
126. El-Ejmi, A. A.; Huglin, M. B. Characterization of *N,N*-Dimethylacrylamide/2-Methoxyethylacrylate Copolymers and Phase Behaviour of their Thermotropic Aqueous Solutions. *Polym. Int.* **1996,** *39* (2), 113–119.

127. Rebizant, V.; et al. Chemistry and Mechanical Properties of Epoxy-Based Thermosets Reinforced by Reactive and Nonreactive SBMX Block Copolymers. *Macromolecules* **2004,** *37* (21), 8017–8027.
128. Robeson, L. M. *Polymer Blends*. Hanser: Munich, 2007; pp 24–149.
129. Flory, P. J. Thermodynamics of High Polymer Solutions. *J. Chem. Phys.* **1941,** *9* (8), 660–660.
130. Flory, P. J. Thermodynamics of High Polymer Solutions. *J. Chem. Phys.* **1942,** *10* (1), 51–61.
131. Huggins, M. L. Solutions of Long Chain Compounds. *J. Chem. Phys.* **1941,** *9* (5), 440–440.
132. Huggins, M. L. Some Properties of Solutions of Long-chain Compounds. *J. Phys. Chem.* **1942,** *46* (1), 151–158.
133. George, S. M.; Puglia, D.; Kenny, J. M.; Parameswaranpillai, J.; Pionteck, J.; Thomas, S. Volume Shrinkage and Rheological Studies of Epoxidised and Unepoxidised Poly(Styrene-Block-Butadiene-Block-Styrene) Triblock Copolymer Modified Epoxy Resin–Diamino Diphenyl Methane Nanostructured Blend Systems. *Phys. Chem. Chem. Phys.* **2015,** *17* (19), 12760–12770.
134. Lü, H.; Zheng, S. Miscibility and Phase Behavior in Thermosetting Blends of Polybenzoxazine and Poly(Ethylene Oxide). *Polymer* **2003,** *44* (16), 4689–4698.
135. Serrano, E.; et al. Nanostructured Thermosetting Systems by Modification with Epoxidized Styrene–Butadiene Star Block Copolymers. Effect of Epoxidation Degree. *Macromolecules* **2006,** *39* (6), 2254–2261.
136. Riess, G. Micellization of Block Copolymers. *Progr. Polym. Sci.* **2003,** *28* (7), 1107–1170.
137. G. Riess, Hurtrez, G.; Bahadur, P. *Encyclopedia of Polymer Science and Engineering*, second ed.; Wiley: New York, 1985; vol. 2, 324–434.
138. Alexandridis, P.; Lindman, B. *Amphiphilic Block Copolymers: Self-Assembly and Applications*. Elsevier: Amsterdam, 2000.
139. Rosselli, S.; et al. Coil–ring–Coil Block Copolymers as Building Blocks for Supramolecular Hollow Cylindrical Brushes. *Angew. Chem. Int. Ed.* **2001,** *40* (17), 3137–3141.
140. Gan, Y.; Dong, D.; and Hogen-Esch, T. E. Synthesis and Characterization of a Catenated Polystyrene–Poly(2-Vinylpyridine) Block Copolymer. *Macromolecules* **2002,** *35* (18), 6799–6803.
141. Pispas, S.; et al. Effect of Architecture on the Micellization Properties of Block Copolymers: A2B Miktoarm Stars vs AB Diblocks. *Macromolecules* **2000,** *33* (5), 1741–1746.
142. Reutenauer, S., Hurtrez, G.; Dumas, P. A New Route to Model (A2B) and Regular Graft Copolymers. *Macromolecules* **2001,** *34* (4), 755–760.
143. Bae, Y. C.; Faust, R. Living Coupling Reaction in Living Cationic Polymerization. 2. Synthesis and Characterization of Amphiphilic A2B2 Star-Block Copolymer: Poly[Bis(Isobutylene)-Star-Bis(Methyl Vinyl Ether)]. *Macromolecules* **1998,** *31* (8), 2480–2487.
144. Quirk, R. P.; Yoo, T.; Lee, B. Anionic Synthesis of Heteroarm, Star-Branched Polymers. Scope and Limitations. *J. Macromol. Sci.—Pure Appl. Chem.* **1994,** *31* (8), 911–926.
145. Iatrou, H.; Avgeropoulos, A.; Hadjichristidis, N. Synthesis of Model Super H-Shaped Block Copolymers. *Macromolecules* **1994,** *27* (21), 6232–6233.

146. Graf, M.; Müller, A. H. Copolymers from End-Functionalized Polyethers and 1,3-Butadiene. In *IUPAC Int. Symp Ionic Polym.*; *Cret.*; *Prepr.*; 2001.
147. Pascual, S.; Narrainen, A. P.; Haddleton, D. M. Synthesis and Micellization of Diblock, Triblock, and Star Block Methacrylate Copolymers. *Abstracts of Papers of the American Chemical Society*; Amer. Chemical Soc.: Washington, DC, 2001; Vol 221.
148. Yoo, M. K.; et al. Conformation of Amphiphilic 6-Arm Star-Block Copolymers in Solvent Mixtures: Photophysical characterization. *Abstracts of Papers of the American Chemical Society*; American Chemical Society: Washington, DC, 2000; Vol 219.
149. Quirk, Roderic, P.; et al. Applications of 1,1-Diphenylethylene Chemistry in Anionic Synthesis of Polymers with Controlled Structures. *Biopolymers PVA Hydrogels, Anionic Polymerisation Nanocomposites*. Springer: Berlin Heidelberg, 2000; pp 67–162.
150. Leduc, M. R.; et al. Dendritic Initiators for "Living" Radical Polymerizations: A Versatile Approach to the Synthesis of Dendritic-Linear Block Copolymers. *J. Am. Chem. Soc.* **1996**, *118* (45), 11111–11118.
151. Aoi, Keigo, et al. Novel Amphiphilic Linear Polymer/Dendrimer Block Copolymer: Synthesis of Poly(2-Methyl-2-Oxazoline)-Block-Poly(Amido Amine) Dendrimer. *Macromol. Rapid Commun.* **1997**, *18* (10), 945–952.
152. An, S. G.; Cho, C. G. Synthesis of Dendritic Amphiphilic Block Copolymers by ATRP. *Abstracts of Papers of the American Chemical Society*; Amer. Chemical Soc.: Washington, DC, 2000; Vol 220.
153. Gitsov, Ivan. Hybrid Dendritic Capsules: Properties and Binding Capabilities of Amphiphilic Copolymers with Linear Dendritic Architecture. In *ACS Symposium Series*; American Chemical Society: Washington, DC, 1999, 2000; Vol 765.
154. Kwon, Y.; et al. Arborescent Polyisobutylene–Polystyrene Block Copolymers—A New Class of Thermoplastic Elastomers. *Polymer* **2002**, *43* (1), 266–267.
155. Hawker, C. J.; et al. "Living" Free Radical Polymerization of Macromonomers: Preparation of Well Defined Graft Copolymers. *Macromol. Chem. Phys.* **1997**, *198* (1), 155–166.
156. Bütün, V.; et al. Synthesis and Aqueous Solution Properties of Novel Neutral/Acidic Block Copolymers. *Polymer* **2000**, *41* (9), 3173–3182.
157. Truelsen, J. H.; Kops, J.; Armes, S. P. Amphiphilic Block Copolymers Synthesized by Aqueous ATRP. In *IUPAC Int'l Symp Ionic Polymerization, Creta, Prepr.* 2001.
158. Guo, Q.; Ed. *Thermosets: Structure, Properties and Applications*. Elsevier: Amsterdam, 2012.
159. Hu, D.; Zheng, S. Reaction-Induced Microphase Separation in Epoxy Resin Containing Polystyrene-Block-Poly(Ethylene Oxide) Alternating Multiblock Copolymer. *Eur. Polym. J.* **2009**, *45* (12), 3326–3338.
160. Hu, D.; Zheng, S. Reaction-Induced Microphase Separation in Polybenzoxazine Thermosets Containing Poly(N-Vinyl Pyrrolidone)-Block-Polystyrene Diblock Copolymer. *Polymer* **2010**, *51* (26), 6346–6354.
161. Aizenberg, J.; Weaver, J. C.; Thanawala, M. S.; Sundar, V. C.; Morse, D. E.; Fratzl, P. Skeleton of *Euplectella* sp.: Structural Hierarchy from the Nanoscale to the Macroscale. *Science* **2005**, *309* (5732), 275–278.
162. Volcani, B. E. Cell Wall Formation in Diatoms: Morphogenesis and Biochemistry. In *Silicon and Siliceous Structures in Biological Systems*; Springer: New York, 1981; pp 157–200.
163. Kresge, C. T.; Leonowicz, M. E.; Roth, W. J.; Vartuli, J. C.; Beck, J. S. Ordered Mesoporous Molecular Sieves Synthesized by a Liquid-Crystal Template Mechanism. *Nature* **1992**, *359* (6397), 710–712.

164. Monnier, A.; Schüth, F.; Huo, Q.; Kumar, D.; Margolese, D.; Maxwell, R. S.; Janicke, M. Cooperative Formation of Inorganic–Organic Interfaces in the Synthesis of Silicate Mesostructures. *Science* **1993**, *261* (5126), 1299–1303.
165. Soler-Illia, G. J. D. A.; Sanchez, C.; Lebeau, B.; Patarin, J. Chemical Strategies to Design Textured Materials: From Microporous and Mesoporous Oxides to Nanonetworks and Hierarchical Structures. *Chem. Rev.* **2002**, *102* (11), 4093–4138.
166. Shenhar, R.; Norsten, T. B.; Rotello, V. M. Polymer-Mediated Nanoparticle Assembly: Structural Control and Applications. *Adv. Mater.* **2005**, *17* (6), 657–669.
167. Busch, O. M.; Hoffmann, C.; Johann, T. R.; Schmidt, H. W.; Strehlau, W.; Schüth, F. Application of a New Color Detection Based Method for the Fast Parallel Screening of DeNO × Catalysts. *J. Am. Chem. Soc.* **2002**, *124* (45), 13527–13532.
168. Kosonen, H.; Ruokolainen, J.; Torkkeli, M.; Serimaa, R.; Nyholm, P.; Ikkala, O. Micro- and Macrophase Separation in Phenolic Resol Resin/PEO–PPO–PEO Block Copolymer Blends: Effect of Hydrogen-Bonded PEO Length. *Macromol. Chem. Phys.* **2002**, *203* (2), 388–392.
169. Templin, M.; Franck, A.; Du Chesne, A.; Leist, H.; Zhang, Y.; Ulrich, R.; Wiesner, U. Organically Modified Aluminosilicate Mesostructures from Block Copolymer Phases. *Science* **1997**, *278* (5344), 1795–1798.
170. Zhao, D.; Feng, J.; Huo, Q.; Melosh, N.; Fredrickson, G. H.; Chmelka, B. F.; Stucky, G. D. Triblock Copolymer Syntheses of Mesoporous Silica with Periodic 50 to 300 Angstrom Pores. *Science* **1998**, *279* (5350), 548–552.
171. Hsiue, G.-H.; Yang, J.-M. Epoxidation of Styrene–Butadiene–Styrene Block Copolymer and Use for Gas Permeation. *J. Polym. Sci., A: Polym. Chem.* **1990**, *28* (13), 3761–3773.
172. Ocando, Connie, et al. Micro- and Macrophase Separation of Thermosetting Systems Modified with Epoxidized Styrene-Block-Butadiene-Block-Styrene Linear Triblock Copolymers and their Influence on Final Mechanical Properties. *Polym. Int.* **2008**, *57* (12), 1333–1342.
173. Pandit, R.; et al. Synthesis and Characterization of Nanostructured Blends of Epoxy Resin and Block Copolymers. *Nepal J. Sci. Technol.* **2013**, *13* (1), 81–88.
174. Ocando, C.; Tercjak, A.; Mondragon, I. Nanostructured Systems Based on SBS Epoxidized Triblock Copolymers and Well-Dispersed Alumina/Epoxy Matrix Composites. *Compos. Sci. Technol.* **2010**, *70* (7), 1106–1112.
175. Serrano, E.; et al. Synthesis and Characterization of Epoxidized Styrene-Butadiene Block Copolymers as Templates for Nanostructured Thermosets. *Macromol. Chem. Phys.* **2004**, *205* (7), 987–996.
176. Rozaini, M. Z. Temperature Dependent Phase Behavior of Pseudo-Ternary Thiourea X-100 Surfactant + 1-Hexanol/Oil/Water Systems. *Open J. Phys. Chem.* **2012**, *2* (03), 169.
177. Stroem, P.; Anderson, D. M. The Cubic Phase Region in the System Didodecyldimethyl ammonium Bromide–Water–Styrene. *Langmuir* **1992**, *8* (2), 691–709.
178. Zhu, X. X.; Banana, K.; Yen, R. Pore Size Control in Cross-Linked Polymer Resins by Reverse Micellar Imprinting. *Macromolecules* **1997**, *30* (10), 3031–3035.
179. Laversanne, R. Polymerization of Acrylamide in Lamellar, Hexagonal, and Cubic Lyotropic Phases. *Macromolecules* **1992**, *25* (1), 489–491.
180. Elliniadis, S.; Higgins, J. S.; Choudhery, R. A.; Jenkins, S. D. Phase Separation in Thermoplastic–Thermoset Polymer Blends. *Macromol. Symp.* **1996**, *112* (1), 55–61.
181. Zhang, J.; Zhang, H.; Yan, D.; Zhou, H.; Yang, Y. Reaction-Induced Phase Separation in Rubber-Modified Epoxy Resin. *Sci. China Ser.; B: Chem.* **1997**, *40* (1), 15–23.

182. Kim, B. S.; Chiba, T.; Inoue, T. Morphology Development via Reaction-Induced Phase Separation in Epoxy/Poly(Ether Sulfone) Blends: Morphology Control Using Poly(Ether Sulfone) with Functional End-Groups. *Polymer* **1995**, *36* (1), 43–47.
183. Boots, H. M. J.; Kloosterboer, J. G.; Serbutoviez, C.; Touwslager, F. J. Polymerization-Induced Phase Separation. 1. Conversion-Phase Diagrams. *Macromolecules* **1996**, *29* (24), 7683–7689.
184. Bogaerts, K.; et al. Curing Kinetics and Morphology of a Nanovesicular Epoxy/Stearyl-Block-Poly(Ethylene Oxide) Surfactant System. *Soft Matter* **2015**, *11* (31), 6212–6222.
185. Guo, Q.; et al. Nanostructures, Semicrytalline Morphology, and Nanoscale Confinement Effect on the Crystallization Kinetics in Self-Organized Block Copolymer/Thermoset Blends. *Macromolecules* **2003**, *36* (10), 3635–3645.
186. Guo, Q.; Wang, K.; Chen, L.; Zheng, S.; Halley, P. J. *J. Polym. Sci., B: Polym. Phys.* **2006**, *44* (6), 975–985.
187. Mijovic, J.; Shen, M.; Sy, J. W.; Mondragon, I. *Macromolecules* **2000**, *33* (14), 5235–5244.
188. Wu, J.; Thio, Y. S.; Bates, F. S. Structure and Properties of PBO–PEO Diblock Copolymer Modified Epoxy. *J. Polym. Sci., B: Polym. Phys.* **2005**, *43* (15), 1950–1965.
189. Lipic, P. M.; Bates, F. S.; Hillmyer, M. A. *J. Am. Chem. Soc.* **1998**, *120* (35), 8963–8970.
190. Dean, J. M.; Lipic, P. M.; Grubbs, R. B.; Cook, R. F.; Bates, F. S. *J. Polym. Sci., B: Polym. Phys.* **2001**, *39* (23), 2996–3010.
191. Guo, Q.; et al. Phase Behavior, Crystallization, and Hierarchical Nanostructures in Self-Organized Thermoset Blends of Epoxy Resin and Amphiphilic Poly(Ethylene Oxide)-Block-Poly(Propylene Oxide)-Block-Poly(Ethylene Oxide) Triblock Copolymers." *Macromolecules* **2002**, *35* (8), 3133–3144.
192. Larrañaga, M.; et al. Towards Microphase Separation in Epoxy Systems Containing PEO/PPO/PEO Block Copolymers by Controlling Cure Conditions and Molar Ratios Between Blocks. Part 2. Structural Characterization. *Colloid Polym. Sci.* **2006**, *284* (12), 1419–1430.
193. Larranaga, M.; Gabilondo, N.; Kortaberria, G.; Serrano, E.; Remiro, P.; Riccardi, C. C.; et al. *Polymer* **2005**, *46* (18), 7082–7093.
194. Larranaga, M.; et al. Cure Kinetics of Epoxy Systems Modified with Block Copolymers. *Polym. Int.* **2004**, *53* (10), 1495–1502.
195. Sun, P.; et al. Mobility, Miscibility, and Microdomain Structure in Nanostructured Thermoset Blends of Epoxy Resin and Amphiphilic Poly(Ethylene Oxide)-block-Poly(Propylene Oxide)-Block-Poly(Ethylene Oxide) Triblock Copolymers Characterized by Solid-State NMR. *Macromolecules* **2005**, *38* (13), 5654–5667.
196. Meng, F.; et al. Formation of Ordered Nanostructures in Epoxy Thermosets: A Mechanism of Reaction-Induced Microphase Separation. *Macromolecules* **2006**, *39* (15), 5072–5080.
197. Argon, A. S.; Cohen, R. E.; Mower, T. M. Mechanisms of Toughening Brittle Polymers. *Mater. Sci. Eng.: A* **1994**, *176* (1–2), 79–90.
198. Micelles, C. Liu, J. Toughening of Epoxies Based on Self-Assembly of Nano-Sized Amphiphilic Block. *Doctoral Dissertation*, 2010.
199. Pearson, R. A.; Yee, A. F. Toughening Mechanisms in Elastomer-Modified Epoxies. *J. Mater. Sci.* **1989**, *24* (7), 2571–2580.
200. Yee, A. F.; Pearson, R. A. Toughening Mechanisms in Elastomer-Modified Epoxies. *J. Mater. Sci.* **1986**, *21* (7), 2462–2474.

201. Jones, R. A.; Richards, R. W. *Polymers at Surfaces and Interfaces*; Cambridge University Press: Cambridge, 1999.
202. Angell, C. A.; Ngai, K. L.; McKenna, G. B.; McMillan, P. F.; Martin, S. W. Relaxation in Glassforming Liquids and Amorphous Solids. *J. Appl. Phys.* **2000**, *88* (6), 3113–3157.
203. Torre, R.; Bartolini, P.; Righini, R. Structural Relaxation in Supercooled Water by Time-Resolved Spectroscopy. *Nature* **2004**, *428* (6980), 296–299.
204. Dinsmore, A. D.; Weeks, E. R.; Prasad, V.; Levitt, A. C.; Weitz, D. A. Three-Dimensional Confocal Microscopy of Colloids. *Appl. Opt.* **2001**, *40* (24), 4152–4159.
205. Debenedetti, P. G.; Stillinger, F. H. Supercooled Liquids and the Glass Transition. *Nature* **2001**, *410* (6825), 259–267.
206. Kennedy, J. E.; Higginbotham, C. L. *Synthesis and Characterisation of Styrene Butadiene Styrene Based Grafted Copolymers for Use in Potential Biomedical Applications*; INTECH Open Access Publisher, 2011.
207. Pedroni, L. G.; et al. Nanocomposites based on MWCNT and Styrene–Butadiene–Styrene Block Copolymers: Effect of the Preparation Method on Dispersion and Polymer–Filler Interactions. *Compos. Sci. Technol.* **2012**, *72* (13), 1487–1492.
208. Sedransk, Kyra, L.; et al. The Metathetic Degradation of Polyisoprene and Polybutadiene in Block Copolymers Using Grubbs Second Generation Catalyst. *Polym. Degrad. Stab.* **2011**, *96* (6), 1074–1080.
209. Zhao, Y.; et al. Largely Improved Mechanical Properties of a Poly(Styrene-*b*-Isoprene-*b*-Styrene) Thermoplastic Elastomer Prepared under Dynamic-Packing Injection Molding. *Ind. Eng. Chem. Res.* **2014**, *53* (39), 15287–15295.
210. Choi, S.; Han, C. D. Molecular Weight Dependence of Zero-Shear Viscosity of Block Copolymers in the Disordered State. *Macromolecules* **2004**, *37* (1), 215–225.
211. Pryamitsyn, V.; et al. Curvature Modification of Block Copolymer Microdomains Using Blends of Block Copolymers with Hydrogen Bonding Interactions. *Macromolecules* **2012**, *45* (21), 8729–8742.
212. Garate, H.; et al. Exploring Microphase Separation Behavior of Epoxidized Poly(Styrene-*b*-Isoprene-*b*-Styrene) Block Copolymer inside Thin Epoxy Coatings. *Macromolecules* **2013**, *46* (6), 2182–2187.
213. Paradise, M.; Goswami, T. Carbon Nanotubes—Production and Industrial Applications. *Mater. Des.* **2007**, *28* (5), 1477–1489.
214. Martone, A.; et al. Reinforcement Efficiency of Multi-Walled Carbon Nanotube/Epoxy Nano Composites. *Compos. Sci. Technol.* **2010**, *70* (7), 1154–1160.
215. Zilli, D.; et al. Comparative Analysis of Electric, Magnetic, and Mechanical Properties of Epoxy Matrix Composites with Different Contents of Multiple Walled Carbon Nanotubes. *Polym. Compos.* **2007**, *28* (5), 612–617.
216. Park, I.; et al. Selective Sequestering of Multi-Walled Carbon Nanotubes in Self-Assembled Block Copolymer. *Sens. Actuat.; B: Chem.* **2007**, *126* (1), 301–305.
217. Müller, A. J.; et al. Super-Nucleation in Nanocomposites and Confinement Effects on the Crystallizable Components within Block Copolymers, Miktoarm Star Copolymers and Nanocomposites. *Eur. Polym. J.* **2011**, *47* (4), 614–629.
218. Garate, H.; et al. Surfactant-Aided Dispersion of Polystyrene-Functionalized Carbon Nanotubes in a Nanostructured Poly(Styrene-*b*-Isoprene-*b*-Styrene) Block Copolymer. *Polymer* **2011**, *52* (10), 2214–2220.

CHAPTER 8

XANTHENE DYE-DOPED PVA-BASED THIN-FILM OPTICAL FILTER CHARACTERISTICS AND ITS GREEN LASER BEAM BLOCKING

R. RENJINI[*], GEORGE MITTY, V. P. N. NAMPOORI, and S. MATHEW

International School of Photonics, Cochin University of Science and Technology, Kochi, India

[*]Corresponding author. E-mail: renjiniretnamayi@gmail.com

ABSTRACT

Polyvinyl alcohol (PVA)-based thin films have wide range of technological applications like lithography, sensors, organic light-emitting diodes, coatings, and electrochemical cells. PVA-based thin films exhibit good optical properties and mechanical strength and are light in weight and relatively low cost in manufacture. The optical properties of thin films can be changed as per the specific requirement with the addition of suitable dopant materials and organic dyes. In this work, the green laser beam-blocking property and ultraviolet (UV)-blocking property of PVA-based erythrosine films were studied. The transmission spectrum shows a 0% transmission for a band of 50 nm (500–550 nm); hence, it can block green laser, and in the UV range also, the film exhibits 0% transmittance, so it can act as a UV protective film. Because of the narrow band rejection in its transmission spectrum, it will not degrade the color recognition. The pass band centered at 400 nm can be utilized for band pass applications of optical filters.

8.1 INTRODUCTION

Applications such as fluorescence microscopy and Raman spectroscopy need an optical filter to selectively transmit desired wavelengths of light while blocking unwanted light. In these applications, there are two distinct types of beams: the illumination (or excitation) beam and the signal (or emission) beam. The signal beam is weaker than the illumination beam. So, filter should have a transition from deep blocking to high transmission over a very short wavelength range. Thus, the filter shows steep and deep spectral edges.[1] Strong absorption and transmission properties at certain bands of the PVA-based films which are doped with a xanthene dye known as erythrosine make them suitable for optical band pass or band reject filter. Optical filters can transmit or absorb specific wavelengths and split an image into two identical images with controlled brightness levels relative to each other. To design an optical filter for a suitable application, certain characteristic features of optical filters have to be considered.[1,2] Following are some optical filter parameters.

- **Central wavelength:** Center wavelength is the midpoint between the wavelengths where transmittance is 50% of the specified minimum transmission.
- **Bandwidth:** Bandwidth is a range of wavelength used to denote a specific band of the spectrum that passes incident energy through a filter. It is also referred to as full-width half maximum.
- **Blocking range:** Blocking range is a range of wavelength used to denote the band of energy that is attenuated by the filter. The degree of its blocking can be specified in terms of its optical density (OD).
- **Optical density:** OD describes how much power the film can be blocked and it indicates the amount of energy transmitted through it. A high OD value indicates very low transmission, and low OD indicates high transmission as shown in Table 8.1.

Metal nanoparticles, such as silver and gold, exhibit interesting optical and electronic phenomena due to the excitation of surface plasmon resonances. When nanoparticles are added in closely packed polymer thin films, the interaction of the particle's local electric fields causes significant changes in their optical properties. These types of doped thin films have several applications in optical and biomedical fields.[6,7] PVA with molecular weight of 10,000 g/mol is used as the polymeric material. Polyvinyl alcohol is a water soluble synthetic polymer and has the idealized formula $[CH_2CH(OH)]_n$. Polyvinyl alcohol has excellent film forming, emulsifying and adhesive properties.[4] It decomposes rapidly above 200°C. Since high intensity laser beams are involved in nonlinear optical studies, the laser damage threshold (LDT) of the film is an important parameter. In the present study, an actively Q-switched diode array side pumped Nd:Yag laser is used for finding the LDT of the PVA thin film.

TABLE 8.1 Optical Density of Ag-Doped PVA-Based Erythrosine Film for Various Colors.

Transmittance (%)	Colors	Optical density
0	Green	$\to \infty$
50	Violet	0.301
50	Indigo	0.301
50	Blue	0.301
90	Yellow	0.045
90	Orange	0.045
90	Red	0.045

8.2 MATERIALS AND METHODS

Xanthene dye-doped PVA-free standing films of various thicknesses can be prepared using tape-casting technique. Erythrosine and PVA are water soluble, so it is easy to prepare PVA-based erythrosine films.[5] PVA solutions (5%) were prepared by dissolving 5 g of PVA in 95 mL of distilled water under constant stirring for 2 h at a temperature of 40°C. Erythrosine dye was then added to the prepared PVA solution in desired concentration. The mix prepared was cast into thin films on a glass plate using double doctor blade tape-casting technique. The prepared mix was poured through the gap between the glass plate and the blade. The thickness of the film can be varied by varying the gap between blade and glass bed and the casting speed. To study the effect of Ag nanoparticle on the optical properties of erythrosine, Ag nanoparticle doped erythrosine films can also be prepared by adding colloidal Ag nanoparticles. Then, the film is kept in the heater for 2 h under 80°C and allowed to dry for 24 h. Finally, the films were carefully separated from the glass plate. The absorption and transmission spectrum of prepared films were characterized using ultra violet (UV)/VIS/NIR spectrophotometer.

8.3 EXPERIMENTAL SETUP

The performance of optical filters is determined by their spectral characteristics, like transmission efficiency of the signal and attenuation (or blocking) of unwanted wavelengths. Generally, spectrophotometers are used to measure the transmission and OD spectral performance of optical filters. Even though these instruments can have significant limitations when the optical filters have high edge steepness and very deep blocking, spectrophotometers are best suited for plotting the transmission and absorption spectrum of thin-film optical filters.

Green laser beam blocking ability of dye-doped PVA films was analyzed using Nd:Yag laser pulses with varying input power. A photodetector placed after the PVA film was used to record the output power of the laser beam which is coming out from the film. The input power levels were recorded using photodetector without having the film in the path of laser beam as in Figure 8.1. The output power levels were recorded using photodetector for the corresponding input power levels by keeping the film in the path of laser beam. To find the laser damage of the film, Nd:Yag laser beam pulses with 7-nm width, 532-nm wavelength, and 6-mm diameter are allowed to fall on the film. The beam spot size on the film is also 6 mm and then varied the input energy level of the pulses. The transmittance of the film after every laser beam irradiation for various time durations is given in as shown in Table 8.3.

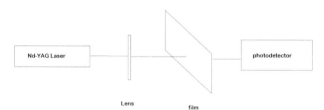

FIGURE 8.1 Experimental setup for determining laser-blocking capability of films.

8.4 RESULTS AND DISCUSSION

Figure 8.2 shows the linear absorption spectrums of PVA-based erythrosine films with various concentrations of erythrosine. From the figure, it is clear that PVA-based erythrosine film with erythrosine concentration 10^{-2} M has maximum absorption in the UV range and in the 500–550-nm range. Figure 8.3 illustrates the transmission spectrum of PVA-based erythrosine film (10^{-2} M). With the addition of colloidal silver, nanoparticles to these films can enhance the transmission efficiency of its pass band. The narrow rejection band (nearly 50 nm) does not degrade the color vision of the film in the visible band. So, it can be adhesively applied to glass or clear plastic to form UV/green laser protective goggles and blocking screens for UV rays and for green laser beam. Another attraction of this film is its ability to maintain color discrimination. Laser-blocking capability of the films was using 532-nm Nd:Yag laser as shown in Table 8.2. The film exhibited excellent laser beam (532 nm) blocking for different input power levels. Due to high-absorption around 500 nm, when excited using 532-nm laser beam, the nonradiative components are predominant than the radiative components in the emission spectrum, which results quenching of fluorescence as shown in Figure 8.4 (Table 8.3).

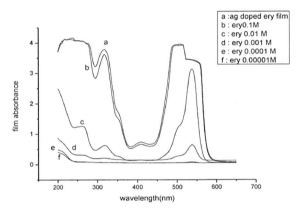

FIGURE 8.2 Absorption spectrum of erythrosine films.

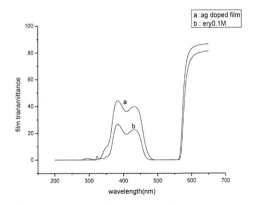

FIGURE 8.3 Transmittance spectrum of erythrosine films.

TABLE 8.2 Blocking Capability of Ag-doped PVA-Based Erythrosine Film for Various Laser Input Power Levels.

Input power (P_{in}) (mJ)	Output power (P_{out}) (mJ)	Percentage power blocked $\{(P_{in}) - (P_{out})\}/(P_{in})$ (%)
0.00185	0	100
0.00726	0	100
0.576	0	100
2.689	0	100
3.550	0	100
11	0	100
14.5	0	100

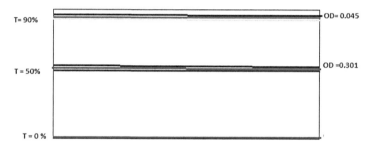

FIGURE 8.4 OD plot for Ag nanoparticle-doped PVA-based erythrosine film in the visible band.

TABLE 8.3 Transmittance of (500–550 nm range) Ag-Doped Erythrosine Film (0.01 M) after Laser Irradiation for Various Time Durations.

i/p Power (mJ)	Time duration (s)	Total number of pulses the film can withstand	Transmittance (%)
3.45	30	300	0
	60	600	0
	300	3000	2
	600	6000	5
7.5	30	300	0
	60	600	2
	300	3000	42
	600	6000	95

8.5 CONCLUSIONS

To summarize, we observed the UV protection, a narrow band pass, a narrow band reject as well as green laser beam-blocking property of xanthenes dye erythrosine-doped PVA films. From the transmission characteristics, it has been observed that the UV-blocking property and the transmittance in the pass band also increase with the addition of Ag nanoparticles for erythrosine films. Experimental results obtained from the response of films for various laser input power level blocking suggest that these films are best suited for practical applications in optical devices like green laser protective goggles. The prepared thin film can withstand 600 pulses with input energy level 3.45 mJ and 300 pulses with input energy level 7.5 mJ without any damage.

KEYWORDS

- **transmission spectrum**
- **rejection band**
- **beam blocking**
- **optical density**
- **PVA**

REFERENCES

1. www.semrock.com/measurement-of-optical-filter-spectra.aspx.
2. Jabbar, W. A.; Habubi, N. F.; Chiad, S. S. Optical Characterization of Silver Doped Poly(Vinyl Alcohol) Films. *J. Arkansas Acad. Sci.* **2010**, *64*, Article 21. Available at: http://scholarworks.uark.edu/jaas/vol64/iss1/21.
3. Kramadhati, S.; Thyagarajan, K. Optical Properties of Pure and Doped (KnO_3 and $MgCl_2$) Polyvinyl Alcohol Polymer Thin Films. *Int. J. Eng. Res. Develop.* **2013**, *6* (8), 15–18. ISSN: 2278-067X, p-ISSN: 2278-800X. www.ijerd.com.
4. Esfahani, Z. H.; Ghanipour, M. Effect of Dye Concentration on the Optical Properties of Red-BS Dye-Doped PVA Film. *J. Theor. Appl. Phys.* **2014**, *8*, 117–121.
5. Sasidharan, S.; Nityaja, B.; Swain D.; Nampoori, V. P. N.; Radhakrishnan, P.; Venugopal Rao, S. Nonlinear Optical Studies of DNA Doped Rhodamine 6G-PVA Films Using Picosecond Pulses. *Optics Photon. J.* **2012**, *2*, 135–139. http://dx.doi.org/10.4236/opj.2012.23019 (published online September 2012) (http://www.SciRP.org/journal/opj).

6. Nithyaja, B.; Misha, H.; Nampoori, V. P. N. Selective Mode Excitation in Dye-Doped DNA Polyvinyl Alcohol Thin Film. *J. Appl. Optics* **2009,** *48* (19), 3521.
7. Bardhan, R.; Grady, N. K.; Cole, J. R.; Joshi, A.; Halas, N. J. Fluorescence Enhancement by Au Nanostructures. *ACS Nano* **2009,** *3* (3), 744–752.

CHAPTER 9

FABRICATION AND FRACTURE TOUGHNESS PROPERTIES OF CASHEW NUT SHELL LIQUID RESIN-BASED GLASS FABRIC COMPOSITES

SAI NAGA SRI HARSHA CH.[1*] and K. PADMANABHAN[2*]

[1]*Srujana-Innovation Center, L. V. Prasad Eye Institute, Hyderabad, India*

[2]*Centre for Excellence in Nano-Composites, School of Mechanical and Building Sciences, VIT University, Vellore, India*

*Corresponding author. E-mail: sainagasriharsha@gmail.com; padmanabhan.k@vit.ac.in

ABSTRACT

Cashew nut shell liquid (CNSL) is found to be a promising natural biopolymer which could be polymerized to get hard plastics. The available literature suggests different ways for polymerization of CNSL but nowhere a proper process for composite fabrication is given. This chapter deals with the synthesis and fabrication of a CNSL-based glass fabric composite material. The composite is observed to have promising mechanical properties. Tensile strength, fracture toughness in mode I single edge notch test, and strain energy release rate of the composite are evaluated experimentally and reported. These values are compared with those of glass/epoxy and carbon/epoxy composites and the possibilities of increasing the fracture toughness properties of CNSL composites are discussed in this chapter.

9.1 INTRODUCTION

The literature on the chemical properties of cashew nut shell liquid (CNSL) gives the principal constituents of natural CNSL as anacardic acid (64.9%), monophenol (20.3%), and Cardol (11.3%).[1,2] The phenol-based monomers present in CNSL when added with formaldehyde cause the polymerization of the resin.[3] The biopolymers synthesized from CNSL are essentially thermosetting polymers which need a combination of addition, condensation, and thermal polymerization. Glass fibers are most extensively used as reinforcements for both thermosetting and thermoplastic matrix composites. This chapter focuses on the fabrication of a glass fabric composite material with CNSL as the matrix and evaluates the mechanical properties like tensile strength and fracture toughness in mode I.

9.2 EXPERIMENTAL DETAILS

The resin mixture consisted of CNSL (100 parts by weight) and 5% sulfuric acid solution (5 parts by weight) is heat treated. The CNSL is mixed thoroughly with toluene to prevent undesirable self-polymerization of the CNSL fluid up to 110°C and keep the viscosity low. The mixture is then placed in a vacuum chamber at about 50 mmHg to remove trapped gas bubbles. Composite material sheets are fabricated through a general hand lay-up process. Initially, CNSL and formaldehyde are applied on a 260-g/m^2 glass fabric sheet of 37 cm × 37 cm. This way, eight layers are laid up to make the composite laminate. The laid composite sheets contain 20% glass fibers by volume (i.e., $V_f = 0.2$). To attain uniform properties, a load of 22 kg is applied by placing two thick-mild steel sheets. For curing, the lay-up is heated in a hot air oven up to a temperature of 120°C for 2 h. Then, the composite is left idle for 2 days with the loads, for ageing. Tensile and single edge notch (SEN) test specimens were prepared from the laminate as per requirements for the tensile test and the mode I fracture toughness test, respectively.[4,5] Tests carried out on an electronic tensometer give the data of load applied and the displacement in the specimen when tested in the opening mode (Fig. 9.1). Tensile specimens are loaded until the failure of the specimen, whereas the SEN specimens are loaded until the precrack starts propagating consistently.

Fabrication and Fracture Toughness Properties

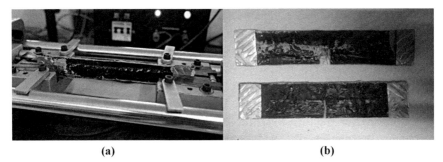

FIGURE 9.1 (a) Mounting of tensile specimen and (b) failure of SEN specimens.

9.3 RESULTS AND DISCUSSION

Tensile properties of the composite are calculated based on the failure load of the specimens as shown in Figure 9.2 and Table 9.1.

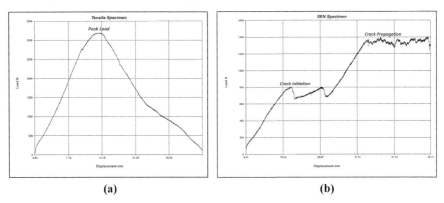

FIGURE 9.2 Load, displacement plots of (a) a tensile specimen and (b) an SEN specimen.

TABLE 9.1 Tensile Test Results.

Specimen no.	Load (N)	Cross sec. area (mm^2)	Stress (MPa)	Strain
1	2004.4	77.5	25.863	0.171
2	3559.9	77.5	45.935	0.184
3	3187.3	77.5	41.126	0.177
4	3805.1	77.5	49.098	0.162
5	3285.3	77.5	42.391	0.195

The tensile strength of the composite is observed to be lower than that of a glass fabric/epoxy composite laminate with the same volume fraction of resin. The specimens show moderate interfacial strength between the resin and the fiber. To increase the interfacial strength, different ways of alkali and acid curing could be attempted. The toughness properties are good as presented in Tables 9.2 and 9.3 but lower than that of glass/epoxy or metal modified glass/epoxy composites whose fracture toughness and the strain energy release rate values lie in the region of 20–40 MPa $\sqrt{m}$ and 2–8 kJ/m^2, respectively.[6,7] A higher volume fraction of glass fabric is also expected to ensure higher fracture toughness values. Efforts are on to improve the mechanical properties of CNSL composites through chemistry and fabrication techniques.

TABLE 9.2 Crack Initiation Properties.

Spec. no.	Load (N)	Stress (MPa)	Fracture toughness (MPa $\sqrt{m}$)	Strain energy release rate (J/m^2)
1	872.8	11.262	8.922	692.57
2	794.4	10.25	8.12	573.69
3	867.5	11.194	8.869	684.23
4	810.3	10.455	8.282	596.87

TABLE 9.3 Crack Propagation Properties.

Spec. no.	Load (N)	Stress (MPa)	Fracture toughness (MPa $\sqrt{m}$)	Strain energy release rate (J/m^2)
1	1371.3	17.6942	14.0173	1709.61
2	1316.4	16.9858	13.4562	1575.45
3	1157.2	14.9316	11.8288	1217.43
4	1412.2	18.2477	14.4558	1818.23

9.4 CONCLUSIONS

The tensile properties, fracture toughness in mode I, and its strain energy release rate were evaluated for a CNSL/glass fabric composite and reported.

ACKNOWLEDGMENT

Dr. G. Bhanu Kiran, assistant professor of GITAM University is gratefully acknowledged for his suggestions.

KEYWORDS

- **CNSL resin**
- **glass fabric**
- **natural resin composites**
- **tensile strength**
- **single edge notch test**
- **mode I fracture**
- **fracture toughness**
- **strain energy release rate**

REFERENCES

1. Gedam, P. H.; Sampathkumaran, P. S. *Progr. Organ. Coat.* **1986,** *14,* 115–157.
2. Menon, A. R. R.; Sudha, J. D.; Pillai, C. K. S.; Mathew, A. G. *J. Sci. Ind. Res.* **1985,** *44,* 324.
3. Papadopoulou, E.; Chrysalis, K. *Thermochim. Acta* **2011,** *512,* 105–109.
4. ASTM. *ASTM 3039, Tensile Properties of Reinforced Plastics,* ASTM: West Conshohocken, PA, 2010.
5. Prakash, V.; Aditya, R.; Kurup, A. L.; Padmanabhan, K. *Int. J. Contemp. Eng. Sci. Technol.* **2010,** *1* (1), 55.
6. Kumar, V.; Balachandran, V.; Sushen, V.; Padmanabhan, K. *Int. J. Contemp. Eng. Sci. Technol.* **2010,** *1* (1), 77.
7. Kumaran, D.; Narayanan, R.; Reddy, A.; Padmanabhan, K. *Int. J. Manuf. Sci. Eng.* **2011,** *2* (2), 97.

CHAPTER 10

PHYTOSYNTHESIS OF Cu AND Fe NANOPARTICLES USING AQUEOUS PLANT EXTRACTS

SUBRAMANIAN L.[1], OBEY KOSHY[2*], and SABU THOMAS[2]

[1]Department of Chemistry, Amrita School of Arts and Sciences, Amrita Vishwa Vidyapeetham, Amrita University, Clappana PO Kollam 690525, Kerala, India

[2]International and Inter University Centre for Nanoscience and Nanotechnology, Mahatma Gandhi University, P.D. Hills, Kottayam 686560, Kerala, India

*Corresponding author. E-mail: Obey.Koshy@gmail.com

ABSTRACT

The synthesis of metal nanoparticles has taken a new route through the emergence of green chemistry. We propose the synthesis of two significant and important metal nanoparticles through simple biosynthetic methods. Metal nanoparticles prepared are characterized by TEM, UV–vis, and FTIR spectroscopy. TEM image shows the iron nanoparticles are nearly spherical in shape of size about 35 nm. UV–vis spectra show distinct peaks of copper at around 420 nm corresponding to the surface plasmon peak of copper nanoparticles. The results obtained from UV–vis, FTIR, and TEM support the fact that the synthesized particles are of nanorange.

10.1 INTRODUCTION

Metal nanoparticles have received a lot of attention in the recent decades due to their unique ability to alter optical, electrical, and biological properties. Nanoparticles are also interesting because of their huge increase in surface area and a surge in conductivity compared to their metal counterparts. Moreover, metal nanoparticles also show surface plasmon resonance. These properties make them central to many applications such as bio-sensing, imaging, drug delivery, HIV treatment, cancer treatment, optical spectroscopy, and Raman scattering. The study of nanoparticles dates back to ancient times. The discovery of Lycurgus cup in 1958 which is a dichroic glass led to the understanding that nanoparticles such as colloidal gold and silver were made thousands of years ago. The Lycurgus cup present in British museum displays different colors depending on direction from which light is allowed to fall on it. Stabilized gold and silver nanoparticles are present in medieval Indian glass paintings also, where they are used as colloids. The study on nanoparticles in modern science starts in 1857 when Michael Faraday gave a description about gold and silver metal particles showing enormous change upon heating above 500°C. However, Richard Feymann's pathbreaking talk "There Is Plenty of Room at the Bottom" in 1959 inspired scientists around the world to think about nanoscience. The discovery of Buckminsterfullerene in 1985 and discoveries of scanning and tunneling microscopies in the starting 1980s led to the further emergence and advancement of nanoscience and nanotechnology. Thus, metallic nanoparticles where synthesized during 1985. The first reports on synthesis of different metallic nanoparticles came in 1984. The term "nano" was first used by the Japanese in Tokyo University of Science. Thus, from 1985, the scientific works, discoveries, and technologies in nanoscience and nanotechnology started to grow and flourish. Scientific research papers were published during the last phase of 20th century. Scientists also thought of blending nanoscience and organic polymers which led to a new field of science known as "Advanced Materials," "Material Science," and "Nanocomposites."

There are basically two methods to make a material on nanoscale. They are top-down and bottom-up approaches. The top-down approaches deal with reducing the material to nanoscale from any region above it. Here, materials which are of macro sizes (milli range and above) and micro sizes are reduced to nanoscale. The second approach is known as bottom-up approach. Here, materials which are below nanoscale (angstrom) are increased to nanoscale. There are numerous methods to synthesize nanoparticles. These include physical methods such as exploding wire technique,[1] plasma,[2] chemical vapor deposition,[3] microwave irradiation,[4] pulsed laser ablation,[5] supercritical fluids,[6] sonochemical reduction,[7] gamma radiation,[7] and vapor liquid solid growth method.[8] The chemical methods used to prepare nanoparticles include chemical reduction of metal salts,[1] microemulsions,[2] thermal decomposition of metal salts,[3] and electrochemical synthesis.[4] In 1951, scientist

Turkvic presented a new chemical method for the synthesis of gold and silver nanoparticles which later became a standard protocol for synthesizing gold nanoparticles. Later, in 1990, another scientist named Brust presented a novel chemical method for synthesizing gold and silver nanoparticles which too went on to become a standard protocol for gold and silver nanoparticle synthesis. These methods are chemical reduction methods from chemical salts and are still considered as the standard chemical methods for metal nanoparticle synthesis resulting good yield. Thus, these methods are used by scientists for synthesizing metal nanoparticles controlling their size and shape. Synthesizing nanoparticles also include controlling and manipulating metal particles in nanoscale. These particles have great applications in electronics, sensing, and drug delivery. The field of electronics has a great obligation to nanoscience and nanotechnology. Moore's law is an apt example for this. Engineers and technologists use nanoparticles to incorporate in their transistors, capacitors, diodes, integrated circuits, etc., to improve their efficiency. Nanoparticles are also used in LCD displays and LEDs. The influence of nanotechnology is very visible in the field of medicine as well. Nanomedicine is an emerging area of medicine which focuses on curing affected cells, terminating microbes, destroying disease affected cells at the nanolevel. This is a great leap in the history of modern medicine and drugs. So, drug delivery, nuclear oncology, and stem cell research have also got something to do with nanoscience. Thus, nanotechnology and nanomaterials have shaped the world and the current era is known as "Nano Era" as the field has made inroads into all walks of human life.

Green approach to nanoscience dates back to the last years of 20th century. Green approach to nanoscience and nanotechnology came into the forefront with the entry of green chemistry. The new stream of study came into existence in the last years of 1980s. Green chemistry is not a specific area of research or study but is a tag name given to all novel, modern methods of understanding, synthesizing, characterizing techniques using ecological friendly, pollution-free methods. Green chemistry involves all those chemical reactions which use toxic-free, pollution-free methods. It simply comprises all biological methods which are free of highly carcinogenic hydrocarbons. This concept of green chemistry came also due to the impact of environmental crisis faced by the modern world. The 12 pillars of green chemistry involve preventing waste than to treat[1] or clean up waste after its formation,[2] designing synthetic methods for maximum incorporation of final product,[3] synthesizing synthetic materials which poses zero toxicity to human health and the environment,[4] designing chemical products which preserve efficacy of function, also reducing toxicity,[5] use of renewable material feedstocks.[6]

The use of renewable and inexhaustible materials for all chemical synthesis processes.[7] The use of environment friendly, sustainable materials for all chemical synthesis processes,[8] reducing derivatives while synthesizing new chemicals especially organic hydrocarbons.[9] Chemical products should be designed in such a way that at the end of their function, they do not persist in the environment and break down into innocuous degradation products.[10] Development of modern analytical methodologies to allow real time processing and controlling prior to the formation of hazardous substances.[11] Chemical substances and the form of a substance used in chemical process should be chosen such a way that it minimizes potential for chemical accidents including releases, explosion, and fires.[12] Finally, the use of safe, environment benign substances including solvents whenever possible. Thus, green chemistry gives pathway for all modern green technologies in all fields of science. This new thinking of sustainable science is also highly applicable to other fields of science including chemical engineering, modern chemical technologies, process chemistry, process engineering, and food technology.

The application of green chemistry into the area of nanoscience and nanotechnology involves inventing new technologies, synthesizing nanoparticles and materials using novel methods, etc. This largely includes synthesizing different nanoparticles using plant extracts, leaf broths, and inventing greener technologies for making products. Green synthesis of nanoparticles includes using aqueous, green plant extracts for synthesizing metal nanoparticles such as gold, silver, zinc, iron, copper, etc. Metal nanoparticles are of major importance and primary significance as they are having great applications. Synthesizing metal nanoparticles using green methods is of great importance. The major reasons for choosing greener synthesis of metal nanoparticles include that all the chemical methods used for synthesizing nanoparticles are found to be toxic in nature. Using toxic chemicals such as $NaBH_4$ and TOAB are highly dangerous and morbid. These organic hydrocarbons are also highly carcinogenic in nature. Green synthesis of nanoparticles is of great interest also due to its method of synthesis. Using greener methods makes it sustainable, renewable, and also inexhaustible. Other interesting aspect of greener method is that it is cost effective and cheap. It is highly economical and cheaper compared to chemical and physical methods. Physical methods require highly sophisticated instruments and are of high cost. Chemical methods of nanoparticle preparation lead to usage of highly cancerous and toxic materials. Thus, greener method is a better solution for nanoparticle synthesis and usage.

Copper nanoparticles are of great significance in photovoltaics, catalysis, energy applications, etc. They have also great application in metallic conductors[15] as copper itself is a conducting metal. Although the synthesis of AgNPs and AuNPs has been widely reported, the synthesis of FeNPs hasn't been reported. This is due to the fact that FeNPs are highly unstable and Fe readily forms oxides. FeNPs show magnetic behavior and properties. However, these particles are very reactive and consequently form aggregates losing their fundamental properties. This leads to the phenomenon called agglomeration. To deal with this challenge, passivating agents such as polyvinylpyrrolidone, polyvinyl alcohol, hyperbranched polyurethane, and polyacronitrile with reducing agents such as sodium borohydride, hydrazine, glycerol, etc., have been used successfully in the synthesis and stabilization of nanoparticles. The use of these passivating and reducing agents introduced the challenge of high cost and toxicity, thus threatening environmental sustainability and limiting the biological application of these noble materials. This crisis can be solved by using eco-friendly, environmental benign methods for the synthesis of nanoparticles. This eventually leads to the entry of green methods for synthesizing nanoparticles. Thus, metal nanoparticles are synthesized using varying types of plant extracts, green algae, leaf broths, etc. The conventional methods for the synthesis of copper nanoparticles include use of surfactant template and carbon nanotube scaffolding[16] with citrate with citrate reduction and the $NaBH_4$ reduction of copper in the presence of surfactant. Other major method for the synthesis of copper nanoparticle is the use of copper nitrate as precursor. Copper nitrate is used as precursor and NaOH is added to the precursor. Here, hydrazine is used as the capping agent. The color of the solution first changes from blue to white indicating reduction. It then changes to pink in color. The pink color appears showing the formation of copper oxide nanoparticles. Copper being highly unstable readily forms copper oxide nanoparticles. The color then easily changes to reddish brown showing the formation of copper nanoparticles. Reports indicate using electrochemical means such as electroplating, electroless plating, vacuum plating, etc., for synthesizing copper nanowires. Synthesis of ultrasmall copper nanoparticles using lemon grass tea is also reported.[17] We report the synthesis and studies of synthesizing copper oxide and copper nanoparticles using aqueous green leaf extract of pineapple. Pine apple leaves are abundant in nature and are great sources of starch and cellulose. Thus, we expect and introduce the reduction of copper from copper sulfate solution to ultrasmall copper nanoparticles and copper oxide nanoparticles. The major challenge faced by green chemists for the green synthesis of copper nanoparticles is its unstable nature. Copper being highly unstable in nature readily forms copper oxide. Thus, copper oxide nanoparticles are formed rather faster and easier compared to copper nanoparticles.

The traditional, conventional methods used for synthesis of iron nanoparticles include the use of $NaBH_4$ as capping agent. Iron nanoparticles have a high propensity to aggregate into agglomerates to lower the energy associated with surface area to volume ratio. In spite of these obstacles, it has been observed that amorphous nanoparticles of zerovalent iron[18] can be synthesized using extracts of some kind of tea. These include the usage of green tea,[19] Oolang tea,[20] and black tea. Iron nanoparticles were found forming instantaneously in aqueous tea extracts and the concentration of tea extracts determined the structure, shape, and morphology of FeNPs. The reduced Fe consists of hexagonal metallic iron, amorphous iron, Fe_3O_4. In addition, iron oxide nanoparticles of size 40 and 50 nm where also found to be synthesized. The use of tea for synthesis of iron nanoparticles is found to be less toxic and environment benign method compared to the conventional methods. The use of tea has great interest as they have a large amount of flavonoids in it. The presence of flavonoids with the presence of a potent array of antioxidants such as polyphenols, reducing sugars, nitrogenous bases, and amino acids makes the tea extracts a large area of interest for nanochemists and nanotechnologists. Nanochemists and nanochemists report the use of $FeCl_3$ and $FeSO_4·9H_2O$ as the precursors for the synthesis of FeNPs. In both the cases, Fe metal ion is present in its +2 oxidation state leaving two electrons upon reduction. Here, we report the use of $Fe(NO_3)_3·9H_2O$, where Fe is reduced to FeNPs from its +3 oxidation state. The chemical reaction is represented below.

$$Fe^{3+} + 3e \rightarrow Fe$$

$$Fe + Tea\ flavonoids \rightarrow FeNPs$$

The antioxidants present in tea flavonoids are highly concentrated and act as great reducing agents. They act as stabilizing agents also. Thus, the antioxidants present act as a surfactant to the FeNPs. Although production of iron nanoparticles is reported, this technology needs further improvement.

10.2 EXPERIMENTAL

10.2.1 MATERIALS REQUIRED

$CuSO_4·4H_2O$ was bought from Sigma Aldrich, and leaves of pineapple was obtained from garden. $Fe(NO_3)_3·9H_2O$ was obtained from Kottayam Chemicals. Green tea was bought from local market.

10.2.2 SYNTHESIS TECHNIQUES FOR ULTRASMALL CuNPs

A quantity of 10 g of pineapple leaves were taken and rinsed thoroughly with deionized water. The leaves were cut into small pieces and were allowed to boil in deionized water for 30 min. Now, the leaf broth obtained was allowed to cool. It was filtered out.

To a stirring solution of 30 mL (0.01 M) $CuSO_4$ solution, 2-mL pineapple broth was added and constantly stirred for 30 min. Similarly, to 30 mL (0.01 M) $CuSO_4$ solution (4, 10, and 20 mL), pineapple broth was added and stirred. Stirring was continued for thirty minutes. The change in characteristic properties shown by $CuSO_4$ on changing the concentration of broth from 0.1 to 0.01 N was studied.

10.2.3 SYNTHESIS TECHNIQUES FOR FeNPs

First, 5 g of green tea bought from nearby market. It was thoroughly washed used in deionized, double distilled water. Now, it was made into a solution by adding 83 mL of water. The solution was heated at 80°C for about 1 h. The solution obtained was filtered thoroughly. Ferric nitrate solution of concentration (0.4×10^{-3}) M was taken. Green tea extract was added to the constantly stirring ferric nitrate solution. The colloidal solution was made in the ratio having $FeNO_3$ to tea extract (1:2, 1:3, and 1:4). The colloidal solution was stirred continuously for about 30 min.

10.3 RESULTS AND DISCUSSION

Figure 10.1a shows the TEM image of iron nanoparticles. Nanoparticles of size around 35 nm are seen in the image. The nanoparticles are not distinctly seen as there is agglomeration of the particles. Figure 10.1b shows the SAED pattern of iron nanoparticles. The image shows the spots depicting that the nanoparticles are highly crystalline.

Figure 10.2 shows UV–vis spectra of $CuSO_4$, broth, and copper nanoparticle prepared at different broth concentration. The spectra of copper sulfate show peak at 267 nm. The surface plasmon peak of copper sulfate gets reduced on addition of broth. The shoulders formed at 420 nm indicate the formation of copper nanoparticles.[18]

From the UV–vis data, it is clear that the copper metal particles get reduced from $CuSO_4$ to CuO nanoparticles or ultrasmall copper nanoparticles. The color of the colloidal solution remains yellow even after 3 weeks indicating stable CuNPs. It gets attached to the OH groups and CO groups present in the broth solution. Ultrasmall CuNPs and CuONPs do not generate special feature in UV–vis spectroscopy as shown in Figure 10.3. The small shoulders indicate the formation of CuNPs.

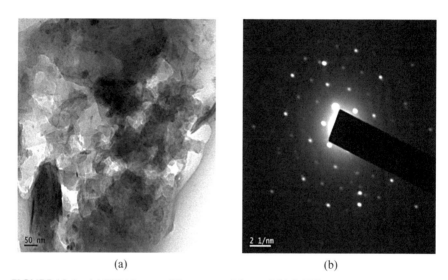

FIGURE 10.1 (a) TEM image of Fe nanoparticles and (b) SAED pattern.

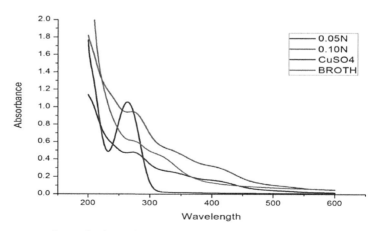

FIGURE 10.2 (See color insert.) UV–vis spectra of $CuSO_4$, broth, and copper nanoparticles prepared at different broth ratio.

Phytosynthesis of Cu and Fe Nanoparticles   149

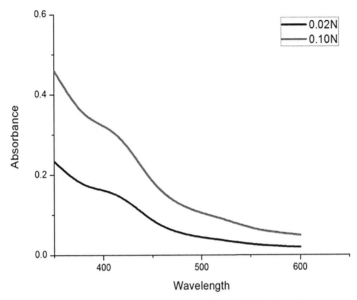

FIGURE 10.3 (See color insert.) UV–vis spectra of $CuSO_4$, broth, and copper nanoparticles prepared at different broth ratio.

The enlarged and magnified view of shoulders present at 420 nm is shown in Figure 10.4.

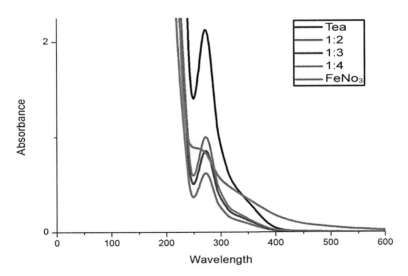

FIGURE 10.4 (See color insert.) UV–vis spectra of $FeNO_3$, tea extract, and iron nanoparticles prepared at different ratios.

Figure 10.4 shows UV–vis spectra of FeNO$_3$, tea extract, and iron nanoparticles prepared at different ratios. The spectra of FeNO$_3$ do not show any distinct peak. The absorbance of green tea extract gets reduced on addition of FeNO$_3$. The absorbance of GT to FeNO$_3$ decreases on addition of FeNO$_3$ solution. The shoulder of FeNO$_3$ gets reduced on addition of GT. This indicates the presence of FeNPs.[21]

The color of ferric nitrate from light brown to dark red indicates the formation of FeNPs. The FeNPs formed will be either ferric oxide nanoparticles or small iron nanoparticles. Color change took place just after 30 min of preparation of colloid. Fe shows no special features in UV–vis but the reduction in the peak of tea upon adding different concentrations of ferric nitrate shows formation of something. From the FTIR data of the tea, it is evident that there are enough flavonoids groups and OH and CO groups, where the Fe$^+$ ions and FeNPs can possibly get attached.

The FTIR peaks (Fig. 10.5) at 3339.69, 2122.87, and 1634.87 cm^{-1} show the presence of OH groups and C=O groups. Thus, the presence of CO groups and OH groups defines the fact that CuO ions get attached to the bonds forming a hydrogen bond.

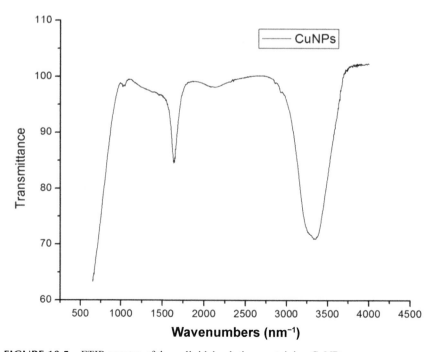

FIGURE 10.5 FTIR spectra of the colloidal solution containing CuNPs.

The FTIR peaks (Fig. 10.6) at 3339.69, 2122.87, and 1634.87 cm^{-1} show the presence of OH groups and C=O groups. Thus, the presence of CO groups and OH groups defines the fact that Fe ions get attached to the bonds forming a hydrogen bond. This later gets converted to FeNPs–OH bonds, which are highly stable.

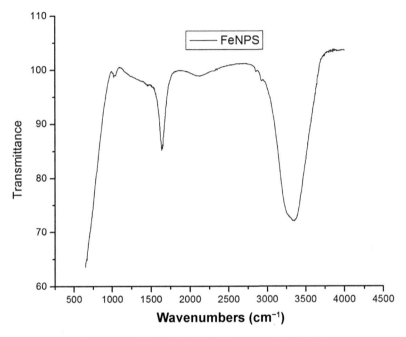

FIGURE 10.6 FTIR spectra of the colloidal solution containing FeNPs.

10.4 CONCLUSION

In summary, we found that metal nanoparticles can be easily synthesized using economically benign methods. Copper and iron nanoparticles were synthesized and characterized simply by using normal sources. Metal nanoparticles can thus be easily synthesized using plant extracts and leaf broths showing high antioxidant and other activities; copper nanoparticles were synthesized using leaf extract of pineapple leaves. Iron nanoparticles were synthesized using green tea. Thus, the area of green chemistry remains wide open and blending green chemistry with other field also remains unexploited. Thus, green-synthesized nanoparticles and materials do have equal advantages and merits when compared to chemically synthesized ones.

KEYWORDS

- metal nanoparticles
- bio-sensing
- imaging
- drug delivery
- optical spectroscopy

REFERENCES

1. Shanker, S.; Rai, A.; Ahmad, A.; Sastry, M. *J. Colloid Interface Sci.* **2004**, *275*, 496–502.
2. Akshaya K Samal, Lakshminarayana Polvarapu, Sergio Rodal Cederia. *Langmuir* **2013**, *29*, 15076–15082.
3. Yang, G.; Xie, J.; Hong, F. *Carbohydr. Polym.* **2012**, *87*, 839–845.
4. Sheny, D. S.; Mathew, J.; Philip, D. *Spectrochim. Acta A* **2011**, *79*, 254–262.
5. Agnihotri, S.; Mukherji, S.; Mukherji, S. *RSC Adv.* **2014**, *4*, 3974–3983.
6. Prakash, P.; Gnanaparakasam, P.; Emmanuel, R. *Colloids Surf., B: Biointerfaces* **2013**, *108*, 255 259.
7. El-Rafie, H. M.; El-Rafie, M. H.; Zahran, M. K. *Carbohydr. Polym.* **2013**, *96*, 403–410.
8. Tian, Y.; Wang, F.; Liu, Y. *Electrochim. Acta* **2014**, *146*, 646–653.
9. Sundari, M. T.; Ramesh, A. *Carbohydr. Polym.* **2012**, *87*, 1701–1705.
10. Hirakawa, T.; Kamat, P. V. *JACS Art.* **2005**, *127*, 3928–3934.
11. Czaja, W.; Krystynowicz, A.; Bielecki, S. *Biomaterials* **2006**, *27*, 145–151.
12. Morones, J. R.; Elechiguerra, J. L.; Camacho, A. *Nanotechnology* **2005**, *16*, 2346–2353.
13. Filippo, E.; Serra, A.; Manno, D. *Sens. Actuators, B* **2009**, *138*, 625–630.
14. Ghosh Chaudhari, R.; Paria, S. *Chem. Rev.* **2012**, *112*, 2373–2433.
15. Brumbaugh, A. D.; Cohen, K. A.; St. Angelo, S. K. *ACS Sustain. Chem. Eng.* **2014**, *2*, 1933–1939.
16. Zhao, J.; Zhang, D.; Zhang, X. *Surf. Interface Anal.* **2015**, *47*, 529–534.
17. Deng, D.; Jin, Y.; Cheng, Y. *ACS Appl. Mater. Interface* **2013**, *5*, 3839–3846.
18. Makarov, V. V.; Makarova, S. S.; Love, A. J. *Langmuir* **2014**, 30, 5982–5988.
19. Huang, L.; Weng, X.; Chen, Z. *Spectrochim. Acta A: Mol. Bilomol. Spectosc.* **2014**, *130*, 295–301.
20. Huang, L.; Weng, X.; Chen, Z. *Ind. Crops Prod.* **2013**, *51*, 342–347.
21. Huang, L.; Weng, X.; Chen, Z. *Spectrochim. Acta, A: Mol. Biomol. Spectrosc.* **2014**, *117*, 801–804.

CHAPTER 11

EFFECT OF CONCENTRATION AND TEMPERATURE ON ZnO NANOPARTICLES PREPARED BY REFLUX METHOD

VAIBHAV KOUTU[1,*], NAJIDHA S.[2], LOKESH SHASTRI[1], and M. M. MALIK[1]

[1]*Department of Physics, Nanoscience and Engineering Center, MANIT, Bhopal, Madhya Pradesh, India*

[2]*Department of Physics, BJM Government College, Kollam, Kerala, India*

[*]*Corresponding author. E-mail: vkoutu@gmail.com*

ABSTRACT

This chapter describes the effect of NaOH concentration and temperature on the crystallinity and size of ZnO nanostructures. Nanosized ZnO nanostructures were prepared using zinc acetate dihydrate and sodium hydroxide as precursors using soft chemical reflux method. Formation of crystalline phases of ZnO was confirmed by X-ray diffraction. Morphology of as-prepared samples was obtained by scanning electron microscope, which confirms the formation of hexagonal whiskers and rod-like structures having average size of 46 nm. The emission spectra of the as-prepared samples were obtained by photoluminescence which depicts the prominent emission in the region around 420–450 nm. These materials may be used in optoelectronic devices and sensing applications.

11.1 INTRODUCTION

ZnO, a II–VI compound n-type semiconductor with hexagonal wurtzite structure,[1] has significant importance in both fundamental research and application due to its wide direct band gap of 3.37 eV and a very high exciton-binding energy of 60 meV at room temperature.[2-5] It is a material with very high thermal and mechanical stability; thus, it possesses unique optical,[2,3,5] acoustical, and electronic properties[3] which stimulate research interest among the scientific community and make ZnO an exceptionally promising material in a wide arena of technological applications.

Various physical and chemical processes have been applied and reported to synthesize ZnO.[5-8] However, preparation via solution route (reflux method)[3,4] provides an easy and convenient method with better yield for the production of nanoparticles. Herein, in this chapter, we report a simple and efficient soft chemical solution technique for the production of ZnO nanoparticles, and the effect of concentration of alkaline precursor and reaction temperature on the final product has been studied. The structure, phase, morphology, and optical properties of the synthesized product were investigated by standard characterization techniques.

11.2 EXPERIMENTAL

Zinc oxide nanostructures were synthesized by using zinc acetate dihydrate $(Zn(Ac)_2)$ (0.1 M, >99% pure) and sodium hydroxide (NaOH) (>99% pure). All the chemicals were purchased from leading suppliers without any further purification. In this experiment, we have prepared four samples—two at varying NaOH concentrations (Series A) and two at different reflux temperatures (Series B), while keeping all the remaining synthesis parameters constant. In Series A, 0.1-M $Zn(Ac)_2$ was dissolved in 50 mL of distilled water in two separate beakers via stirring at 90°C to obtain clear solutions. In one part, 50 mL aqueous solution of 0.1-M NaOH (A-1) and in other part equal volume of 0.4-M NaOH (A-2) were added dropwise. The resulting mixtures were stirred at 90°C for 1 h for homogenous mixing. After that, both the mixtures were transferred in separate round-bottom flasks and refluxed at 90°C for 1 h.

For Series B, two separate 50 mL 0.1-M $Zn(Ac)_2$ solutions were prepared by dissolving zinc precursor at 90 and 100°C, respectively, via stirring. To both of the solutions, 50 mL aqueous solution of 0.1-M NaOH was added dropwise and the resulting mixtures were transferred in separate round-bottom flasks. One mixture was refluxed at 90°C (B-1) and the other was refluxed at 100°C (B-2) for 1 h each.

The mixtures were allowed to cool at room temperature for the precipitate to settle down. The precipitates were filtered, washed with methanol several times, and dried at 60°C to obtain final ZnO powders. The final dried powder was collected and characterized using X-ray diffractrometer (Rigaku Miniflex II), scanning electron microscope (Jeol JSM 6390), and room temperature photoluminescence (Hitachi F-7000 spectrophotometer).

11.3 RESULTS AND DISCUSSION

The method adopted for the preparation of ZnO nanostructures is found to be suitable and an efficient one. Figure 11.1 shows the X-ray diffraction patterns for the as-prepared ZnO samples.

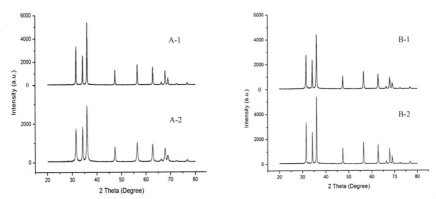

FIGURE 11.1 (a) XRD pattern for ZnO at different NaOH concentrations (Series A). (b) XRD pattern for ZnO at different reflux temperatures (Series B).

X-ray analysis confirms the formation of crystalline ZnO phases having hexagonal wurtzite structures.[6] All the prominent peaks are identified and indexed and are found to be in a good correlation with JCPDS file no. 79-0208 with no traces of impurities.

Figure 11.2a,b shows the scanning electron microscopy images of the as-prepared ZnO nanopowders. The SEM images reveal the formation of hexagonal structures—whiskers-like structures are observed for ZnO prepared via Series B having mean particle size of 48 nm (sample B-1) and 44 nm (sample B-2); and rod-like structures are observed for ZnO prepared via Series A having mean particle size of 49 nm (sample A-1) and 42 nm (sample A-2). The images indicate the growth of the nanostructures along a particular direction.[6,7]

FIGURE 11.2 (a) SEM images of ZnO nanopowders at different NaOH concentrations (Series A). (b) SEM images of ZnO nanopowders at different reflux temperatures (Series B).

Room temperature photoluminescence emission spectra are shown in Figure 11.3. PL studies were done at an excitation wavelength of 320 nm for Series A samples and at 325 nm for Series B samples. Secondary excitation peak at 640 and 650 nm, respectively, are clearly visible in the graphs. It has been reported in many studies that ZnO shows four PL emissions[5,9]—(1) around 390 nm due to free exciton recombination; (2) 460 nm due to intrinsic defects such as oxygen and zinc interstitials; (3) around 540 nm because of deep level emissions due to impurities, structural defects in crystal; and (4) around 630 nm due to oxygen and zinc antisites. The PL spectra for all the samples shown in Figure 11.3a,b reveal the presence of abovementioned peaks. The most prominent peaks are observed at around 420 and 450 nm and are attributed to intrinsic defects such as oxygen and zinc interstitials.[9]

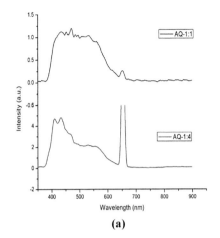 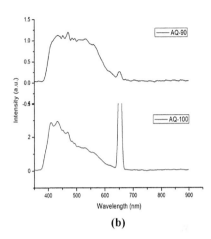

FIGURE 11.3 (a) PL emission spectra of ZnO nanopowders at different NaOH concentrations (Series A). (b) PL emission spectra of ZnO nanopowders at different reflux temperatures (Series B).

11.4 CONCLUSION

In this chapter, ZnO nanoparticles were prepared by soft-chemical reflux method at varying NaOH concentrations and reflux temperatures. Crystallinity of as-prepared ZnO samples shows that the ZnO phases have hexagonal wurtzite structure. Same morphology was also observed in SEM images which show hexagonal whiskers and rod-like structures having average particle size of 46 nm. Room temperature photoluminescence studies show emissions in the visible region. This study reveals that these materials may have sensing applications and can be used in optoelectronic devices.

KEYWORDS

- ZnO nanostructures
- X-ray diffraction
- scanning electron microscope
- optoelectronic devices
- wurtzite structure

REFERENCES

1. Mayekar, J.; Dhar, V.; Srinivasan, R. To Study the Role of Temperature and Sodium Hydroxide Concentration in the Synthesis of Zinc Oxide Nanoparticles. *Int. J. Sci. Res. Publ.* **2013,** 3 (1), 1–5.
2. Ladislav, S.; Praus, P.; Wojtalová, M. Preparation and Properties of ZnO Nanoparticles. In: *NANOCON*, 5th–7th November 2014, Brno, Czech Republic, EU.
3. Wahab, R.; Ansari, S. G.; Kim, Y. S.; Song, M.; Shin, H. The Role of pH Variation on the Growth of Zinc Oxide Nanostructures. *Appl. Surf. Sci.* **2009,** *255*, 4891–4896.
4. Wahab, R.; Hwang, I. H.; Kim, Y.; Shin, H. Photocatalytic Activity of Zinc Oxide Micro-Flowers Synthesized via Solution Method. *Chem. Eng. J.* **2011,** *168*, 359–366.
5. Singh, A. K.; Viswanath, V.; Janu, V. C. Synthesis, Effect of Capping Agents, Structural, Optical and Photoluminescence Properties of ZnO Nanoparticles. *J. Luminesc.* **2009,** *129* (8), 874–878.
6. Gusatti, M.; Rosário, J. A.; Barroso, G. S.; Campos, C. E. M.; Riella, H. G.; Kuhnen, N. C. Synthesis of ZnO Nanostructures in Low Reaction Temperature. *Chem. Eng. Trans.* **2009,** *17*, 1017–1021.
7. Moharram, A. H.; Mansour, S. A.; Hussein, M. A.; Rashad, M. Direct Precipitation and Characterization of ZnO Nanoparticles. *J. Nanomater.* **2014,** *2014*, 20.
8. Conde, M. N.; Dakhsi, K.; Zouihri, H.; Abdelouahdi, K.; Laanab, L.; Benaissa, M.; Jaber, B. Preparation of ZnO Nanoparticles Without Any Annealing and Ripening Treatment. *JMSE A* **2011,** *1*, 985–990.
9. Singh, A. K. Synthesis, Characterization, Electrical and Sensing Properties of ZnO Nanoparticles. *Adv. Powder Technol.* **2010,** *21* (6), 609–613.

CHAPTER 12

GRAPHENE-BASED NANOCOMPOSITE FOR REMEDIATION OF INORGANIC POLLUTANTS IN WATER

VIMLESH CHANDRA*

Department of Chemistry, Dr. Harisingh Gour Central University, Sagar 470003, Madhya Pradesh, India

*E-mail: vchandg@gmail.com

ABSTRACT

Graphene, a one-atom thick planar sheet of sp^2-bonded carbon atoms packed in a honeycomb lattice, is considered to be the mother of all graphitic materials and created tremendous interest to both physicist and chemist. Graphene and its derivative have potential applications in nanoelectronics, supercapacitors, solar cells, batteries, flexible displays, hydrogen storage, waste-water treatment, and sensors. In this chapter, a brief overview on various aspects of graphene such as discovery, synthesis, functionalization, composites, and its applications in removal and detection of aqueous inorganic pollutants had been illustrated.

12.1 INTRODUCTION

Water is the most precious natural resource on earth and supports life for people, animals, and plants. In spite of this, 1.2 billion people lack access to safe drinking water, and millions of people die annually from diseases transmitted through contaminated water.[1] This is due to the presence of trace amount of toxic organic (dyes, pesticides), inorganic (arsenic, uranium, salt ions), and biological (virus, bacteria) pollutants in water. The global population is expected to reach up to ~10 billion in 2020, and because of this, the world will be under great drinking water scarcity. Therefore, the removal of pollutants from wastewater is an urgent need for providing disease-free health across the globe. A number of possible methods have been used for removal of aqueous pollutants. Among these, adsorption, photocatalysis, and nanofiltration turnout to be effective and economical techniques applied for water treatment.[2] At first, activated carbon (powder, granular, pellet) has been used for removal of toxic pollutants, and then it was replaced with low-cost adsorbent.[3,4] From last two decades, nanotechnology has been playing an increasing important role in developing nanomaterials with size ranging from 1 to 100 nm. These nanomaterials display unique properties not found in the bulk-sized materials.[5] Many experiments have been carried out in developing nanomaterials and composites to overcome problems, such as agglomeration, electron–hole recombination, nonmagnetic nature of adsorbent, and weak interaction between adsorbate (pollutants) and adsorbent.[6] Graphene is the world's thinnest, strongest, and most conductive material and has very large number of applications in optoelectronics, solar/fuel cell, drug delivery, desalination, toxic materials removal, supercapacitor, batteries, energy generation/storage, catalyst, and many others.[7,8] Graphene has very high conductivity and two free mesoporous surfaces with theoretically predicted surface area as being >2500 $m^2 g^{-1}$ and experimentally measured to be 400–700 $m^2 g^{-1}$.[9] Graphene is predictably going to arise the notion of "nano" as was dreamed long ago "There's Plenty of Room at the Bottom"—Richard Feynman 1959. The high surface area and large-scale production of graphene-based nanomaterials (adsorbent/photocatalyst/membrane) at low cost makes a potential candidate for treatment of water containing toxic pollutants (organic, inorganic, and biological) and salt.[10–14] This book chapter evaluates the important chemical route for synthesis of graphene and graphene-based composites with metal, oxides, polymers nanoparticles, etc. It highlights the chronic diseases transmitted to human beings on drinking contaminated water and methods for treatment of these toxic inorganic pollutants. Toxicity of graphene has discussed toward soil, environment, and human being for its safe application in water-treatment techniques.

12.2 GRAPHENE

Graphite and diamond are the natural-occurring allotropes of carbon. The graphite was derived from Greek word "graphein" (to write/draw) by A. G. Werner in 1789, for its use in pencils. Graphite is semimetal and has lamellar planer structure and has been used in several industrial and academic researches.

In each layer, the carbon atoms are arranged in hexagonal lattice with separation of 0.142 nm with interlayer distance 0.335 nm. The single graphite crystal shows highly anisotropic nature, known as very good conductors along the graphite planes and very poor one across the planes.[15–17] Krishnan and Ganguli found for the two directions ratio of resisitivities 10^4 or higher.[15] Calculation made by Wallace and Coulson has shown that a graphite crystal along the planes has more conducting than across the plane.[16,17] This anisotropic nature of graphite motivated for isolation of single layer of graphite. Thin graphite layers around 5–10 nm were stripping off from highly ordered graphite crystal with mechanical exfoliation and using adhesion tape.[18] The graphite oxide was prepared several decades ago by Brodie, Staudenmaier, and Hummers et al. using strong acid and oxidizing agent.[19–22] Boehm et al. synthesized single/bilayer of graphite oxide and given abbreviations GO.[23] The reduced forms of graphite oxide were observed with electron microscopy.[24] The single layer of graphite named graphene by IUAPC in 1995, where "–ene" was taken from fused polycyclic aromatic hydrocarbons.[25] Ruoff et al. demonstrated the tailoring of highly oriented pyrolytic graphite to obtain uniformly sized islands.[26] Kim et al. developed a unique micromechanical method to extract extremely thin graphite ranging from 10 to 100 nm size thickness from bulk.[27] In 2004, Geim and Novoselov isolated single layer of graphite (graphene) using scotch tape and received the 2010 Noble Prize in Physics.[7]

12.2.1 SYNTHESIS OF GRAPHENE

12.2.1.1 TOP-DOWN APPROACH

Chemical oxidation reduction routes are very attractive method for large-scale production of graphene from graphite via top-down approach. The graphene oxide, abbreviation GO, was prepared several decades ago by Hummers et al.[22] and recently by many others which is based on the oxidation of graphite with strong oxidizing agents in acidic media. The GO exhibits lamellar structure with randomly distributed nonoxygenated aromatic region (sp^2-carbon atoms), oxygenated six member aliphatic regions (sp^3-carbon atoms) containing hydroxyl, epoxy, and carboxyl functional groups. The epoxy and hydroxyl group lies above and below of each carbon layer and carboxylic groups attached at edges of the layers. The GO is highly water soluble and potentials candidates for many applications (Fig. 12.1). First time, hydrazine was used for the reduction of GO to graphene[28] and after this work, various reducing agents such as sodium borohydride, ascorbic acid, hydroiodic acid, hydroquinone, pyrogallol, hot strong alkaline solutions, hydroxyl amine, urea, and thiourea were used for reducing GO.[29] The thermal,[30] microwave,[31] and photo-irradiation[32] route for reduction of GO to graphene were employed to fabricate graphene in the environment-friendly way. Recently, GO has been transformed to reduced GO by high energy ball milling in inert atmosphere. The process of ball milling introduces defects and removes oxygen functional groups, thereby creating the possibility of fine tuning the bandgap of all intermediate stages of the structural evolution.[33]

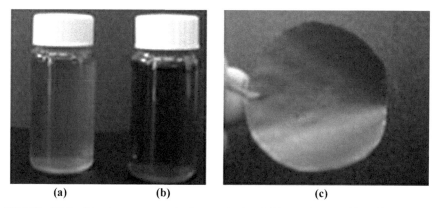

FIGURE 12.1 Top-down approach for synthesis of (a) graphene oxide, (b) graphene aqueous solution, and (c) graphene film.

The sonochemical routes offer an attractive method for synthesizing nanomaterials, polymers, and composites for numerous applications.[34] The intense collapse effects arising due to cavitation enhance the polymerization.[35] The sonochemical method has been used for exfoliating graphene from graphite in different solvents and in presence of surfactants and polymers.[36] The weak van der Waals attractions forces are sufficient to keep stack of graphite layers. The interfacial tension between graphite and liquid plays a vital role in isolating graphene sheets in the liquid and solvents with surface tension $\gamma = 40$ mJ m^{-2} are the best choice.[37]

12.2.1.2 BOTTOM-UP APPROACH

The mechanical exfoliation method yields a very small amount of high-quality graphene with carrier mobility (2.0×10^5 cm^2 (V s)$^{-1}$).[7] Somani and coworkers reported the first successful synthesis of few-layer graphene films via chemical vapor deposition (CVD) using camphor as the precursor and Ni foils as substrate.[38] After this, large-area monolayer graphene films were prepared on Ni foil[39,40] and copper foil.[41] Plasma-enhanced CVD offers another route for fabrication of graphene films on various substrate from a gas mixture of 5–100% CH_4 in H_2 and at high temperature.[42] The few-layer graphene was achieved via annealing of SiC substrate at 1200°C in ultrahigh vacuum.[43] Mullen's group synthesized two-dimension graphene ribbons with the size of 12 nm through the Suzuki–Miyaura coupling of 1,4-diiodo-2,3,5,6-tetraphenylbenzene with 4-bromophenylboronic acid.[44] Jiao and coworkers synthesized graphene sheets via unzipping of carbon nanotube.[45]

12.2.2 CHARACTERIZATION OF GRAPHENE

The formation of graphene can be confirmed by counting the number of graphene layers (N) and by using electron diffraction patterns.[46] Atomic force microscopy (AFM) can be used in the measurement of thickness of graphene and GO of the deposited flakes on substrate such as SiO_2 and mica sheet (Fig. 12.2).[47,48] The number of graphene layers (N) in a sample can be determined by elastic light scattering (Rayleigh) spectroscopy[49] and Raman spectroscopy.[50,51]

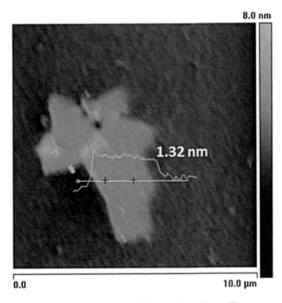

FIGURE 12.2 AFM image of graphene oxide on mica sheet. The cross-sectional analysis shows thickness of 1.32 nm.

12.3 REMEDIATION METHODS

12.3.1 ADSORPTION

Adsorption is process which occurs with the enrichment of one or more components in the interfacial layer. The sorption is combined effect of mass transport into solids and including surface adsorption, absorption by penetration into the solid, and condensation within pores (Fig. 12.3).

Adsorption processes can be divided into the two categories, physical adsorption (physisorption) and chemical adsorption (chemisorption). In physical adsorption, no chemical bonds formation occur; however, attraction between the adsorbate and adsorbent exists by the formation of intermolecular electrostatic, such as London dispersion forces, or van der Waals forces from induced dipole–dipole interactions. The adsorption is an exothermic process since reducing the entropy due to increase in ordering of the adsorbate on the adsorbent surface. Chemical adsorption is due to the chemical bonds formation between adsorbent and adsorbate so regeneration of the adsorbent for subsequent reuse is very difficult.[52,53]

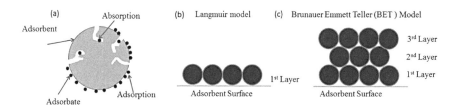

FIGURE 12.3 Sorption process (a) absorption versus adsorption, (b) monolayer (Langmuir), and (c) multilayer adsorption (Brunauer–Emmett–Teller, BET, model).

The adsorption governs by different type of adsorption isotherms (Fig. 12.4) and classified as follows:

Type I isotherm	Microporous
Type II isotherm	Nonpowder
Type III isotherm	Macroporous
Type IV isotherm	Mesoporous
Type V isotherm	Porous (weak adsorption interaction)
Type VI isotherm	Energetically uniform surface

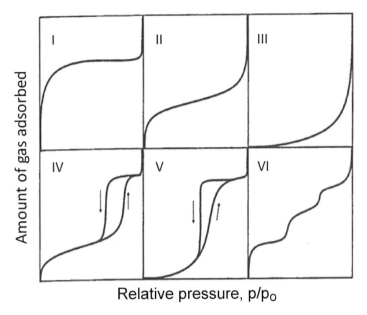

FIGURE 12.4 Pictorial representation of different type adsorption isotherms.

Pores are minute openings in solids that are accessible to vapors and gases. Porosity is the measure of empty space due to these pores and their distribution in the structure of a solid. The porosity plays a key role in physical adsorption and highly porous materials have high surface area. Porosity increases with decreasing temperature or increasing pressure. The porosity can be classified as follows: microporous (pore size < 2 nm); mesoporous (2 < pore size < 50 nm); and macroporous (pore size > 50 nm).

12.3.2 PHOTOCATALYSIS

A photocatalyst is a material that absorbs light quanta in the ultraviolet, visible, or infrared region and produce hole in the valance band and electron in the conduction band (Fig. 12.5). These holes and electrons are responsible for oxidation and reduction of toxic pollutants present in the water.[54] TiO_2-driven photoprocesses were established in 1972 by Fujishima and Honda[55] and Carey et al. employed the principle first for decontamination in 1976.[56] After this, a wide range of toxic pollutants such as dyes, volatile organic compounds, and pesticides[57–59] have been successfully treated in the drinking water. Various toxic anions can be oxidized into harmless or less toxic compounds by using TiO$_2$ as a photocatalyst. For instance, nitrite is oxidized into nitrate, sulfide, sulfite, and thiosulfate that are converted into sulfate, whereas cyanide is converted either into isocyanide or nitrogen or nitrate.[60] The photocatalysis works well by absorbing wavelengths in the near-UV region (i.e., 390 nm), which is about 3% of the solar spectrum. Current efforts in the field are being devoted to optimize the efficiency of these processes in the visible region (400 < λ > 750 nm) in many ways such as doping with transition metal cations and nonmetal anions to produce intermediates states in the bandgap,[61] *spectral sensitization* using dye or polymer[62] and attaching photosensitizers such as another small bandgap semiconductor or organic compounds which absorb visible light.[63] The photocatalytic activity of semiconductor photocatalyst is limited by the aggregation of nanoparticles and increases on the formation of mesoporous particles.[64] The charge recombination and separation/migration are two important competitive processes inside the semiconductor photocatalyst that largely affect the efficiency of the photocatalytic reaction.[65] The enhancement of photocatalytic efficiency can be achieved by reducing electron–hole recombination and aggregation through attachment of graphene.

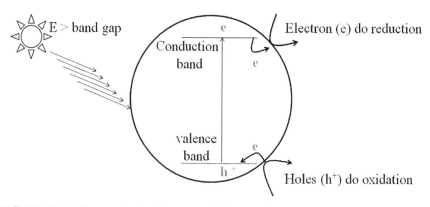

FIGURE 12.5 Photocatalysis using energy higher than bandgap.

12.4 INORGANIC POLLUTANTS

Inorganic pollutants in the water are due to erosion of natural deposits, and manmade activities, such as industrial, agricultural, and domestic wastes. Inorganic pollutants are not biodegradable and they remain in water, indefinitely for this reason World Health Organization (WHO) and Environmental Protection Agency (EPA) has recommended guidelines for heavy metals in drinking water based on toxicity data (Table 12.1).[66,67] On drinking water of more than permissible limit of concentration (*WHO/EPA*) can cause severe health problem in human and effect skin, bone, teeth; nervous system; and kidneys/liver.[68–70] The analysis of trace amount (ppm ppb^{-1}) of toxic metal ions in drinking water carried out using sophisticated analytical such as atomic absorption spectroscopy (AAS),[71] inductively coupled plasma-mass spectrometry,[72] mass spectroscopy,[73] X-ray fluorescence spectroscopy,[74] and potentiometric methods.[75] The development for low-cost instruments for detections of inorganic toxic pollutants in drinking water based on colorimetry, fluorimeter, and voltammetry to allow the miniaturization and infield applications.

TABLE 12.1 Maximum Contamination Level of Few Inorganic Pollutants in Water.

Metal	WHO (mg L^{-1})	EPA (mg L^{-1})[a]
As	0.010	0.010
Cr	0.050	0.100
Hg	0.001	0.002
Cd	0.003	0.005
Pb	0.010	0.015
U	0.030	0.030
Cu	2.000	1.300
Zn	3.000	5.000
Ni	0.020	0.040

[a]1 mg L^{-1} = 1 ppm and 1 µg L^{-1} = 1 ppb.

12.4.1 ARSENIC

Arsenic is a semimetal element and used in paints, dyes, metals, drugs, soaps, and semiconductors. It enters drinking water supplies from erosion of natural deposits or from agricultural and industrial practices such as copper smelting, mining, and coal burning.[76] It is odorless and tasteless and higher levels of arsenic found in ground water sources than in surface water sources (i.e., lakes and rivers).[77] Arsenite (AsO$_3^{3-}$) and arsenate (AsO$_4^{3-}$) are one of the most toxic and carcinogenic chemical ions of arsenic in drinking water. WHO-decided maximum concentration of arsenic in water 0.01 mg L^{-1}. On long term, chronic exposure to arsenic causes partial paralysis, blindness, and cancer.[78] Iron and its oxide nanoparticles have been used for purification of toxic pollutants contaminated water.[79] The nanoparticles of iron, magnetite, and maghemite are magnetic in nature so after adsorption of pollutants, adsorbent can be separated using magnet without filtration.[80] The bare nanoparticle can leach out in flowing water and contaminate the drinking water. The bare nanoparticles have very small surface area so results in weak adsorption capacity. To overcome these issues, pure iron and its derivative have been functionalized with different organic/inorganic materials and showed enhanced removal of arsenic.[81-83] The graphene has very high surface area and mesoporous in nature and functionalized graphene sheets with different type of nanoparticles have been used in decontaminations of drinking water. Chandra et al. developed first time graphene functionalized magnetite nanoparticles for adsorption of As(III) and As(V) from drinking water. The superparamagnetic magnetite-graphene composite was developed via chemical route containing magnetite particle of size 10 nm and have shown nearly complete (over 99.9%) arsenic removal within 1 ppb level.[84] The magnetic nature of adsorbent due to the Fe_3O_4 nanoparticles makes it easily separable using external magnetic fields (Fig. 12.6). Mishra et al. studied electrochemical sorption behavior of arsenic using magnetite graphene nanocomposite electrode and using Langmuir isotherm fitting the maximum adsorption capacities for arsenate and arsenite was nearly 172.1 and 180.3 mg g^{-1}, respectively.[85] Paul et al. reported the humic acid coated graphene–Fe_3O_4 composite and observed almost double the removal efficiency of As(III) and As(V) present in water even earlier studies have shown the negative role of humic acid.[86] Guo et al. synthesized graphene sheets functionalized with polydopamine and Fe_3O_4 nanoparticles

FIGURE 12.6 The separation of magnetic adsorbent after adsorption of toxic pollutants using external magnetic field.

via self-assembly under basic conditions and used for As(III) and As(V) removal from aqueous solutions.[87] Vadahanambi et al. reported synthesis of three-dimensional graphene–carbon nanotube–iron oxide nanostructures for arsenic removal where carbon nanotubes are vertically standing on graphene sheets and iron oxide nanoparticles are decorated on both the graphene and the carbon nanotubes.[88] Wen et al. reported layered double hydroxide nanocomposite exhibited swelling behavior in water and forming a gel. The As(V) adsorption experiment showed a maximum adsorption capacity of 183.11 mg g^{-1}.[89] Wu et al. reported a composite material containing magnetite, graphene, and layered double hydroxides via two-step reaction and have shown enhanced arsenic removal due to incorporation of magnetite particles and graphene provides more active sites for arsenate uptake.[90] Luo et al. developed the nanocomposites of magnetite Fe_3O_4-reduced graphite oxide–MnO_2 for removal of arsenic from water and has shown a high adsorption capacity of 14.04 and 12.22 mg g^{-1} for As(III) and As(V), respectively.[91] Nandi et al. reported synthesis of magnetic manganese incorporated magnetite with high specific surface area (280 m^2 g^{-1}) and pore volume (0.3362 cm^3 g^{-1}) showed higher binding efficiency with As(III) and almost complete (>99.9 %) As(III) removal (≤10 µg L^{-1}) from water.[92] Wang and coworkers studied the adsorption of As(III) and As(V) on nanoscale zero-valent iron decorated on graphene sheets and shown maximum adsorption capacity of 35.83 and 29.04 mg g^{-1}, respectively, and pseudo-second-order adsorption kinetics.[93] Zhu et al. reported magnetic graphene nanopalatelet composite decorated with Fe–Fe_2O_3 nanoparticles and were used for effective removal of arsenic(III) within 1 ppb in the polluted water and results have shown adsorption capacity of 11.34 mg g^{-1}.[94] There are many challenges in making stable dispersion of magnetic nanoparticles and need to rectify for future scope in the drinking water treatment.[95] Zhang et al. developed the cross-linked ferric hydroxide–GO composite via in situ oxidation of ferrous sulfate using hydrogen peroxide for the removal of arsenate and shown maximum efficiency above 95% over the 4–7 pH.[96] Chen et al. showed β-FeOOH@GO–COOH nanocomposite an efficient adsorption medium for the uptake of arsenite and arsenate within a wide range of pH 3–10, providing high adsorption capacities of 77.5 mg g^{-1} for As(III) and 45.7 mg g^{-1} for As(V), respectively.[97] Andjelkovic et al. developed a three-dimensional (3D) graphene–α-FeOOH nanoparticle aerogel composite from natural graphite rocks without the use of harsh chemicals for the efficient removal of arsenic from contaminated water.[98] Luo et al. modified hydrated zirconium oxide $ZrO(OH)_2$ nanoparticles with graphite oxide by hydrothermal coprecipitation reaction and showed good anti-interference ability to co-existing anions, and exhibited excellent recyclability and high adsorption of 95.15 and 84.89 mg g^{-1} for As(III) and As(V), respectively, which are 3.54 and 4.64 times that of ZrO(OH)$_2$ nanoparticles.[99] Cortes et al. studied the adsorption of As(III) onto doped graphene adsorbent on the basis of quantum chemistry calculations and sorted as Al–G > Fe–G ≫ Si–G ≫ GO ≫ G due to chemical and physical interactions with high adsorption energies (>~1 eV).[100]

12.4.2 MERCURY

Mercury is a shiny, silver-white metal and an odorless liquid that can slowly evaporate into the air. Mercury occurs in three major forms: elementary mercury, inorganic mercury compounds, and organomercurials (methyl mercury).[101] Mercury releases into environments through coal combustion, nonferrous metal smelting, and incineration of municipal solid wastes.[102] Fish and shellfish concentrate mercury in their bodies, often in the form of methylmercury (CH$_3$–Hg$^+$), a highly toxic organic compound of mercury. Exposure to mercury leads to severe damage to brains, kidneys, lungs, and other organs.[103] The US EPA has mandated an upper limit of 2 ppb for mercury(II) in drinking water. Materials with sulfur, nitrogen, and oxygen containing functional groups show high binding capacity toward mercury.[104–109] Conducting polymers incorporated with these functional groups show low binding due to the macroporous nature and small surface area and enhanced when functionalized with high surface area materials. Chandra et al. developed polypyrrole functionalized graphene sheet showing mesoporous in nature with surface area 166 m^2 g^{-1} around 35 times higher than pure polypyrrole (6.15 m^2 g^{-1}). The π–π interaction between graphene and polypyrrole forced to grow polypyrrole along graphene sheets and hence adopted surface morphology similar to graphene sheet. The presence of graphene sheets enhanced adsorption capacity 980 mg g^{-1}, and selectivity toward mercuric ions from mixed ions solution of Hg(II), Cu(II), Cd(II), Pb(II), and Zn(II) in the drinking water.[110] Li et al. synthesized polyaniline-reduced GO composite for removal of Hg(II) in drinking water and showed increased capacity from 515.46 to 1000.00 mg g^{-1} at pH 4.[111] Thakur et al. developed sulfur/reduced graphene-oxide nanohybrid having sulfur nanoparticles of average size ~20 nm. The nanohybrid demonstrated a fast and efficient Hg^{2+} removal at around pH 6–8, pseudo-second-order kinetics, and endothermic and spontaneous adsorption process.[112] Kumar et al. functionalized GO with L-cystine and characterized by different techniques and studied interaction of thiol functional group with Hg(II). The adsorption experiment showed maximum adsorption capacity of 79.36 mg g^{-1} and adsorbent was regenerated using thiourea.[113] Diagboya et al. synthesized GO–Fe$_3$O$_4$ nanocomposite via attaching magnetite nanoparticles to 3-aminopropyltriethoxysilane GO and used for removal Hg^{2+} from aqueous solution. The adsorbent showed adsorption capacity of 16.6 mg g^{-1} and kinetics fitted well Elovich model.[114] To avoid contamination of water from elemental mercury (Hg^0) emitted from combustion flue gas, Liu et al. developed Ag-nanoparticle-decorated GO that showed enhanced Hg^0 removal compared to the pure GO.[115] Bao et al. developed thiol-functionalized magnetite/GO hybrid and it exhibited adsorption capacity 289.9 mg g^{-1} in a solution with an initial Hg^{2+} concentration of 100 mg L^{-1} and adsorbent was regenerated using H$^+$.[116]

12.4.3 CHROMIUM

Chromium is an odorless and tasteless metallic element and has been used in making steel and other alloys. Chromium exists in nature as hexavalent chromium, Cr(VI), trivalent chromium, and Cr(III) and found in rocks, plants, soil and volcanic dust, humans and animals. Chromium(III) is an essential human dietary element and occurs naturally in many vegetables, fruits, meats, grains, and yeast. Chromium released in water due to erosion of natural chromium deposits and inadequate industrial waste disposal practices. Chromium(VI) is highly toxic and mobile compared to chromium(III) and results in the severe health problems such as liver damage, pulmonary congestion, and ulcer, lung cancer when consumed more than permissible limit (<0.1 mg L^{-1}, 100 ppb).[117–119] The chromium pollution arises mainly from the industries involved in mining, leather tanning, cement, dye, electroplating, steel, metal alloys, photographic material, and metal corrosion inhibition.[120] Humera et al. developed iron nanoparticles decorated graphene sheets for mitigation of hexavalent chromium ions.[121] Wang et al. synthesized a novel ternary magnetic polypyrrole–Fe$_3$O$_4$–graphene nanocomposite via two step chemical route with specific surface area 80.53 m^2 g^{-1} and showed adsorption capacity for Cr(VI) 293.3 mg g^{-1} much higher than that of Fe$_3$O$_4$/graphene composite. XPS analysis revealed that the Cr(VI) was reduced to the low-toxicity Cr(III) by the nitrogen species of polypyrrole and iron nanoparticles.[121,122] Liu et al. used a facile self-assembly approach for fabrication of Fe$_3$O$_4$ hollow microspheres/GO composite driven by the mutual electrostatic interactions. The presence of GO reduces agglomeration of nanoparticles and increases dispersion in polluted water. The composite shows saturation magnetization of 37.8 emu g^{-1} and sorption capacity for Cr(VI) is 32.33 mg g^{-1}.[123] Li et al. developed three-dimensional magnetic GO foam/Fe$_3$O$_4$ which exhibited a very large surface area 574.2 m^2 g^{-1}, high saturation magnetization of 40.2 emu g^{-1} and high Cr(VI) adsorption capacity 258.6 mg g^{-1}.[124] Zhou et al. fabricated reduced GO (RGO)–Fe$_3$O$_4$ composite having size larger than 100 nm via solvothermal process and using nontoxic and cost-effective precursors for removal of Cr(VI). The Cr(VI) adsorption studies shows adsorption capacity 500 μg L^{-1} at neutral pH and kinetics follows pseudo-second-order and the large-saturation magnetization of 41.12 emu g^{-1} allows fast separation of the adsorbent from water.[125] Yao et al. synthesized the graphene/Fe$_3$O$_4$@polypyrrole hybrid nanocomposites with saturation magnetization of 12.4 emu g^{-1} which allows fast and economic separation of adsorbent from solution after adsorption of Cr(VI) of 348.4 mg g^{-1}. The adsorbent is stable and recyclable and adsorption isotherms and kinetics are Langmuir and pseudo-second-order models, respectively.[126] Zhu et al. decorated graphene sheets with core@double–shell nanoparticles, where the core is crystalline iron nanoparticle, the outer shell iron oxide, and the external shell amorphous Si–S–O. The material exhibited almost complete removal of Cr(VI) from wastewater within 5 min and could be separated from solution using an external magnetic field.[127] Cong et al. reported the metal ions induced self-assembly of α-FeOOH–RGO hydrogels by the NH$_4$OH reduction of FeSO$_4$ and GO and shown to be effective in the removal of Cr(VI) polluted water.[128] Dinda et al. demonstrated the adsorption of chromium using sulfuric acid doped poly diaminopyridine/graphene composite to remove high concentration of toxic Cr(VI).[129] Kumar et al. synthesized a novel trioctylamine impregnated GO adsorbent for the removal of hexavalent chromium through cation-pi, lone pair-pi, and electrostatic interactions with adsorption capacity of 232.55 mg g^{-1}. The kinetics follows second order and exothermic in nature and regeneration of adsorbent was achieved using ammonium hydroxide solution.[130] Zhang et al. reported hierarchical nanocomposites of polyaniline nanorods array on GO nanosheets exhibit excellent water treatment performance with a superb removal capacity of 1149.4 mg g^{-1} for Cr(VI).[131] Ge et al. studied on chromium(VI) adsorption on triethylenetetramine modified GO/chitosan composite synthesized via microwave irradiation method and shows higher yield and uptake for Cr(VI) 219.5 mg g^{-1} at pH 2 compared to the conventional heating method.[132] Zhang et al. prepared α-MnO$_2$–NH$_2$–RGO hybrid and characterized and used in removing hexavalent chromium ions (Cr^{6+}) from aqueous solutions. The adsorption equilibrium data were best described by the Freundlich isothermal model and the maximum sorption capacity toward Cr^{6+} was 371 mg g^{-1}. The kinetic adsorption was fitted to the pseudo-second-order kinetics model and governed by physical or chemical sorption on heterogeneous materials with exothermic nature.[133] Dubey et al. developed graphene–sand composite with surface area 100 m^2 g^{-1} and 157 m^2 g^{-1} before and after chemical activation, respectively, with adsorption capacity of Cr(VI) 2859.38 mg g^{-1}.[134] Chen et al. synthesized polyethyleneimine (PEI)–GO composite via amidation reaction between amine groups of the PEI and the carboxyl groups of the GO. The adsorption of Cr(VI) on this composite was completed within 1 h with uptake capacity of 539.53 mg g^{-1}.[135] Dinda et al. synthesized UV-active 2,6-diamino pyridine-reduced GO composite which showed Cr(VI) removal capacity 500 mg L^{-1} in 3 h only. The presence of an extra pyridinic-nitrogen lone pair facilitates strong interaction and hence enhances removal capacity of toxic Cr(VI).[136] Kan et al. synthesized TiO$_2$–graphene composites via reduction of GO by TiO$_2$ and composite was used for reductive mitigation of chromium(VI).[137]

12.4.4 CADMIUM

Cadmium has no known beneficial function in the human body. The major sources of cadmium in drinking water are corrosion of galvanized pipes, erosion of natural deposits, discharge from metal refineries, and runoff from waste batteries and paints. Cadmium causes cancer, birth defects, and genetic mutations and maximum amount are found in kidneys and the liver. EPA has set an enforceable regulation for cadmium, called a maximum contaminant level at 0.005 mg L^{-1} or 5 ppb.[138] Deng et al. synthesized magnetic GO and used as an adsorbent for removal of Cd(II). The kinetic data followed a pseudo-second-order model and equilibrium data were well fitted by the Langmuir model and the maximum sorption capacity for aqueous Cd(II) was 91.29 mg g^{-1}.[139] Yang et al. decorated GO nanosheets with a cysteine-rich metal-binding protein, cyanobacterium metallothionein (SmtA) and characterized using FT-IR, AFM, and TGA. In comparison with bare GO, the functionalized GO shows a 3.3-fold improvement over the binding capacity of cadmium, that is, 7.70 mg g^{-1} for SmtA–GO@cytopore compared to 2.34 mg g^{-1} for that by GO@cytopore.[140] Liu et al. grafted Fe$_3$O$_4$–graphene composite functionalized with thiol groups using (3-mercaptopropyl) trimethoxysilane. The Langmuir isotherm shows 125 mg g^{-1} maximum Cd(II) adsorption.[141]

12.4.5 LEAD

Lead(II) is a significant environmental health threat due to its nonbiodegradability, toxicity, wide-spread presence, and tendency to accumulate in living organisms.[142] It can cause nervous system damage, renal kidney disease, mental retardation, cancer, and anemia in humans.[143,144] It is introduced into water systems in various ways, including mining, painting and printing processes, plumbing, automobile batteries, and petroleum industries.[143] As a priority contaminant, the EPA has set a permissible limit of 50 ppb in drinking water.[145] Polymer-functionalized graphene has been used for Pb(II) adsorption from drinking water. Yang et al. developed poly(acrylamide) polymer brushes on RGO sheets by in situ free-radical polymerization. The experimental data for Pb(II) adsorption followed the Langmuir isotherm and display adsorption capacities as high as 1000 mg g^{-1}.[146] In situ polymerization of 1,5-diaminonaphthalene and 1,4-diaminoanthraquinone was carried on the surface of reduced graphite oxide by Olanipekun and coworkers.[147] The adsorption experiments were performed using 100 ppm aqueous solution of Pb^{2+} ions and concentration was investigated by AAS which shown highest absorptivity for RGO-P15DAN. Luo et al. prepared an oligomer-linked GO composite through simple cross-linking reactions between GO sheets and poly-3-aminopropyltriethoxysilane and in this way introduced a large amount of amino functional groups. The adsorption studies exhibited with the maximum adsorption capacity of 312.5 mg g^{-1} Pb(II) at 303 K and adsorption capacity increased with increasing temperature.[148] Song et al. used tea polyphenols as a reductant and functionalization reagent of GO and the resulting functionalized graphene nanosheets have mostly single-layer structure, are stable, and have very good water dispersibility.[149] Gui et al. synthesized a sandwich-like magnesium silicate/reduced GO nanocomposite by a hydrothermal approach exhibits a high specific surface area of 450 m^2 g^{-1} and shown the maximum adsorption capacities for lead ion of 416 mg g^{-1}.[150] Fan et al. fabricated magnetic chitosan/GO materials and studied the effects of pH, contact time, and concentration on Pb(II) ions sorption. The equilibrium studies showed Pb(II) adsorption followed the Langmuir model with adsorption capacity 76.94 mg g^{-1}.[151] Jabeen et al. reported synthesis of nano zerovalent iron nanoparticles decorated on graphene via a sodium borohydride reduction of GO and iron chloride under an argon atmosphere. TEM and XRD analysis shows the formation of ~10 nm particles with *bcc*–Fe phase. Adsorption experiments show a maximum Pb(II) adsorption capacity for the 6 wt% GO loading and X-ray photoelectron spectroscopy analysis confirmed the composite's ability to adsorb and immobilize lead.

The adsorption of Pb(II) ions fit a pseudo-second-order kinetic model, and adsorption isotherms described by Freundlich equations.[152] Hao et al. reported that synthesis of SiO_2/graphene composite via chemical reduction route and composite shows high efficiency and high selectivity toward Pb(II) ion. The maximum adsorption capacity of SiO_2/graphene composite for Pb(II) ion was found to be 113.6 mg g^{-1}, which was much higher than that of bare SiO_2 nanoparticles.[153]

12.4.6 COPPER

Copper is a reddish metal that occurs naturally in rock, soil, water, sediment, and air. It has many practical uses in our society and is commonly found in coins, electrical wiring, and pipes. It is an essential element for living organisms needed to prevent anemia and keep the skeletal, reproductive, and nervous systems healthy. However, too much copper can cause adverse health effects like gastrointestinal distress, and liver or kidney damage. EPA and WHO has recommended a maximum of limit for copper 1.3 mg L^{-1} or 1.3 ppm and 2.0 mg L^{-1}, respectively. The major sources of copper in drinking water are erosion of natural deposits, leaches out plumbing and industrial waste.[154] Singh et al. reported a simple eco-friendly method to obtain reduced GO using polyethylene glycol-400 as a reducing agent at 100°C in a low-power microwave. The reduced GO formed shows high specific area ranging from 644 to 1275 m^2 g^{-1} with increase in absorption capacity of Cu^{2+} ions.[155] Xing et al. dispersed and cross-linked GO nanosheets with a water-soluble polymer, namely poly(allylamine hydrochloride) and the resulting composite was used for removal of Cu(II) ions from aqueous solutions and shown maximum adsorption capacity about 349.03 mg g^{-1}.[156] Yang et al. studied the effects of humic acid on Cu(II) adsorption onto few-layer of graphene. Humic acid was adsorbed on the surface of graphene through π–π interaction and increases electron density and O-containing functional groups. EXAFS results suggested that Cu(II) was adsorbed on graphene and form graphene–humic acid–Cu ternary surface complexes, whereas in case of GO, humic acid showed precipitation of Cu(II).[157] Zhao et al. synthesized S-doped graphene sponge via hydrothermal route for the removal of Cu^{2+} with a huge adsorption capacity of 228 mg g^{-1}, 40 times higher than that of active carbon. The thermodynamics study showed that the adsorption was spontaneous, physisorption, and endothermic in nature.[158] Li et al. functionalized GO–Fe_3O_4 composite with fulvic acid and used for Cu(II) sorption and dominated by innersphere surface complexation.[159]

12.4.7 RADIONUCLIDE

Radionuclides are radioactive isotopes occur naturally (80%) or result from manmade sources (20%) and are colorless, odorless, and tasteless when dissolved in water. A radionuclide is an unstable nucleus and emits energy in the form of rays/particles to become more stable. The three basic types of radiation are positively charged alpha particles (+ve charge), negatively charged beta particles (−ve charge), and charge less gamma rays. Radiation exposure can occur by ingesting, inhaling, injecting, or absorbing radioactive materials. The amount of radiation exposure is usually expressed in a unit called millirem (mR or mrem), which is a measure of energy deposited in human tissue and its ability to produce biological damage.[160] Radioactive elements are naturally present in a wide range of concentrations in soil and rock and water which decay slowly and produce other radioactive elements known as daughter element. Fission products from manmade nuclear reactions, particularly radioactive cesium and iodine and natural occurring present in water and its effect on human health have become a major environmental concern.[161] The radioactivity measurements of water are expressed in picocuries per liter (pCi L^{-1}). Uranium is huge problem around the thermal power plant area and in the naturally deposited region. Uranium occurs naturally in the +2, +3, +4, +5, and +6 valence states, but it is most commonly found in the hexavalent form. In nature, hexavalent uranium is commonly associated with oxygen as the uranyl ion, UO_2^{2+}. Uranium(VI) compounds are more toxic than Uranium(IV) due to their high water solubility and mobility.[162] On the other hand, uranium(IV) is insoluble and immobile under ambient condition. The WHO has determined that safe drinking water should contain less than 30 parts per billion (ppb) uranium. The uranium content in the waters of Malwa region of Punjab (India) has 50% above the permissible WHO limit. Li et al. studied U(VI) sorption as a function of solution pH, ionic strength, and initial concentration of U(VI) on single-layered GO prepared by the Hummers method. The maximum sorption capacity of GO for U(VI) was evaluated to be 299 mg g^{-1} at pH = 4 due to formation of inner-sphere surface complexes.[163] In further work, Li et al. developed zero-valent iron nanoparticle and its graphene composites and studied uranium removal of 24 ppm U(VI) aqueous solution. Experimental data showed complete removal even in the presence of NaHCO$_3$, humic acid, and for a change of solution pH from 5 to 9. The maximum sorption capacity was found 8173 mg g^{-1}.[164] Sun et al. also developed the reduced GO-supported nanoscale zero-valent iron nanoparticles and observed enhanced capacity due to presence of hydroxyl functional groups. The fitting of EXAFS spectra showed the UC (at ~2.9 Å) and UFe (at ~3.2 Å) shells indicating the formation of inner-sphere surface complexes.[165] Tan et al. studied adsorption of uranium(VI) from aqueous solution by Fe_3O_4@TiO_2 composites. The adsorption of uranium(VI) could be well described by Langmuir with maximum sorption capacity 118.8 mg g^{-1} at pH 6.0. Thermodynamic parameters showed endothermic and spontaneous process.[166] Zhao et al. synthesized amidoximated magnetite/GO composites to adsorb uranium(VI) from aqueous solutions. Effects of pH, ionic strength, and coexisted ions on the sorption of U(VI) on composites were investigated and sorption isotherm agreed well with the Langmuir model having a maximum sorption capacity of 1.197 mmol g^{-1} at pH = 5.0 and T = 298 K.[167] Cheng et al. reported adsorptive behavior of uranium from aqueous solution on GO supported on sepiolite composites as a function of pH, ionic strength, temperature, and initial uranium concentration. The uptake equilibrium is best described by Langmuir adsorption isotherm, and the maximum adsorption capacity at pH 5.0 were calculated to be 161.29 mg g^{-1}.[168] Zhou et al. developed a novel core–shell Fe_3O_4@titanate nanocomposites as efficient adsorbents with sorption capacity 118.4 mg g^{-1} for Ba^{2+} ions. The results demonstrate that the Fe_3O_4@titanate nanomaterials can be used as a promising emergency radioactive adsorbent after a nuclear leakage.[169] Yang et al. reported a simple procedure to prepare magnetic Prussian blue/GO nanocomposites with efficient removal capacity of Cs^+ (55.56 mg g^{-1}) and adsorption mechanism occurs via H^+-exchange and/or ion trapping.[170] Romanchuk et al. have shown the efficacy of GO with actinides including Am(III), Th(IV), Pu(IV), Np(V), U(VI), and typical fission products Sr(II), Eu(III), and Tc(VII) from contaminated water in acidic solutions (pH < 2). The results indicate formation of nanoparticle aggregates of GO sheets.[171] Pan and coworkers prepared GO by modified method and investigated the removal of Th^{4+} ions from aqueous solutions. The sorbent provided significant Th^{4+} removal (>98.7%) at pH 3.0 and the adsorption equilibrium was achieved after only 10 min and a maximum adsorption capacity of 411 mg g^{-1} after 2 h.[172]

12.4.8 OTHERS IONS

Vasudevan et al. did systematic study for adsorption of *phosphate* on graphene by varying pH, ionic strength, and temperature and observed adsorption capacity of 89.37 mg g^{-1} and kinetic follow second-order model with spontaneous and endothermic process.[173] Tran et al. presented a new method for removal of *phosphate* using three-dimensional graphene aerogels decorated with goethite (α-FeOOH) and magnetite (Fe_3O_4) nanoparticles and showed adsorption capacity of 350 mg g^{-1} at an initial phosphate concentration of 200 mg L^{-1} in water.[174] Lakshmi et al. investigate the removal of perchlorate (ClO_4^-) from water by graphene synthesized via facile liquid-phase exfoliation and showed 99.2% efficiency. The kinetics studied shows second-order kinetics model and equilibrium data were well described by the typical Langmuir adsorption isotherm with maximum adsorption capacity of 0.024 mg g^{-1}.[175] Li and colleagues studied adsorption of fluoride from aqueous solution by graphene at different initial pH, contact time, and temperature. The experimental results showed maximum adsorption capacity of fluoride ions 35.6 mg g^{-1} at pH = 7 and 298 K.[176] Nandi et al. exploited reduced-graphene-oxide-based superparamagnetic nanocomposite for immobilization of Ni(II) in an aqueous solution by the fluorescent sensor platform. The results were explored at varying pH, doses, contact times, and temperatures and observed pseudo-second-order kinetics with monolayer sorption capacity of 228 mg g^{-1} at 300 K.[177] Selenium ions are toxic at concentrations of >40 ppb (40 μg L^{-1}) which has been a very challenging environmental issue. Fu et al. reported synthesis of functionalized water-dispersible magnetic nanoparticle–GO composites for removal of Se(IV) and Se(VI) in aqueous system. The composite showed removal percentage of >99.9% for Se(IV) and ~80% for Se(VI) from water (pH 6–7) within 10 s.[178] Liu et al. synthesized polyacrylamide-modified GO composites for adsorption of strontium(II) from aqueous solutions and experimental results showed adsorption capacity of 2.11 mmol g^{-1} at pH = 8.5.[179]

12.5 DESALINATION

Nanofiltration has received increasing attention as a cost- and energy-effective technique for removing dissolved salts from water, and thus producing clean water from seawater. The materials used in membrane are polymeric materials such as cellulose, nylon, and acetate, and nonpolymeric materials such as ceramics, metals, and composites.[180] The nanofiltration membrane works similar to reverse osmosis and is very common method for removing salts (desalination) from water. However, it requires extremely high pressure to force water through thick membrane which are about a thousand times thicker than graphene hence need a lot of energy.[181] Graphene and its derivatives have offered a novel class of mechanically robust, one atom thick, high-flux, high-selectivity, and fouling resistant separation membranes that provide opportunities to advance water desalination technologies. The new graphene system operates at much lower pressure due to the large pore size of membrane.[182] Qui et al. suggested that graphene sheets suspended in water are corrugated can be controlled at the nanometer scale by hydrothermal treatment. Corrugation makes graphene as the potential materials in application for nanofiltration.[183] Tanugi et al. observed nanometer-scale pores in single-layer freestanding graphene for effectively filter of NaCl salt from water and performance depends critically on pore diameter. The role of chemical functionalization on graphene showed presence of hydroxyl groups roughly double the water flux.[184] Konatham et al. employed molecular dynamics simulations to study the transport of water and ions through pores created on the basal plane of one graphene sheet ranged from 7.5 to 14.5 Å. The results indicate that effective ion exclusion using nonfunctionalized (pristine) pores of diameter ~7.5 Å.[185] Surwade et al. created nanometer-sized pores in a graphene monolayer using an oxygen plasma etching process and experimental data showed salt rejection rate of nearly 100% and water fluxes of up to 10^6 g m^{-2} s^{-1} at 40°C and using pressure difference while water fluxes measured using osmotic pressure as a driving force did not exceeded 70 g m^{-2} s^{-1} atm^{-1}.[186] Russo et al. demonstrated a scalable method for creating extremely small structures in graphene with atomic precision by inducing defect nucleation centers with energetic ions, followed by edge-selective electron recoil sputtering.[187] Mahmoud et al. reviewed the literature on application of graphene membrane in desalination and concluded that membranes formed from polyamide–graphene composite will be very effective for water remediation.[188] Bano et al. observed that polyamide nanofiltration membranes embedded with various GO contents showed 12-fold increase in water flux, with a negligible change in salt rejection. The presence of GO also improved the antifouling property due to an increase in the hydrophilicity of the membrane.[189]

12.6 TOXICITY OF GRAPHENE

The emergence of graphene as a single-atom-thick carbon nanosheet with great mechanical strength, high surface area and fascinating optoelectronic properties attracted great interest as a promising nanomaterial for a variety of applications.[9] However, the safety concerns about graphene and its derivatives are questionable. The graphene is light weight, can be inhale, and might be transported deep within the lungs. The toxicity of graphene-based derivative in biological systems such as bacteria, mammalian cells, and animal models is are serious issues for safer design and manufacturing of graphene and its derivatives for future application.[190] Akhavan et al. studied bacterial toxicity of graphene nanosheets for both Gram-positive and Gram-negative models of bacteria and found that the GO reduced by hydrazine was more toxic to the bacteria than the unreduced GO.[191] Vallabani et al. characterized the GO using dynamic light scattering along with the toxicological aspects related to cytotoxicity and apoptosis in normal human lung cells. A significant concentration and time dependent decrease in cell viability was observed at different concentrations (10–100 μg mL^{-1}) by the MTT assay after 24 and 48 h of exposure and significant increase of early and late apoptotic cells was observed as compared to control cells which demonstrates that GO induces cytotoxicity and apoptosis in human lung cells.[192] Yang et al. observed that the physicochemical properties such as surface functional groups, charges, coatings, sizes, and structural defects of graphene may affect its in vitro/in vivo behavior as well as its toxicity in biological systems.[193] Schinwald et al. derived respirable graphene nanoplatelets 25 μm in diameter and deposited beyond the ciliated airways following inhalation and utilized models of pharyngeal aspiration. In vitro tests showed that these particles trigger the inflammatory response in lung cells and those found in the pleural space. Intriguingly, the immune response is not seen with nanoparticulate carbon black.[194] Chong and colleagues studied the interaction between serum proteins and graphene derivatives using molecular dynamic simulations which reveal that the protein-coated GO resulted in a markedly less cytotoxicity than pristine graphene.[195] Hu et al. studied the toxicity of GO with protozoa *Euglena gracilis* with exposer to GO ranging from 0.5 mg L^{-1} to 5 mg L^{-1} for 10 days. Results showed that the 96 h EC50 value of GO in *E. gracilis* was 3.76 ± 0.74 mg L^{-1}. GO at a concentration of 2.5 mg L^{-1} exerted significant ($P < 0.01$) adverse effects on the organism.[196] Seabra et al. observed that there is no evidence that particles below 100 nm show any drastic change in their hazard compare to the conventional particle.[197] Li et al. observed that edges of graphene platelets were capable of piercing human lung tissue.[198] Ahmad et al. investigate the acute toxicity effect of GO on the biological wastewater treatment process in concentrations between 50 and 300 mg L^{-1}. Microscopic techniques confirmed penetration and accumulation of GO inside the activated sludge floc matrix and produced significant amount of reactive oxygen species which is responsible for the toxic effect of

GO.[199] Sanchez and colleagues reviewed the papers published on graphene's and suggested that in vitro toxicity of graphene derivative is toxic to cells and depending on layer number, lateral size, stiffness, hydrophobicity, surface functionalization, and dose.[200] Guo et al. summarized the recent findings on the toxicological effects and the potential toxicity mechanisms of graphene-family nanomaterials in bacteria, mammalian cells, and animal models. Graphene, GO, and reduced GO elicit toxic effects both in vitro and in vivo, whereas surface modifications can significantly reduce their toxic interactions with living systems.[201] Bianco et al. critically discussed in the review articles that in vitro and in vivo studies clearly showed no particular risks while others have indicated health hazards.[202]

12.7 CONCLUSION AND OUTLOOK

The presence of inorganic toxic pollutants such as arsenic, mercury, chromium, lead, cadmium, uranium, etc. in the water is the most challenging problem of the 21st century. The conventional remediation technology will be failed in providing drinkable water to 10 billion populations across globe in 2020. There is urgent need to find out eco-friendly, low-cost, and easily handling technique for purifying contaminated water. Graphene, a unique two-dimensional single-atom-thin nanomaterial with exceptional structural, mechanical, and electronic properties, has spurred an enormous interest in waste-water treatment. The graphene-based adsorbent, photocatalyst, and membranes have been developed for potential applications in toxic pollutants remediation and desalination of water. The large production of graphene in the environment-friendly route is still a challenge and at the same time it ignites a growing concern on its biosafety and potential cytotoxicity to human and animal cells. The risk presented by graphene on soil, environment, and aquatic organisms needs to be understood and identified for future application of graphene to make clean water more accessible around the globe.

ACKNOWLEDGMENT

V. Chandra acknowledges the financial support from University Grants Commission under the scheme Start-Up Research Grant for Newly Recruited Faculty.

KEYWORDS

- water
- dyes
- pesticides
- arsenic
- graphene

REFERENCES

1. Shannon, M. A.; Bohn, P. W.; Elimelech, M.; Georgiadis, J. G.; Marinas, B. J.; Mayes, A. M. Science and Technology for Water Purification in the Coming Decades. *Nature* **2008**, *452*, 301–310.
2. Malato, S.; Fernández-Ibáñez, P.; Maldonado; M. I.; Blanco, J.; Gernjak, W. Decontamination and Disinfection of Water by Solar Photocatalysis: Recent Overview and Trends. *Catal. Today* **2009**, *147*, 1–59.
3. Mohan, D.; Pittman Jr., C. U. Activated Carbons and Low Cost Adsorbents for Remediation of Tri- and Hexavalent Chromium from Water. *J. Hazard. Mater. B* **2007**, *137*, 762–811.
4. Mauter, M. S.; Elimelech, M. Environmental Applications of Carbon-Based Nanomaterials. *Environ. Sci. Technol.* **2008**, *42*, 5843–5859.
5. Stone, V.; Nowack, B.; Baun, A.; van den Brink, N.; von der Kammer, F.; Dusinska, M.; Nanomaterials for Environmental Studies: Classification, Reference Material Issues, and Strategies for Physico-Chemical Characterization. *Sci. Total Environ.* **2010**, *408*, 1745–1754.
6. Ali, I. New Generation Adsorbents for Water Treatment. *Chem. Rev.* **2012**, *112*, 5073–5091.
7. Novoselov, K. S.; Geim, A. K.; Morozov, S. V.; Jiang, D.; Zhang, Y.; Dubonos, S. V.; Grigorieva, I. V.; Firsov, A. A. Electric Field Effect in Atomically Thin Carbon Films. *Science* **2004**, *306*, 666–669.
8. Rao, C. N. R.; Sood, A. K.; Subrahmanyam, K. S.; Govindaraj, A. Graphene: The New Two-Dimensional Nanomaterial. *Angew. Chem.* **2009**, *48*, 7752–7777.
9. Georgakilas, V.; Otyepka, M.; Bourlinos, A. B.; Chandra, V.; Kim, N.; Hobza, P.; Zboril, R.; Kim, K. S. Functionalization of Graphene: Covalent and Non-Covalent Approaches, Derivatives and Applications. *Chem. Rev.* **2012**, *112*, 6156–6214.
10. Sreeprasad, T. S.; Maliyekkal, S. M.; Lisha, K. P.; Pradeep, T. Reduced Graphene Oxide–Metal/Metal Oxide Composites: Facile Synthesis and Application in Water Purification. *J. Hazard. Mater.* **2011**, *186*, 921–931.
11. Kemp, K. C.; Seema, H.; Saleh, M.; Le, N. H.; Mahesh, K.; Chandra, V.; Kim, K. S. Environmental Applications Using Graphene Composites: Water Remediation and Gas Adsorption. *Nanoscale* **2013**, *5*, 3149–3171.

12. Chowdhury, S.; Balasubramanian, R. Recent Advances in the Use of Graphene-Family Nanoadsorbents for Removal of Toxic Pollutants from Wastewater. *Adv. Colloid Interface Sci.* **2014**, *204*, 35–56.
13. Perreault, F.; de Faria, A. F.; Elimelech, M. Environmental Applications of Graphene-Based Nanomaterials. *Chem. Soc. Rev.* **2015**, *44*, 5861–5896 .
14. Yusuf, M.; Elfghi, F. M.; Zaidi, S. A.; Abdullah, E. C.; Ali, K. M. Applications of Graphene and Its Derivatives as an Adsorbent for Heavy Metals and Dyes Removal: A Systematic and Comprehensive Overview. *RSC Adv.* **2015**, *5*, 50392–50420.
15. Krishnan, K. S.; Ganguli, N. Large Anisotropy of the Electrical Conductivity of Graphite. *Nature* **1939**, *144*, 667–667.
16. Coulson, C. A. Energy Bands in Graphite. *Nature* **1947**, *159*, 265–266.
17. Wallace, P. R. *Band Theory Graphite Phys. Rev.* **1947**, *71*, 622–634.
18. Moran, H. F. Single Crystals of Graphite and Mica as Specimen Support for Electron Microscopy. *J. Appl. Phys.* **1960**, *31*, 1844–1846.
19. Brodie, B. C. On the Atomic Weight of Graphite. *Philos. Trans. R. Soc. Lond.* **1859**, *149*, 249–259.
20. Brodie, B. C. Researches on the Atomic Weight of Graphite. *Q. J. Chem. Soc.* **1860**, *12*, 261–268.
21. Staudenmaier, L. Verfahren zur Darstellung der Graphitsäure. *Ber. Dtsch. Chem. Ges.* **1898**, *31*, 1481–1487.
22. Hummers Jr., W. S.; Offeman, R. E. Preparation of Graphitic Oxide. *J. Am. Chem. Soc.* **1958**, *80*, 1339–1339.
23. Clauss, A.; Plass, R.; Boehm, H. P.; Hofmann, U. Z. Untersuchungen zur Struktur des Graphitoxyds *Anorg. Allg. Chem.* **1957**, *291*, 205–220.
24. Boehm, H. P.; Clauss, A.; Hofmann, U.; Fischer, G. O. Dunnste Kohlenstoff-Folien, *Zeitsch. Naturforsch., B—Chem. Biochem. Biophys. Biol. Verwandt. Geb.* **1962**, *B17*, 150.
25. Fitzer, E.; Kochling, K.-H.; Boehm, H. P.; Marsh, H. Recommended Terminology for the Description of Carbon as a Solid. *Pure Appl. Chem.* **1995**, *67*, 473–506.
26. Lu, X.; Yu, M.; Huang, H.; Ruoff, R. S. Tailoring Graphite with the Goal of Achieving Single Sheets. *Nanotechnology* **1999**, *10*, 269–273.
27. Zhang, Y.; Small, J. P.; Pontius, W. V.; Kim, P. Fabrication and Electric-Field-Dependent Transport Measurements of Mesoscopic Graphite Devices. *Appl. Phys. Lett.* **2005**, *86*, 073104–073106.
28. Stankovich, S.; Dikin, D. A.; Piner, R. D.; Kohlhaas, K. A.; Kleinhammes, A.; Jia, Y.; Wu, Y.; Nguyen, S. B. T.; Ruoff, R. S. Synthesis of Graphene-Based Nanosheets *via* Chemical Reduction of Exfoliated Graphite Oxide. *Carbon* **2007**, *45*, 1558–1565.
29. Pei, S.; Cheng, H. M. The Reduction of Graphene Oxide. *Carbon* **2012**, *50*, 3210–3228.
30. Schniepp, H. C.; Li, J.-L.; McAllister, M. J.; Sai, H.; Alonso, M. H.; Adamson, D. H.; Prud'homme, R. K.; Car, R.; Saville, D. A.; Aksay, I. A. Functionalized Single Graphene Sheets Derived from Splitting Graphite Oxide. *J. Phys. Chem. B* **2006**, *110*, 8535–8539.
31. Hassan, H. M. A.; Abdelsayed, V.; Khder, A. E. R. S.; Zeid, K. M. A.; Terner, J.; El-Shall, M. S.; Al-Resayes, S. I.; El-Azhary, A. A. Microwave Synthesis of Graphene Sheets Supporting Metal Nanocrystals in Aqueous and Organic Media. *J. Mater. Chem.* **2009**, *19*, 3832–3837.
32. Cote, L. J.; Cruz-Silva, R.; Huang, J. Flash Reduction and Patterning of Graphite Oxide and Its Polymer Composite. *J. Am. Chem. Soc.* **2009**, *131*, 11027–11032.

33. Mondal, O.; Mitra, S.; Pal, M.; Datta, A.; Dhara, S.; Chakravorty, D. Reduced Graphene Oxide Synthesis by High Energy Ball Milling. *Mater. Chem. Phys.* **2015**, *161*, 123–129.
34. Mohan, M.; Chandra, V.; Manoharan, S. S. Nano Body-Centered Cubic CoFe$_2$ Alloy Precursor for Cobalt Ferrite via Sonoreduction Process. *J. Mater. Res.* **2008**, *23*, 1849–1853.
35. Sivakumar, M.; Gedanken, A. A Sonochemical Method for the Synthesis of Polyaniline and Au–Polyaniline Composites Using H$_2$O$_2$ for Enhancing Rate and Yield. *Synthetic Metals* **2005**, *148*, 301–306.
36. Ciesielski, A.; Samorì, P. Graphene *via* Sonication Assisted Liquid-Phase Exfoliation. *Chem. Soc. Rev.* **2014**, *43*, 381–398.
37. Hernandez, Y.; Nicolosi, V.; Lotya, M.; Blighe, F. M.; Sun, Z.; De, S.; McGovern, I. T.; Holland, B.; Byrne, M.; Gun'Ko, Y. K.; Boland, J. J.; Niraj, P.; Duesberg, G.; Krishnamurthy, S.; Goodhue, R.; Hutchison, J.; Scardaci, V.; Ferrari, A. C.; Coleman, J. N. High-Yield Production of Graphene by Liquid-Phase Exfoliation of Graphite. *Nat. Nanotechnol.* **2008**, *3*, 563–568.
38. Somani, P. R.; Somani, S. P.; Umeno, M. Planer Nano-Graphenes from Camphor by CVD. *Chem. Phys. Lett.* **2006**, *430*, 56–59.
39. Kim, K. S.; Zhao, Y.; Jang, H.; Lee, S. Y.; Kim, J. M.; Kim, K. S.; Ahn, J.-H.; Kim, P.; Choi, J.-Y.; Hong, B. H. Large-Scale Pattern Growth of Graphene Films for Stretchable Transparent Electrodes. *Nature* **2009**, *457*, 706–710.
40. Reina, A.; Jia, X.; Ho, J.; Nezich, D.; Son, H.; Bulovic, V.; Dresselhaus, M. S.; Kong, J. Large Area, Few-Layer Graphene Films on Arbitrary Substrates by Chemical Vapor Deposition. *Nano Lett.* **2009**, *9*, 30–35.
41. Li, X.; Cai, W.; An, J.; Kim, S.; Nah, J.; Yang, D.; Piner, R.; Velamakanni, A.; Jung, I.; Tutuc, E.; Banerjee, S. K.; Colombo, L.; Ruoff, R. S. Large-Area Synthesis of High-Quality and Uniform Graphene Films on Copper Foils. *Science* **2009**, *324*, 1312–1314.
42. Wang, J. J.; Zhu, M. Y.; Outlaw, R. A.; Zhao, X.; Manos, D. M.; Holoway, B. C. Synthesis of Carbon Nanosheets by Inductively Coupled Radio-Frequency Plasma Enhanced Chemical Vapor Deposition. *Carbon* **2004**, *42*, 2867–2872.
43. de Heer, W. A.; Berger, C.; Wu, X.; First, P. N.; Conrad, E. H.; Li, X. Epitaxial Graphene. *Solid State Commun.* **2007**, *143*, 92–100.
44. Yang, X.; Dou, X.; Rouhanipour, A.; Zhi, L.; Rader, H. J.; Mullen, K. Two-Dimensional Graphene Nanoribbons. *J. Am. Chem. Soc.* **2008**, *130*, 4216–4217.
45. Jiao, L.; Zhang, L.; Wang, X.; Diankov, G.; Dai, H. Narrow Graphene Nanoribbons from Carbon Nanotubes. *Nature* **2009**, *458*, 877–880.
46. Ferrari, A. C.; Meyer, J. C.; Scardaci, V.; Casiraghi, C.; Lazzeri, M.; Mauri, F.; Piscanec, S.; Jiang, D.; Novoselov, K. S.; Roth, S.; Geim, A. K. Raman Spectrum of Graphene and Graphene Layers. *Phys. Rev. Lett.* **2006**, *97*, 187401.
47. Novoselov, K. S.; Jiang, D.; Schedin, F.; Booth, T. J.; Khotkevich, V. V.; Morozov, S. V.; Geim, A. K. Two-Dimensional Atomic Crystals. *Proc. Natl. Acad. Sci. U. S. A.* **2005**, *102*, 10451–10453.
48. Valles, C.; Drummond, C.; Saadaoui, H.; Furtado, C. A.; He, M.; Roubeau, O.; Ortolani, L.; Monthioux, M.; Penicaud, A. Solutions of Negatively Charged Graphene Sheets and Ribbons. *J. Am. Chem. Soc.* **2008**, *130*, 15802–15804.
49. Casiraghi, C.; Hartschuh, A.; Lidorikis, E.; Qian, H.; Harutyunyan, H.; Gokus, T.; Novoselov, K. S.; Ferrari, A. C. Rayleigh Imaging of Graphene and Graphene Layers. *Nano Lett.* **2007**, *7*, 2711–2717.

50. Ferrari, A. C.; Basko, D. M. Raman Spectroscopy as a Versatile Tool for Studying the Properties of Graphene. *Nat. Nanotechnol.* **2013**, *8*, 235–246.
51. Malard, L. M.; Pimenta, M. A.; Dresselhaus, G; Dresselhaus, M. S. Raman Spectroscopy in Graphene. *Phys. Rep.* **2009**, *473*, 51–87.
52. Dabrowski, A. Adsorption—From Theory to Practice. *Adv. Colloid Interface Sci.* **2001**, *93*, 135–224.
53. Kadirvelu, K.; Thamaraiselvi, K.; Namasivayam, C. Removal of Heavy Metals from Industrial Wastewaters by Adsorption onto Activated Carbon Prepared from an Agricultural Solid Waste. *Bioresour. Technol.* **2001**, *76*, 63–65.
54. Hoffmann, M. R.; Martin, S. T.; Choi, W. Y.; Bahnemann, D. W. Environmental Applications of Semiconductor Photocatalysis. *Chem. Rev.* **1995**, *95*, 69–96.
55. Fujishima, A.; Honda, K. Electrochemical Photolysis of Water at a Semiconductor Electrode *Nature* **1972**, *238*, 37–38.
56. Carey, J. H.; Lawrence, J.; Tosine, H. M. Photodechlorination of PCB's in the Presence of Titanium Dioxide in Aqueous Suspensions. *Bull. Environ. Contam. Toxicol.* **1976**, *16*, 697–701.
57. Lachhe, H.; Puzenat, E.; Houas, A.; Ksibi, M.; Elaloui, E.; Guillard, C.; Herrmann, J.-M. Photocatalytic Degradation of Various Types of Dyes (Alizarin S, Crocein Orange G, Methyl Red, Congo Red, Methylene Blue) in Water by UV-Irradiated Titania. *Appl. Catal., B: Environ.* **2002**, *39*, 75–90.
58. Mo, J. H.; Zhang, Y. P.; Xu, Q. J.; Lamson, J. J.; Zhao, R. Y. Photocatalytic Purification of Volatile Organic Compounds in Indoor Air: A Literature Review. *Atmos. Environ.* **2009**, *43*, 2229–2246.
59. Konstantinou, I. K.; Albanis, T. A. Photocatalytic Transformation of Pesticides in Aqueous Titanium Dioxide Suspensions Using Artificial and Solar Light: Intermediates and Degradation Pathways. *Appl. Catal. B* **2002**, *1310*, 1–17.
60. Herrmann, J.-M. Heterogeneous Photocatalysis: Fundamentals and Applications to the Removal of Various Types of Aqueous Pollutants. *Catal. Today* **1999**, *53*, 115–129.
61. Asahi, R.; Morikawa, T.; Ohwaki, T.; Aoki, K.; Taga, Y. Visible-Light Photocatalysis in Nitrogen-Doped Titanium Oxides. *Science* **2001**, *293*, 269–271.
62. *Ehret, A.;* Stuhl, L.; Spitler, M. T. Spectral Sensitization of TiO_2 Nanocrystalline Electrodes with Aggregated Cyanine Dyes. *J. Phys. Chem. B* **2001**, *105*, 9960–9965.
63. Khanchandani, S.; Kundu, S.; Patra, A.; Ganguli, A. K. Band Gap Tuning of ZnO/In_2S_3 Core/Shell Nanorod Arrays for Enhanced Visible-Light-Driven Photocatalysis. *J. Phys. Chem. C* **2013**, *117*, 5558–5567.
64. Shiraishi, Y.; Morishita, M.; Hirai, T. Adsorption-Driven Photocatalytic Activity of Mesoporous Titanium Dioxide. *J. Am. Chem. Soc.* **2005**, *127*, 12820–12822.
65. Chen, X.; Shen, S.; Guo, L.; Mao, S. S. Semiconductor-Based Photocatalytic Hydrogen Generation. *Chem. Rev.* **2010**, *110*, 6503–6570.
66. Kumar, M.; Puri, A. A Review of Permissible Limits of Drinking Water. *Indian J. Occup., Enviorn. Med.* **2012**, *16*, 40–44.
67. http://www.lenntech.com/who-eu-water-standards.htm.
68. Hamilton, J. W.; Kaltreider, R. C.; Bajenova, O. V.; Ihnat, M. A.; McCaffrey, J.; Turpie, B. W.; Rowell, E. E.; Oh, J.; Nemeth, M. J.; Pesce, C. A.; Lariviere, J. P. Molecular Basis for Effects of Carcinogenic Heavy Metals on Inducible Gene Expression. *Environ. Health Perspect.* **1998**, *106*, 1005–1015.
69. Vallee, B. L.; Ulmer, D. D. Biochemical Effects of Mercury, Cadmium, and Lead. *Annu. Rev. Biochem.* **1972**, *41*, 91–128.

70. Partanen, T.; Heikkila, P.; Hernberg, S.; Kauppinen, T.; Moneta, G.; Ojajarvi, A. Renal Cell Cancer and Occupational Exposure to Chemical Agents. *Scand. J. Work. Environ. Health* **1991**, *17*, 231–239.
71. Gasparik, J.; Vladarova, D.; Capcarova, M.; Smehyl, P.; Slamecka, J.; Garaj, P.; Stawarz, R.; Massanyi, P. *J. Environ. Sci. Health, A* **2010**, *45*, 818–823.
72. Caroli, S.; Forte, G.; Iamiceli, A. L.; Galoppi, B. Determination of Essential and Potentially Toxic Trace Elements in Honey by Inductively Coupled Plasma-Based Techniques. *Talanta* **1999**, *50*, 327–336.
73. Flamini, R.; Panighel, A. Mass Spectrometry in Grape and Wine Chemistry. Part II: The Consumer Protection. *Mass Spectrom. Rev.* **2006**, *25*, 741–774.
74. Potts, P. J.; Webb, P. C. X-Ray Fluorescence Spectrometry. *J. Geochem. Explor.* **1992**, *44*, 251–296.
75. Mimendia, A.; Legin, A.; Merkoc-i, A.; del Valle, M. Use of Sequential Injection Analysis to Construct a Potentiometric Electronic Tongue: Application to the Multidetermination of Heavy Metals. *Sens. Actuators, B* **2010**, *146*, 420–426.
76. Mohan, D.; Pittman, C. U., Jr. Arsenic Removal from Water/Wastewater Using Adsorbents: A Critical Review. *J. Hazard. Mater.* **2007**, *142*, 1–53.
77. Polizzotto, M. L.; Kocar, B. D.; Benner, S. G.; Sampson, M.; Fendorf, S. Near-Surface Wetland Sediments as a Source of Arsenic Release to Ground Water in Asia. *Nature* **2008**, *454*, 505–509.
78. Meharg, A. Earth, V. *Venomous Earth: How Arsenic Caused the World's Worst Mass Poisoning*; Macmillan: Houndsmill, England, 2005.
79. Zhong, L.-S. ; Hu, J.-S.; Liang, H.-P.; Cao, A.-M.; Song, W.-G.; Wan, L.-J. Self-Assembled 3D Flowerlike Iron Oxide Nanostructures and Their Application in Water Treatment *Adv. Mater.* **2006**, *18*, 2426–2431.
80. Yavuz, C. T.; Mayo, J. T.; Yu, W. W.; Prakash, A.; Falkner, J. C.; Yean, S.; Cong, L.; Shipley, H. J.; Kan, A.; Tomson, M.; Natelson, D.; Colvin, V. L. Low-Field Magnetic Separation of Monodisperse Fe_3O_4 Nanocrystals. *Science* **2006**, *314*, 964–967.
81. Gu, Z.; Fang, J.; Deng, B. Preparation and Evaluation of GAC-Based Iron-Containing Adsorbents for Arsenic Removal. *Environ. Sci. Technol.* 2005, *39*, 3833–3843.
82. Mayo, J. T.; Yavuz, C.; Yean, S.; Cong, L.; Shipley, H.; Yu, W.; Falkner, J.; Kan, A.; Tomson, M.; Colvin, V. L. The Effect of Nanocrystalline Magnetite Size on Arsenic Removal. *Sci. Technol. Adv. Mater.* 2007, *8*, 71–75.
83. Cumbal, L.; Gupta, A. K. S. Arsenic Removal Using Polymer-Supported Hydrated Iron(III) Oxide Nanoparticles: Role of Donnan Membrane Effect. *Environ. Sci. Technol.* **2005**, *39*, 6508–6515.
84. Chandra, V.; Park, J.; Chun, Y.; Lee, J. W.; Hwang, I.-C.; Kim, K. S. Water-Dispersible Magnetite-Reduced Graphene Oxide Composites for Arsenic Removal. *ACS Nano* **2010**, *4*, 3979–3986.
85. Mishra, A. K.; Ramaprabhu, S. Ultrahigh Arsenic Sorption Using Iron Oxide–Graphene Nanocomposite Supercapacitor Assembly. *J. Appl. Phys.* **2012**, *112*, 104315.
86. Paul, B.; Parashar, V.; Mishra, A. Graphene in the Fe_3O_4 Nano-Composite Switching the Negative Influence of Humic Acid Coating into an Enhancing Effect in the Removal of Arsenic from Water. *Environ. Sci.: Water Res. Technol.* **2015**, *1*, 77–83.
87. Guo, L; Ye, P; Wang, J; Fu, F; Wu, Z. Three-Dimensional Fe_3O_4-Graphene Macroscopic Composites for Arsenic and Arsenate Removal. *J. Hazard. Mater.* **2015**, *298*, 28–35.

88. Vadahanambi, S.; Lee, S. H.; Kim, W. J.; Oh, I. K. Arsenic Removal from Contaminated Water Using Three-Dimensional Graphene–Carbon Nanotube–Iron Oxide Nanostructures. *Environ. Sci. Technol.* **2013**, *47*, 10510–10517.
89. Wen, T.; Wu, X.; Tan, X.; Wang, X.; Xu, A. One-Pot Synthesis of Water-Swellable Mg–Al Layered Double Hydroxides and Graphene Oxide Nanocomposites for Efficient Removal of As(V) from Aqueous Solutions. *ACS Appl. Mater. Interfaces* **2013**, *5*, 3304–3311.
90. Wu, X.-L.; Wang, L.; Chen, C.-L.; Xu, A.-W.; Wang, X.-K. Water-Dispersible Magnetite–graphene–LDH Composites for Efficient Arsenate Removal. *J. Mater. Chem.* **2011**, *21*, 17353–17359.
91. Luo, X.; Wang, C.; Luo, S.; Dong, R.; Tu, X.; Zeng, G. Adsorption of As(III) and As(V) from Water Using Magnetite Fe_3O_4-Reduced Graphite Oxide–MnO_2 Nanocomposites. *Chem. Eng. J.* **2012**, *187*, 45–52.
92. Nandi, D.; Gupta, K.; Ghosh, A. K.; De, A.; Banerjee, S.; Ghosh, U. C. Manganese-Incorporated Iron(III) Oxide–Graphene Magnetic Nanocomposite: Synthesis, Characterization, and Application for the Arsenic (III)-Sorption from Aqueous Solution. *J. Nanopart. Res.* **2012**, *14*, 1272.
93. Wang, C.; Luo, H.; Zhang, Z.; Wu, Y.; Zhang, J.; Wu, Y.; Zhang, J.; Chen, S. Removal of As(III) and As(V) from Aqueous Solutions Using Nanoscale Zero Valent Iron-Reduced Graphite Oxide Modified Composites. *J. Hazard. Mater.* **2014**, *268*, 124–131.
94. Zhu, J.; Sadu, R.; Wei, S.; Haldolaarachchige, N.; Luo, Z.; Young, D. P.; Guo, Z. Magnetic Graphene Nanoplatelet Composites Toward Arsenic Removal. *ECS J. Solid State Sci. Technol.* **2012**, *1*, M1–M5.
95. Mandel, K.; Hutter, F. The Magnetic Nanoparticle Separation Problem. *Nano Today* **2012**, *7*, 485–487.
96. Zhang, K.; Dwivedi, V.; Chi, C.; Wu, J. Graphene Oxide/Ferric Hydroxide Composites for Efficient Arsenate Removal from Drinking Water. *J. Hazard. Mater.* **2010**, *182*, 162–168.
97. Chen, M. L.; Sun, Y.; Huo, C. B.; Liu, C.; Wang, J. H. Akaganeite Decorated Graphene Oxide Composite for Arsenic Adsorption/Removal and It's Proconcentration at Ultra-Trace Level. *Chemosphere* **2015**, *130*, 52–59.
98. Andjelkovic, I.; Tran, D. N. H.; Kabiri, S.; Azari, S.; Markovic, M.; Losic, D. Graphene Aerogels Decorated with α-FeOOH Nanoparticles for Efficient Adsorption of Arsenic from Contaminated Waters. *ACS Appl. Mater. Interfaces* **2015**, *7*, 9758–9766.
99. Luo, X.; Wang, C.; Wang, L.; Deng, F.; Luo, S.; Tu, X.; Au, C. Nanocomposites of Graphene Oxide-Hydrated Zirconium Oxide for Simultaneous Removal of As(III) and As(V) from Water. *Chem. Eng. J.* **2013**, *220*, 98–106.
100. Cortés-Arriagada, D.; Toro-Labbé, A. Improving As(III) Adsorption on Graphene Based Surfaces: Impact of Chemical Doping. *Phys. Chem. Chem. Phys.* **2015**, *17*, 12056–12064.
101. Harris, H. H.; Pickering, I. J.; George, G. N. The Chemical Form of Mercury in Fish. *Science* **2003**, *301*, 1203–1203.
102. Tchounwou, P. B.; Ayensu, W. K.; Ninashvili, N.; Sutton, D. Environmental Exposure to Mercury and its Toxicopathologic Implications for Public Health. *Environ. Toxicol.* **2003**, *18*, 149–175.
103. Chiarle, S; Ratto, M; Rovatti, M. Mercury Removal from Water by Ion Exchange Resins Adsorption. *Water Res.* **2000**, *34*, 2971–2978.

104. Zhu, J.; Yang, J.; Deng, B. Enhanced Mercury Ion Adsorption by Amine-Modified Activated Carbon. *J. Hazard. Mater* **2009**, *166*, 866–872.
105. Cai, J. H.; Jia, C. Q. Mercury Removal from Aqueous Solution Using Coke-Derived Sulfur-Impregnated Activated Carbons. *Ind. Eng. Chem. Res.* **2010**, *49*, 2716–272.
106. Brown, J.; Mercier, L.; Pinnavaia, T. J. Selective Adsorption of Hg^{2+} by Thiol Functionalized Nanoporous Silica. *Chem. Commun.* **1999**, 69–70.
107. Shin, S.; Jang, J. Thiol Containing Polymer Encapsulated Magnetic Nanoparticles as Reusable and Efficiently Separable Adsorbent for Heavy Metal Ions. *Chem. Commun.* **2007**, 4230–4232.
108. Wang, J.; Deng, B. L.; Chen, H. Removal of Aqueous Hg(II) by Polyaniline: Sorption Characteristics and Mechanisms. *Environ. Sci. Technol.* **2009**, *43*, 5223–5228.
109. Lin, Y.; Cui, X.; Bontha, J. Electrically Controlled Anion Exchange Based on Polypyrrole and Carbon Nanotubes Nanocomposite for Perchlorate Removal. *Environ. Sci. Technol.* **2006**, *40*, 4004–4009.
110. Chandra, V.; Kim, K. S. Highly Selective Adsorption of Hg^{2+} by a Polypyrrole-Reduced Graphene Oxide Composite. *Chem. Commun.* **2011**, *47*, 3942–3944.
111. Li, R.; Liu, L.; Yang, F. Preparation of Polyaniline/Reduced Graphene Oxide Nanocomposite and Its Application in Adsorption of Aqueous Hg(II). *Chem. Eng. J.* **2013**, *229*, 460–468.
112. Thakur, S.; Das, G.; Raul, P. K.; Karak, N. Green One-Step Approach to Prepare Sulfur/Reduced Graphene Oxide Nanohybrid for Effective Mercury Ions Removal. *J. Phys. Chem. C* **2013**, *117*, 7636–7642.
113. Kumar, A. S. K.; Jiang, S. J. Preparation and Characterization of Exfoliated Graphene Oxide-l-Cystine as an Effective Adsorbent of Hg(II) Adsorption. *RSC Adv.* **2015**, *5*, 6294–6304.
114. Diagboya, P. N.; Olu-Owolabi, B. I.; Adebowale, K. O. Synthesis of Covalently Bonded Graphene Oxide–Iron Magnetic Nanoparticles and the Kinetics of Mercury Removal. *RSC Adv.* **2015**, *5*, 2536–2542.
115. Liu, Y.; Tian, C.; Yan, B.; Lu, Q.; Xie, Y.; Chen, J.; Gupta, R.; Xu, Z.; Kuznicki, S. M.; Liu, Q.; Zeng, H. Nanocomposites of Graphene Oxide, Ag Nanoparticles, and Magnetic Ferrite Nanoparticles for Elemental Mercury (Hg0) Removal. *RSC Adv.* **2015**, *5*, 15634–15640.
116. Bao, J.; Fu, Y.; Bao, Z. Thiol-Functionalized Magnetite/Graphene Oxide Hybrid as a Reusable Adsorbent for Hg^{2+} Removal. *Nanoscale Res. Lett.* **2013**, *8*, 486.
117. Mohan, D.; Pittman Jr.; C. U. Activated Carbons and Low Cost Adsorbents for Remediation of Tri- and Hexavalent Chromium from Water. *J. Hazard. Mater. B* **2006**, *137*, 762–811.
118. Hsu, N. H.; Wang, S. L.; Lin, Y. C.; Sheng, G. D.; Lee, J. F. Reduction of Cr(VI) by Cropresidue-Derived Black Carbon. *Environ. Sci. Technol.* **2009**, *43*, 8801–8806.
119. Legrand, L.; Figuigui, A. E.; Mercier, F.; Chausse, A. Reduction of Aqueous Chromate by Fe(II)/Fe(III) Carbonate Green Rust: Kinetic and Mechanistic Studies. *Environ. Sci. Technol.* **2004**, *38*, 4587–4595.
120. Ramos, R. L.; Martinez, A. J.; Coronado, R. M. G. Adsorption of Chromium(VI) from Aqueous Solutions on Activated Carbon. *Water Sci. Technol.* **1994**, *30*, 191–197.
121. Humera, J.; Chandra, V.; Jung, S.; Lee, J. W.; Kim, K. S.; Kim, S. B. Enhanced Cr(VI) Removal Using Iron Nanoparticle Decorated graphene. *Nanoscale* **2011**, *3*, 3583–3585.
122. Wang, H.; Yuan, X.; Wu, Y.; Chen, X.; Leng, L.; Wang, H.; Li, H.; Zeng, G. Facile Synthesis of Polypyrrole Decorated Reduced Graphene Oxide–Fe$_3$O$_4$ Magnetic

Composites and Its Application for the Cr(VI) Removal. *Chem. Eng. J.* **2015**, *262*, 597–606.
123. Liu, M.; Wen, T.; Wu, X.; Chen, C.; Hu, J.; Li, J.; Wang, X. Synthesis of Porous Fe_3O_4 Hollow Microspheres/Graphene Oxide Composite for Cr(VI) Removal. ***Dalton Trans.*** **2013**, *42*, 14710–14717.
124. Lei, Y.; Chen, F.; Luo, Y.; Zhang, L. Three-Dimensional Magnetic Graphene Oxide Foam/Fe_3O_4 Nanocomposite as an Efficient Absorbent for Cr(VI) Removal. *J. Mater. Sci.* **2014**, *49*, 4236–4245.
125. Zhou, L.; Deng, H.; Wan, J.; Shi, J.; Su, T. A Solvothermal Method to Produce RGO-Fe_3O_4 Hybrid Composite for Fast Chromium Removal from Aqueous Solution. *Appl. Surf. Sci.* **2013**, *283*, 1024–1031.
126. Yao, W.; Ni, T.; Chen, S.; Li, H.; Lu, Y. Graphene/Fe_3O_4@Polypyrrole Nanocomposites as a Synergistic Adsorbent for Cr(VI) Ion Removal Composites. *Sci. Technol.* **2014**, *99*, 15–22.
127. Zhu, J.; Wei, S.; Gu, H.; Rapole, S. B.; Wang, Q.; Luo, Z.; Haldolaarachchige, N.; Young, D. P. ; Guo, Z. One-Pot Synthesis of Magnetic Graphene Nanocomposites Decorated with Core@Double-Shell Nanoparticles for Fast Chromium Removal. *Environ. Sci. Technol.* **2012**, *46*, 977–985.
128. Cong, H.-P.; Ren, X.-C.; Wang, P.; Yu, S.-H. Macroscopic Multifunctional Graphene-Based Hydrogels and Aerogels by a Metal Ion Induced Self-Assembly Process. *ACS Nano* **2012**, *6*, 2693–2703.
129. Dinda, D.; Saha, S. K. Sulfuric Acid Doped Poly-Diaminopyridine/Graphene Composite to Remove High Concentration of Toxic Cr(VI). *J. Hazard. Mater.* **2015**, *291*, 93–101.
130. Kumar, S. K.; Kakan, S. S.; Rajesh, N. A Novel Amine Impregnated Graphene Oxide Adsorbent for the Removal of Hexavalent Chromium. *Chem. Eng. J.* **2013**, *230*, 328–337.
131. Zhang, S.; Zeng, M.; Xu, W.; Li, J.; Li, J.; Xu, J.; Wang, X. Polyaniline Nanorods Dotted on Graphene Oxide Nanosheets as a Novel Super Adsorbent for Cr(VI). *Dalton Trans.* **2013**, *42*, 7854–7858.
132. Ge, H.; Ma, Z. Microwave Preparation of Triethylenetetramine Modified Graphene Oxide/Chitosan Composite for Adsorption of Cr(VI). *Carbohydr. Polym.* **2015**, *131*, 280–287.
133. Zhang, L.; Tian, Y.; Guo, Y.; Gao, H.; Li, H.; Yan, S. Introduction of α-MnO_2 Nanosheets to NH_2 Graphene to Remove Cr^{6+} from Aqueous Solutions. *RSC Adv.* **2015**, *5*, 44096–44106.
134. Dubey, R; Bajpai, J.; Bajpai, A. K. Green Synthesis of Graphene Sand Composite (GSC) as Novel Adsorbent for Efficient Removal of Cr(VI) Ions From Aqueous Solution. *J. Water Process Eng.* **2015**, *5*, 83–94.
135. Chen, J. H.; Xing, H. T.; Guo, H. X.; Weng, W.; Hu, S. R.; Li, S. X.; Huang, Y. H.; Sun, X.; Su, Z. B. Investigation on the Adsorption Properties of Cr(VI) Ions on a Novel Graphene Oxide (GO) Based Composite Adsorbent. *J. Mater. Chem. A*, **2014**, *2*, 12561–12570.
136. Dinda, D.; Gupta, A.; Saha, S. K. Removal of toxic Cr(VI) by UV-Active Functionalized Graphene Oxide for Water Purification. *J. Mater. Chem. A* **2013**, *1*, 11221–11228.
137. Kan, Z.; Kemp, K. C.; Chandra, V. Homogeneous Anchoring of TiO_2 Nanoparticles on Graphene Sheets for Waste Water Treatment. *Mater. Lett.* **2012**, *81*, 127–130.
138. Kumar, R.; Chawla, J.; Kaur, I. Removal of Cadmium Ion from Wastewater by Carbon-Based Nanosorbents: A Review. *J. Water. Health* **2015**, *13*, 18–33.

139. Deng, J. H.; Zhang, X. R.; Zeng, G. M.; Gong, J. L.; Niu, Q. Y.; Liang, J. Simultaneous Removal of Cd(II) and Ionic Dyes from Aqueous Solution Using Magnetic Graphene Oxide Nanocomposite as an Adsorbent. *Chem. Eng. J.* **2013**, *226*, 189–200.
140. Yang, T.; Liu, L.; Liu, J.; Chen, M. L.; Wang, J. H. *Cyanobacterium metallothionein* Decorated Graphene Oxide Nanosheets for Highly Selective Adsorption of Ultra-Trace Cadmium. *J. Mater. Chem.* **2012**, *22*, 21909–21916.
141. Liu, J.; Du, H.; Yuan, S.; He, W.; Liu, Z. Synthesis of Thiol-Functionalized Magnetic Graphene as Adsorbent for Cd(II) Removal from Aqueous Systems. *J. Environ. Chem. Eng.* **2015**, *3*, 617–621.
142. Xu, D.; Tan, X.; Chen, C.; Wang, X. Removal of Pb(II) from Aqueous Solution by Oxidized Multiwalled Carbon Nano-Tubes. *J. Hazard. Mater.* **2008,** *154*, 407- 416.
143. Zhang, X.; Lin, S.; Lu, X.-Q.; Chen, Z.-L. Removal of Pb(II) from Water Using Synthesized Kaolin Supported Nanoscale Zero-Valent Iron. *Chem. Eng. J.* **2010**, *163*, 243–248.
144. Li, Y. H.; Di, Z.; Ding, J.; Wu, D.; Luan, Z.; Zhu, Y. Adsorption Thermodynamic, Kinetic and Desorption Studies of Pb^{2+} on Carbon Nanotubes. *Water Res.* **2005**, *39*, 605–609.
145. http://www.epa.gov/ogwdw/lead/.
146. Yang, Y.; Xie, Y.; Pang, L.; Li, M.; Song, X.; Wen, J.; Zhao, H. Preparation of Reduced Graphene Oxide/Poly(acrylamide) Nanocomposite and its Adsorption of Pb(II) and Methylene Blue. *Langmuir* **2013**, *29*, 10727–10736.
147. Olanipekun, O.; Oyefusi, A.; Neelgund, G. M.; Oki, A. Synthesis and Characterization of Reduced Graphite Oxide-Polymer Composites and Their Application in Adsorption of Lead. *Spectrochim. Acta, A: Mol. Biomol. Spectrosc.* **2015**, *149*, 991–996.
148. Luo, S.; Xu, X.; Zhou, G.; Liu, C.; Tang, Y.; Liu, Y. Amino Siloxane Oligomer-Linked Graphene Oxide as an Efficient Adsorbent for Removal of Pb(II) from Wastewater. *J. Hazard Mater.* **2014**, *274*, 145–155.
149. Song, H.; Hao, L.; Tian, Y.; Wan, X.; Zhang, L.; Lv, Y. Stable and Water-Dispersible Graphene Nanosheets: Sustainable Preparation, Functionalization, and High-Performance Adsorbents for Pb^{2+}. *Chem. Plus Chem.* **2012**, *77*, 379–386.
150. Gui, C.-X.; Wang, Q.-Q.; Hao, S.-M.; Qu, J.; Huang, P.-P.; Cao, C.-Y.; Song, W.-G.; Yu, Z.-Z. Sandwichlike Magnesium Silicate/Reduced Graphene Oxide Nanocomposite for Enhanced Pb^{2+} and Methylene Blue Adsorption. *ACS Appl. Mater. Interfaces* **2014**, *6*, 14653–14659.
151. Fan, L.; Luo, C.; Sun, M.; Li, X.; Qiu, H. Highly Selective Adsorption of Lead Ions by Water-dispersible Magnetic Chitosan/Graphene Oxide Composites. *Colloids Surf B: Biointerfaces* **2013**, *103*, 523–531.
152. Jabeen, H.; Kemp, K. C.; Chandra, V. Synthesis of Nano Zerovalent Iron Nanoparticles–Graphene Composite for the Treatment of Lead Contaminated Water. *J. Environ. Manage.* **2013**, *130*, 429–435.
153. Hao, L.; Song, H.; Zhang, L.; Wan, X.; Tang, Y.; Lv, Y. SiO$_2$/Graphene Composite for Highly Selective Adsorption of Pb(II) Ion. *J. Colloid Interface Sci.* **2012**, *369*, 381–387.
154. Potera, C. Copper in Drinking Water: Using Symptoms of Exposure to Define Safety. *Environ. Health Perspect.* **2004**, *112*, A568–A569.
155. Singh, V. K.; Han, S. H.; Yeon, J. R.; Kim, J. H.; Shin, K. A Simple Green Synthesis Route for the Reduction of Graphene Oxide and Its Application to Cu^{2+} Removal. *Sci. Adv. Mater.* **2013**, *5*, 566–574.
156. Xing, H. T.; Chen, J. H.; Sun, X.; Huang, Y. H.; Su, Z. B.; Hu, S. R.; Weng, W.; Li, S. X.; Guo, H. X.; Wu, W. B.; He, Y. S.; Li, F. M.; Huang, Y. NH$_2$-Rich Polymer/Graphene

Oxide Use as a Novel Adsorbent for Removal of Cu(II) from Aqueous Solution. *Chem. Eng. J.* **2015**, *263*, 280–289.
157. Yang, S.; Li, L.; Pei, Z.; Li, C.; Shan, X.; Wen, B.; Zhang, S.; Zheng, L.; Zhang, J.; Xie, Y.; Huang, R. Effects of Humic Acid on Copper Adsorption onto Few-layer Reduced Graphene Oxide and Few-layer Graphene Oxide. *Carbon* **2014**, *75*, 227–235.
158. Zhao, L.; Yu, B.; Xue, F.; Xie, J.; Zhang, X.; Wu, R.; Wang, R.; Hu, Z.; Yang, S. T.; Luo, J. Facile Hydrothermal Preparation of Recyclable S-Doped Graphene Sponge for Cu^{2+} Adsorption. *J. Hazard. Mater.* **2015**, *286*, 449–456.
159. Li, J.; Zhang, S.; Chen, C.; Zhao, G.; Yang, X.; Li, J.; Wang, X. Removal of Cu(II) and Fulvic Acid by Graphene Oxide Nanosheets Decorated with Fe_3O_4 Nanoparticles *ACS Appl. Mater. Interfaces* **2012**, *4*, 4991–5000.
160. Arnikar, H. J. *Essentials of Nuclear Chemistry*, 4th revised ed.; New Age International Publishers, New Delhi, India, 2011.
161. Morton, R. J.; Straub, C. P. Removal of Radionuclides from Water by Water Treatment Processes. *J. Am. Water Works Assoc.* **1956**, *48*, 545–558.
162. Crane, R. A.; Dickinson, M.; Popescu, I. C.; Scott, T. B. Magnetite and Zero-Valent Iron Nanoparticles for the Remediation of Uranium Contaminated Environmental Water. *Water Res.* **2011**, *45*, 2931–2942.
163. Li, Z.; Chen, F.; Yuan, L.; Liu, Y.; Zhao, Y.; Chai, Z.; Shi, W. Uranium (VI) Adsorption On Graphene Oxide Nanosheets From Aqueous Solutions. *Chem. Eng. J.* **2012**, *210*, 539–546.
164. Li, Z. J.; Wang, L.; Yuan, L. Y.; Xiao, C. L.; Mei, L.; Zheng, L. R.; Zhang, J.; Yang, J. H.; Zhao, Y. L.; Zhu, Z. T.; Chai, Z. F.; Shi, W. Q. Efficient Removal of Uranium from Aqueous Solution by Zero-Valent Iron Nanoparticle and Its Graphene Composite. *J. Hazard. Mater.* **2015**, *290*, 26–33.
165. Sun, Y.; Ding, C.; Cheng, W.; Wang, X. Simultaneous Adsorption and Reduction of U(VI) on Reduced Graphene Oxide-Supported Nanoscale Zerovalent Iron. *J. Hazard. Mater.* **2014**, *280*, 399–408.
166. Tan, L.; Zhang, X.; Liu, Q.; Jing, X.; Liu, J.; Song, D.; Hu, H.; Liu, L.; Wang, J. Synthesis of Fe_3O_4@TiO_2 Core–Shell Magnetic Composites for Highly Efficient Sorption of Uranium(VI). *Colloids Surf., A: Physicochem. Eng. Aspects* **2015**, *469*, 279–286.
167. Zhao, Y.; Li, J.; Zhang, S.; Chen, H.; Shao, D. Efficient Enrichment of Uranium(VI) on Amidoximated Magnetite/Graphene Oxide Composites. *RSC Adv.* **2013**, *3*, 18952–18959.
168. Cheng, H.; Zeng, K.; Yu, J. Adsorption of Uranium from Aqueous Solution by Graphene Oxide Nanosheets Supported on Sepiolite. *J. Radioanal. Nucl. Chem.* **2013**, *298*, 599–603.
169. Zhou, L.; Xu, M.; Wei, G.; Li, L.; Chubik, M. V.; Chubik, M. P.; Gromov, A. A.; Han, W. Fe_3O_4@Titanate Nanocomposites: Novel Reclaimable Adsorbents for Removing Radioactive Ions from Wastewater. *J. Mater. Sci.: Mater. Electron.* **2015**, *26*, 2742–2747.
170. Yang, H.; Sun, L.; Zhai, J.; Li, H.; Zhao, Y.; Yu, H. In Situ Controllable Synthesis of Magnetic Prussian Blue/Graphene Oxide Nanocomposites for Removal of Radioactive Cesium in Water. *J. Mater. Chem. A* **2014**, *2*, 326–332.
171. Romanchuk, A. Y.; Slesarev, A. S.; Kalmykov, S. N.; Kosynkin, D. V.; Tour, J. M. Graphene Oxide for Effective Radionuclide Removal. *Phys. Chem. Chem. Phys.* **2013**, *15*, 2321–2327.
172. Pan, N.; Guan, D.; He, T.; Wang, R.; Wyman, I.; Jin, Y.; Xia, C. Removal of Th^{4+} Ions from Aqueous Solutions by Graphene Oxide. *J. Radioanal. Nucl. Chem.* **2013**, *298*, 1999–2008.

173. Vasudevan, S.; Lakshmi, J. The Adsorption of Phosphate by Graphene from Aqueous Solution. *RSC Adv.* 2012, *2*, 5234–5242.
174. Tran, D. N. H.; Kabiri, S.; Wang, L.; Losic, D. Engineered Graphene–Nanoparticle Aerogel Composites for Efficient Removal of Phosphate from Water. *J. Mater. Chem. A* 2015, *3*, 6844–6852.
175. Lakshmi, J.; Vasudevan, S. Graphene—A Promising Material for Removal of Perchlorate (ClO_4^-) from Water. *Environ. Sci. Pollut. Res.* 2013, *20*, 5114–5124.
176. Li, Y.; Zhang, P.; Du, Q.; Peng, X.; Liu, T.; Wang, Z.; Xia, Y.; Zhang, W.; Wang, K.; Zhu, H.; Wu, D. Adsorption of Fluoride from Aqueous Solution by Graphene. *J. Colloid Interface Sci.* 2011, *363*, 348–354.
177. Nandi, D.; Saha, I.; Ray, S. S.; Maity, A. Development of a Reduced-Graphene-Oxide Based Superparamagnetic Nanocomposite for the Removal of Nickel(II) from an Aqueous Medium via a Fluorescence Sensor Platform. *J. Colloid Interface Sci.* 2015, *454*, 69–79.
178. Fu, Y.; Wang, J.; Liu, Q.; Zeng, H. Water-Dispersible Magnetic Nanoparticle–Graphene Oxide Composites for Selenium Removal. *Carbon* 2014, *77*, 710–721.
179. Qi, H.; Liu, H.; Gao, Y. Removal of Sr(II) from Aqueous Solutions Using Polyacrylamide Modified Graphene Oxide Composites. *J. Mol. Liq.* 2015, *208*, 394–401.
180. Shannon, M. A.; Bohn, P. W.; Elimelech, M.; Georgiadis, J. G.; Marinas, B. J.; Mayes, A. M. Science and Technology for Water Purification in the Coming Decades. *Nature* 2008, *452*, 301–310.
181. Elimelech, M.; Phillip, W. A. The Future of Seawater Desalination: Energy, Technology, and the Environment. *Science* 2011, *333*, 712–717.
182. Han, Y.; Xu, Z.; Gao, C. Ultrathin Graphene Nanofiltration Membrane for Water Purification. *Adv. Funct. Mater.* 2013, *23*, 3693–3700.
183. Qiu, L.; Zhang, X.; Yang, W.; Wang, Y.; Simona, G. P.; Li, D. Controllable Corrugation of Chemically Converted Graphene Sheets in Water and Potential Application for Nanofiltration. *Chem. Commun.* 2011, *47*, 5810–5812.
184. Tanugi, D. C.; Grossman, J. C. Water Desalination across Nanoporous Graphene. *Nano Lett.* 2012, *12*, 3602–3608.
185. Konatham, D.; Yu, J.; Ho, T. A.; Striolo, A. Simulation Insights for Graphene-Based Water Desalination Membranes. *Langmuir* 2013, *29*, 11884–11897.
186. Surwade, S. P.; Smirnov, S. N. ; Vlassiouk, I. V.; Unocic, R. R.; Veith, G. M.; Dai, S.; Mahurin, S. M. Water Desalination Using Nanoporous Single-Layer Graphene. *Nat. Nanotechnol.* 2015, *10*, 459–464.
187. Russo, C. J.; Golovchenko, J. A. Atom-by-Atom Nucleation and Growth of Graphene Nanopores. *PNAS* 2012, *109*, 5953–5957.
188. Mahmoud, K. A.; Mansoor, B.; Khraisheh, A. M. M. Functional Graphene Nanosheets: The Next Generation Membranes for Water Desalination. *Desalination* 2015, *356*, 208–225.
189. Bano, S.; Mahmood, A.; Kim, S.-J.; Lee, K. H. Graphene Oxide Modified Polyamide Nanofiltration Membrane with Improved Flux and Antifouling Properties. *J. Mater. Chem. A* 2015, *3*, 2065–2071.
190. Jastrzębska, A. M.; Olszyna, A. R. The Ecotoxicity of Graphene Family Materials: Current Status, Knowledge Gaps and Future Needs. *J. Nanopart. Res.* 2015. DOI:10.1007/s11051-014-2817-0.
191. Akhavan, O.; Ghaderi, E. Toxicity of Graphene and Graphene Oxide Nanowalls Against Bacteria. *ACS Nano* 2010, *4*, 5731–5736.

192. Vallabani, N. V. S.; Mittal, S.; Shukla, R. K.; Pandey, A. K.; Dhakate, S. R.; Pasricha, R.; Dhawan, A. Toxicity of Graphene in Normal Human Lung Cells (BEAS-2B). *J. Biomed. Nanotechnol.* **2011**, *7*, 106–107.
193. Yang, K.; Li, Y.; Tan, X.; Peng, R.; Liu, Z. Behavior and Toxicity of Graphene and Its Functionalized Derivatives in Biological Systems. *Small* **2013**, *9*, 1492–1503.
194. Schinwald, A.; Murphy, F. A.; Jones, A.; MacNee, W.; Donaldson, K. Graphene-Based Nanoplatelets: A New Risk to the Respiratory System as a Consequence of their Unusual Aerodynamic Properties. *ACS Nano* **2012**, *6*, 736–746.
195. Chong, Y.; Ge, C.; Yang, Z.; Garate, J. A.; Gu, Z.; Weber, J. K.; Liu, J.; Zhou, R. Reduced Cytotoxicity of Graphene Nanosheets Mediated by Blood-Protein Coating. *ACS Nano* **2015**, *9*, 5713–5724.
196. Hu, C.; Wang, Q.; Zhao, H.; Wang, L.; Guo, S.; Li, X. Ecotoxicological Effects of Graphene Oxide on the Protozoan *Euglena gracilis*. *Chemosphere* **2015**, *128*, 184–190.
197. Seabra, A. B.; Paula, A. J.; de Lima, R.; Alves, O. L.; Durán, N. Nanotoxicity of Graphene and Graphene Oxide. *Chem. Res. Toxicol.* **2014**, *27*, 159–168.
198. Li, Y.; Yuan, H.; Bussche, A.; Creighton, M.; Hurt, R. H.; Kane, A. B.; Gao, H. Graphene Microsheets Enter Cells through Spontaneous Membrane Penetration at Edge Asperities and Corner Sites. *PNAS* **2013**, *110*, 12295–12300.
199. Ahmed, F.; Rodrigues, D. F. Investigation of Acute Effects of Graphene Oxide on Wastewater Microbial Community: A Case Study. *J. Hazard. Mater.* **2013**, *256–257*, 33–39.
200. Sanchez, V. C.; Jachak, A.; Hurt, R. H.; Kane, A. Biological Interactions of Graphene-Family Nanomaterials—An Interdisciplinary Review. *Chem. Res. Toxicol.* **2012**, *25*, 15–34.
201. Guo, X.; Mei, N. Assessment of the Toxic Potential of Graphene Family Nanomaterials. *J. Food Drug Anal.* **2014**, *22*, 105–115.
202. Bianco, A. Graphene: Safe or Toxic? The Two Faces of the Medal. *Angew. Chem.* **2013**, *52*, 4986–4997.

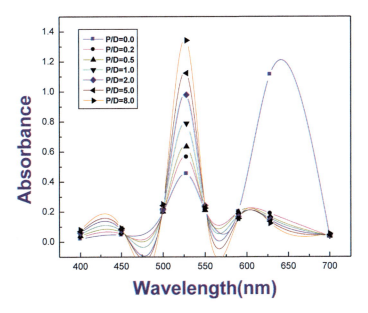

FIGURE 6.1 Absorption spectrum of MB–NaHep at various P/D ratios.

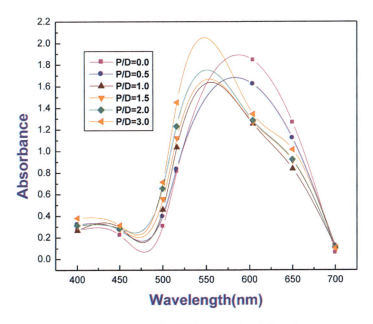

FIGURE 6.2 Absorption spectrum of TB–NaHep at various P/D ratios.

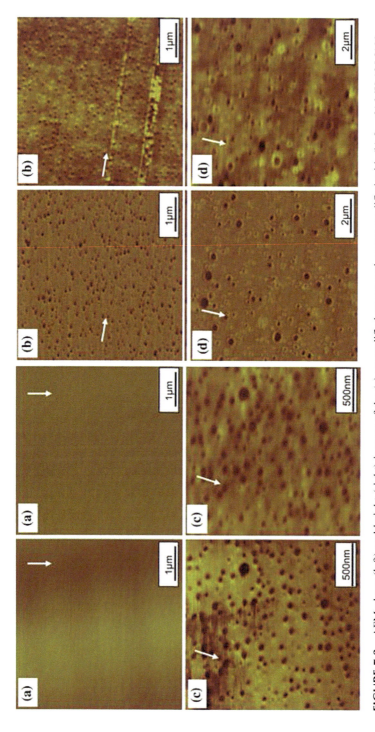

FIGURE 7.3 AFM phase (left) and height (right) images of the (a) unmodified epoxy and epoxy modified with (b) 9 wt% MX 125 [100 nm styrene–butadiene core and poly(methyl methacrylate) (PMMA) shell 85–115 nm], (c) 9 wt% MX 156 (100 nm polybutadiene core and PMMA shell 85–115 nm), and (d) 9 wt% MX 960 (300 nm siloxane and PMMA shell 250–350 nm) (the arrows indicate the cutting direction) (adapted from Ref. [6] with permission).

Polymeric and Nanostructured Materials

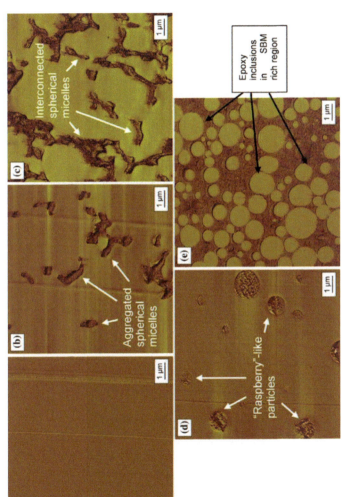

FIGURE 7.6 AFM phase images of (a) unmodified epoxy (appears flat and homogeneous thermoset), (b) 5 wt% E21 grade (SBM) modified epoxies [E21 SBM phase separated as a network of aggregated spherical micelles], (c) 10 wt% E21 grade (SBM)-modified epoxies (spherical micelles become increasingly interconnected at higher E21 content due to reduction of secondary phase separating into individual particles, which is result of increase in viscosity), (d) 5 wt% E41 grade (SBM)-modified epoxies (SBM particles phase separated with a "raspberry"-like microstructure "sphere-on-sphere"), and (e) 10 wt% E41 grade (SBM)-modified epoxies (particle diameter increased and partially phase-inverted microstructure) (adapted from Ref. [12] with permission). [SBM—(tri-block copolymer of poly(styrene)-b-1,4-poly(butadiene)-b-poly(methyl methacrylate modifier, E21 SBM has low polarity higher butadiene content and molecular weight than E41).]

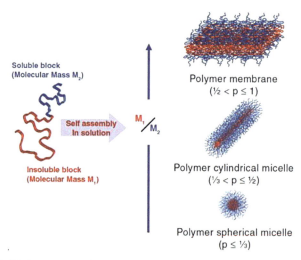

FIGURE 7.7 Various geometries produced by block copolymers in a particular solvent (membranes include two monolayers of soluble block–insoluble block–soluble block copolymers aligned to produce a sandwich-like membrane, and both cylindrical and spherical micelles consist of a nonsoluble core surrounded by a soluble corona) (adapted from Ref. [59] with permission).

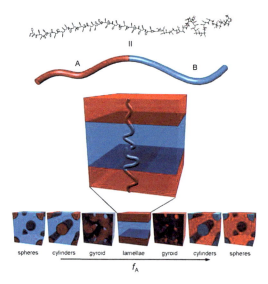

FIGURE 7.8 Schematics representation of thermodynamically stable di-block copolymer phases. The A–B di-block copolymer, such as the PS-*b*-PMMA molecule indicated at the top, is represented as a simple two-color chain. The chains self-organize such that contact between the immiscible blocks is minimized, with the structure determined primarily by the relative lengths of the two polymer blocks (f_A) (adapted from Ref. [91] with permission).

Polymeric and Nanostructured Materials　　　　　　　　　　　　　　　　　　　　E

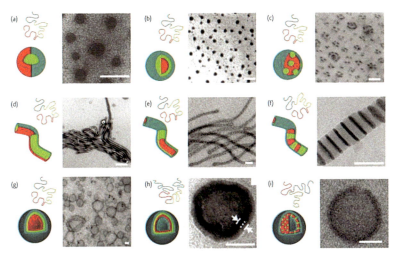

FIGURE 7.9 Assemblies formed in a particular solvent conditions by multiblock copolymers: (a) Janus sphere,[79] (b) core–shell spheres,[80] (c) raspberry-like spheres,[69] (d) Janus cylinders,[81] (e) core–shell cylinder,[82] (f) segmented cylinders,[83] (g) asymmetric (Janus) membrane vesicles,[84] (h) double-layer membrane vesicles, and (i) vesicles with hexagonally packed cylinder.[85] Scale bar 50 nm (adapted from Ref. [66] with permission).

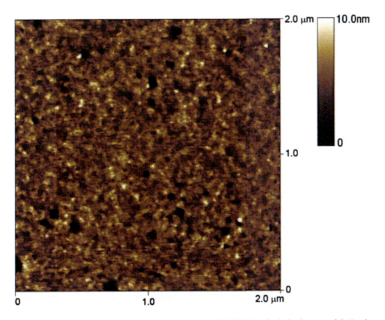

FIGURE 7.15 AFM image of PS (polystyrene)-*alt*-PEO (poly(ethylene oxide)) alternating multiblock copolymer, the microphase separation at the nanometer scale took place and the co-continuous nanophases (dark and light regions) can be attributed to PS and PEO microdomains, separately (adapted from Ref. [159] with permission).

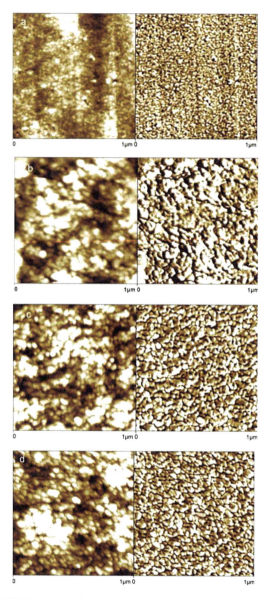

FIGURE 7.16 AFM images of the thermosets containing (a) 10 wt% (small and uniform dispersed PS nanodomains), at higher concentration of PS-*alt*-PEO alternating multiblock copolymer, coagulated and the size of the nanodomains is increased, as shown in (b) 20, (c) 30, and (d) 40 wt% of PS-*alt*-PEO alternating multiblock copolymer, PS-*alt*-PEO multiblock copolymer displayed disordered nanostructures and the formation of the nanostructures depends on the block topology. Left image: topography. Right image: phase contrast images (adapted from Ref. [159] with permission).

Polymeric and Nanostructured Materials G

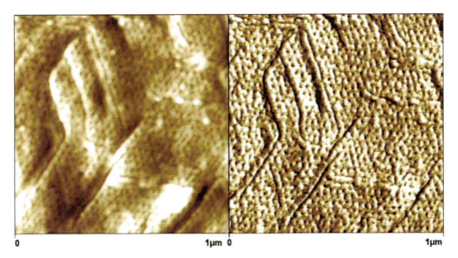

FIGURE 7.17 AFM image of epoxy thermoset containing 40 wt% PS-*b*-PEO di-block copolymer, the formation of the long-ranged ordered nanostructures attributed to the specific topology of the block copolymer (i.e., di-block). The di-block topology offers an ordered arrangement of PS nanodomains in the epoxy matrix during the curing reaction (adapted from Ref. [159] with permission).

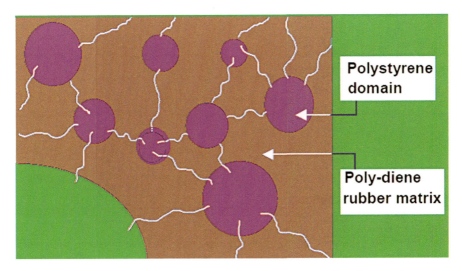

FIGURE 7.28 Schematic representation of tri-block styrene–butadiene and styrene-isoprene block copolymers.

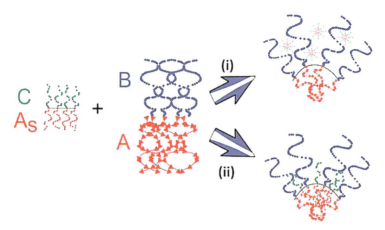

FIGURE 7.29 Two possible mechanisms to explain the influence of polystyrene-block-poly(4-hydroxystyrene) copolymer (A_sC) chains on the morphologies of polystyrene-block-poly(2-vinyl pyridine) copolymer (AB) di-block copolymers and PS (A and A_s) blocks are immiscible with both poly(4-hydroxystyrene) and poly(2-vinyl pyridine) copolymer blocks, (i) the A_sC copolymers dissolve in the B phase and increase the effective volume fraction of the B component; (ii) the A_sC copolymers segregate to the AB interface and modify the curvature through hydrogen bonding interactions. The dissolution of A–C copolymers into the B phase is possible to establish particularly one of the contributions to the mechanisms controlling the phase transformation (adapted from Ref. [211] with permission from American Chemical Society).

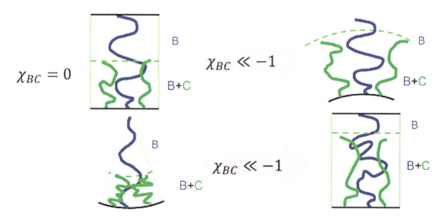

FIGURE 7.30 Schematic of the proposed mechanism for transformation of lamellar morphologies to convex morphologies and concave morphologies to lamellar phases. A pure A–B block copolymer system that is either arranged in lamellar (symmetric) or curved (adapted from Ref. [211] with permission from American Chemical Society).

Polymeric and Nanostructured Materials

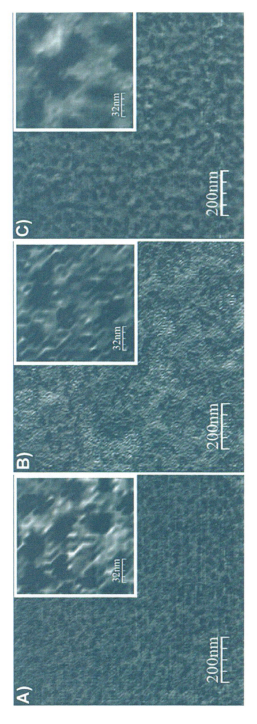

FIGURE 7.32 Tapping mode-AFM phase image of nanostructured epoxy system/SIS85 (85:85% epoxidation degrees) (23 wt%) blends for (a) stage one (before curing, sphere-like nanostructures is observed), (b) stage two (after curing at 80°C for 100 min, less ordered sphere-like nanostructured is observed), and (c) stage three (after curing at 80°C for 180 min, bigger and less organized is observed) (adapted from Ref. [212] with permission from American Chemical Society).

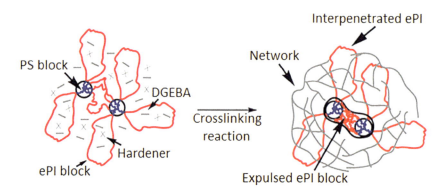

FIGURE 7.33 Uncured (left, PS subchains self-assemble in sphere-like nanodomains (blue)) and cured (right, the interconnection of PS nanodomains) for SIS85/DGEBA/Hardener (adapted from Ref. [212] with permission from American Chemical Society).

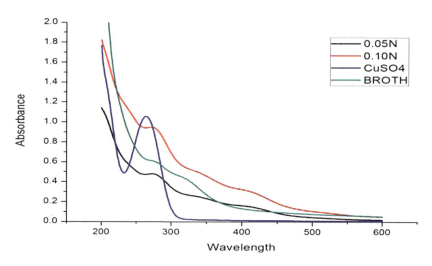

FIGURE 10.2 UV–vis spectra of $CuSO_4$, broth, and copper nanoparticles prepared at different broth ratio.

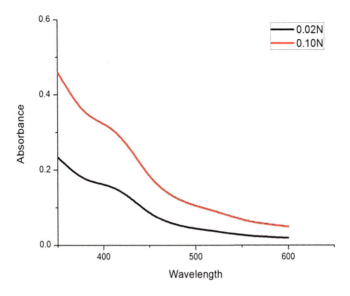

FIGURE 10.3 UV–vis spectra of $CuSO_4$, broth, and copper nanoparticles prepared at different broth ratio.

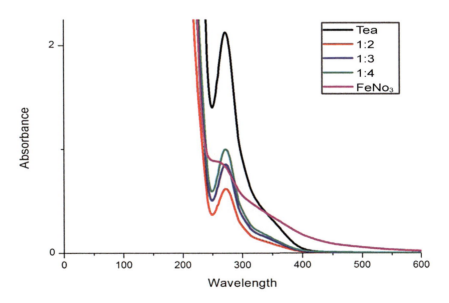

FIGURE 10.4 UV–vis spectra of $FeNO_3$, tea extract, and iron nanoparticles prepared at different ratios.

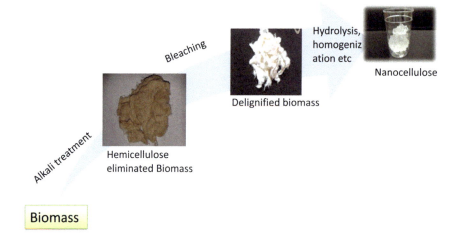

FIGURE 13.4 Extraction of nanocellulose.

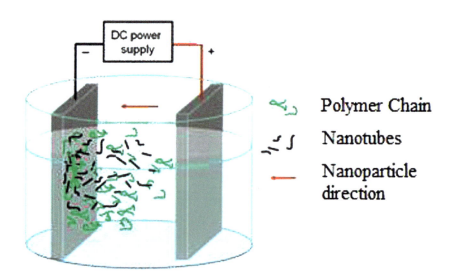

FIGURE 15.5 Electrophoretic deposition (modified from Federica M.; De, R.; Virginia, M. New Method to Obtain Hybrid Conducting Nanocomposites Based on Polyaniline and Carbon Nanotubes. *ENI* **2011**, *6*, 86).

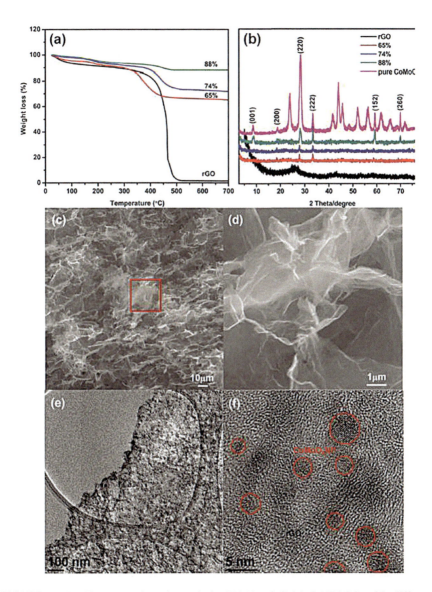

FIGURE 17.1 Thermogravimetric analysis (TGA) of CoMoO$_4$NP/rGO with different (a) CoMoO$_4$ contents and (b) XRD patterns. (c–f) Characterization of the morphology and elemental analysis of CoMoO$_4$NP/rGO (74%) nanocomposite. (c,d) Typical SEM images of the nanocomposite with different magnifications, revealing that the nanoparticles changed the structure of the graphene sheets. (e,f) TEM images of the nanocomposite, showing that the CoMoO$_4$ nanoparticles are confined in the matrix of the GNSs. Copyright reserved to the American Chemical Society.[45]

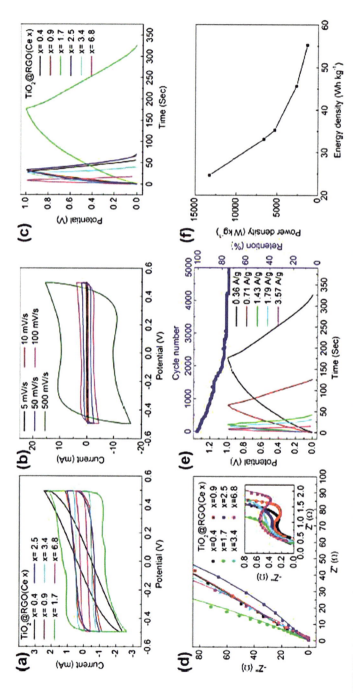

FIGURE 17.5 (a) CV curves for 3D TiO$_2$@RGO composites measures at a scan rate of 50 mV s^{-1}. (b) CV curves collected for a 3D TiO$_2$@RGO(Ce 1.7) composite at different scan rates. (c) GV charge/discharge profiles for 3D TiO$_2$@RGO composites obtained at a current density of 0.36 A g^{-1}. (d) Nyquist plots for 3D TiO$_2$@RGO composites measured over a frequency range of 100 kHz to 0.01 Hz with an AC perturbation of 5 mV. (e) GV charge/discharge profiles collected for the 3D TiO$_2$@RGO composite at different current densities (left Y axis), and the long-term stability measured over nearly 5000 cycles at a current density of 1 A g^{-1} (right Y axis). (f) Relationship between power density and energy density for the 3D TiO$_2$@RGO composite. Copyright reserved to the American Chemical Society.[66]

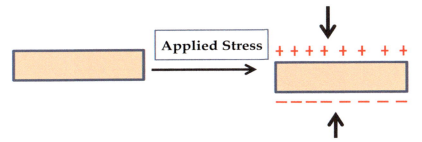

FIGURE 18.1 The diagrammatic representation of a piezoelectric material.

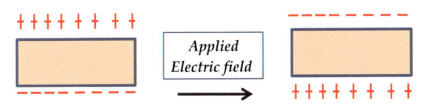

FIGURE 18.2 The diagrammatic representation of a ferroelectric material.

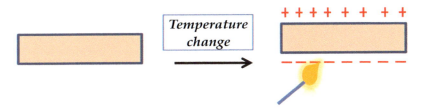

FIGURE 18.3 The diagrammatic representation of pyroelectric materials.

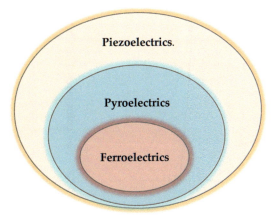

FIGURE 18.4 Relationship between piezoelectric, pyroelectric, and ferroelectric materials.

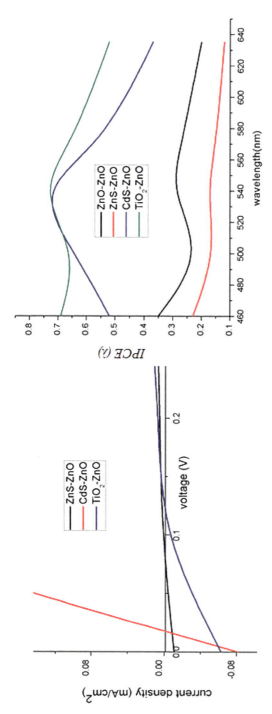

FIGURE 19.4 (See color insert.) (a) Current density against voltage characteristics of different composites and (b) their IPCE (λ).

PART II
Research on Polymer Technology

CHAPTER 13

CELLULOSE: THE POTENTIAL BIOPOLYMER

ANEESA PADINJAKKARA* and SABU THOMAS

International and Inter University Centre for Nanoscience and Nanotechnology, Mahatma Gandhi University, Kottayam 686560, Kerala, India

*Corresponding author. E-mail: anee18p@gmail.com

13.1 INTRODUCTION

Cellulose is the most abundant biopolymer on the earth with the formula $(C_6H_{10}O_5)_n$. It has been estimated that globally about 10^{12} t of cellulose are synthesized and destroyed each year. So the effective utilization of cellulose is required. Cellulose is the major component of plant primary cell walls and also produced by some bacteria. Some species of bacteria secrete it to form biofilms. Cellulose content will vary from plant to plant. Cotton contains more than 90% cellulose (Fig. 13.1).[1]

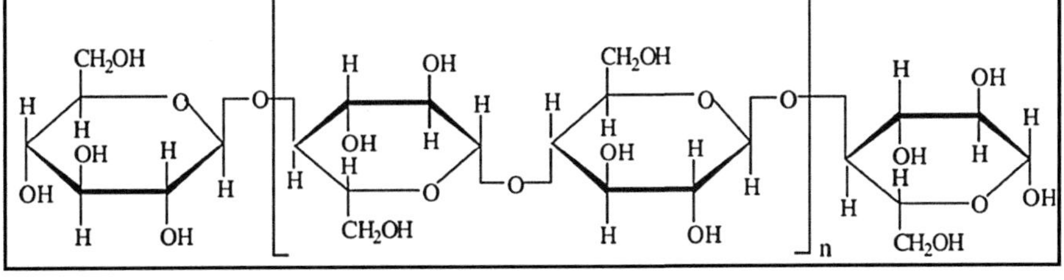

FIGURE 13.1 Structure of cellulose.

Lignocellulosic biomass is an example for natural composite, it comprises mainly cellulose, hemicellulose, and lignin; here, the cellulose act as reinforcing filler or fiber, hemicelluloses and lignin act as matrix. Cellulose is a typical example where the reinforcing filler exist as whiskers-like microfibrils (Fig. 13.2).[2]

FIGURE 13.2 Cotton.

The structure of cellulose consist a linear chain of several hundred to many thousands of β(1 → 4) linked D-glucose units (with a syndiotactic configuration), these chains aggregate to form microfibrils, long thread-like bundles of molecules. Cellulose got rigid rod-like structure due to the strong inter- and intramolecular hydrogen bonds between hydroxyl groups and oxygen of adjacent molecules. (The multiple hydroxyl groups on the glucose from one chain form hydrogen bonds with oxygen atoms on the same or on a neighbor chain, holding the chains firmly together side-by-side and forming microfibrils.) Due to the regular molecular arrangement, cellulose exhibits a crystalline X-ray diffraction pattern.[1]

The stable structure of cellulose leads to outstanding mechanical properties, including a high Young's modulus (138 GPa in the crystal region along the longitudinal direction) and a very low coefficient of thermal expansion. Consequently, cellulose whiskers and fibrils are potential material for the use as reinforcement or filler in composites at both micro- and nanodimensions. In the current world, cellulose in the nanodimension or nanocellulose got very much attraction due to its excellent properties. Production of biofuels (cellulosic ethanol) as an alternative fuel source from cellulose is also under investigation.[1,7]

Cellulose nanofibers got immense attraction as reinforcement for transparent resins also because they are free from light scattering due to their diameters being less than one-tenth of the visible light wavelength. Plant based cellulose nanofibers have the potential to be extracted into fibers thinner than bacterial cellulose; therefore, many researchers have been extensively studying the extraction of nanofibers from wood and other plant fibers. Bacterial cellulose is very expensive and can cause contamination problem.[1]

The development of low-cost, sustainable, and renewable resources is required to overcome the present environmental problems and energy demands. Efficient separation of constitutive components is the one of the considerable hurdles for the efficient utilization of renewable biomass. For the current nanotechnological field, it is very important and interesting that the isolation of nanocellulose in an efficient way.

The separation of the nanocellulose from the biomass is carrying out through two main steps. The first step is different pretreatments for the removal of matrix substances such as lignin and hemicelluloses. Second step is the fibrillation process. The fibrillation of plant fibers was achieved by various ways such as hydrolysis with strong acids (HCl or H_2SO_4), hydrolysis using mild acid (oxalic acid) with steam explosion, mechanical treatments using high-pressure homogenizer, grinder, cryocrushing, etc. Recently, some researchers utilized ultrasonic and enzymatic treatment methods for the fibrillation.

13.2 EXTRACTION OF CELLULOSE AND NANOCELLULOSE

In the current world, many researchers are very much interested on the extraction of cellulose, nanocellulose, and its utilization for different application. Government and industry are providing lot of funds for the utilization of renewable resources.

The extraction of cellulose or purification of cellulose from biomass is mainly through two different chemical treatments such as alkali treatment and bleaching. For the alkali treatment, normally aqueous sodium hydroxide solution was used. Other alkali like KOH also can use for this treatment (Fig. 13.3).[1-6]

Alkaline solution can dissolve the hemicelluloses in the biomass, so the alkali treatment mainly helps to remove the hemicelluloses from the natural fiber or biomass. The alkali solution can also remove the lignin slightly but not completely. Usually, a very low concentration of alkaline solution used for this treatment, for example, 2 wt% NaOH solution in water. The higher concentrated alkali solution may change the crystalline structure of cellulose.

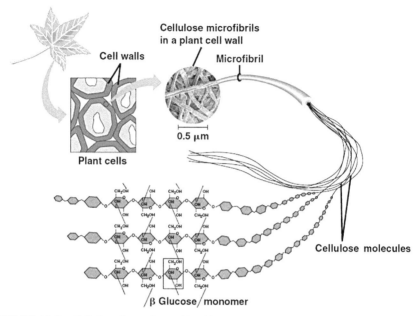

FIGURE 13.3 Cellulose from plant cell wall.

Bleaching treatment is used for the complete removal of lignin. This step is also known as delignification because here the elimination of lignin is occurring. Normally, sodium hypochlorite or chlorine dioxide ($NaClO_2$) is used for the bleaching of alkali-treated biomass. The bleaching treatment normally conduct multiple time up the biomass become white in color and it is indicating that the complete removal of lignin. Alkali treatment and bleaching causes the purification of cellulose.

After the alkali treatment and bleaching, the pure cellulose in the micrometer dimension was obtained.

Different treatments are adopted by the scientist for the isolation of nanocellulose from the purified cellulose.

Acid hydrolysis is the one of the technique for the extraction of nanocellulose from the isolated cellulose (from biomass). Researchers have been used strong acid like sulfuric acid or hydrochloric acid and also mild acid like oxalic acid (with steam explosion) for the acid hydrolysis. Researchers also used diverse mechanical treatment for the individualization of nanocellulose from microcellulose fiber such as disintegrator, high-shear homogenizer, commercial grinder, ball mill, etc. (Fig. 13.4).[1-5]

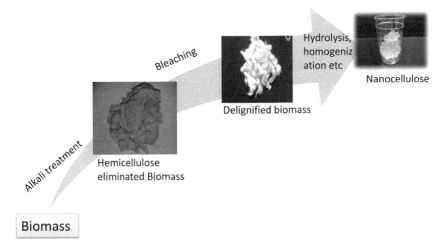

FIGURE 13.4 (See color insert.) Extraction of nanocellulose.

13.3 APPLICATION OF CELLULOSE AND NANOCELLULOSE

Cellulose is using in the textile industry, medical field, cosmetic field, etc. The chemically modified cellulose such as cellulose acetate and cellulose nitrate using to make film for various applications as dialysis membrane, tape, etc. Nanocellulose got more application compared to normal or micro-sized cellulose. Currently, nanocellulose mainly used for the reinforcement of the nanocomposite. It also used in water purification membrane, transparent display, in the medical scaffolds, high gas barrier film, etc.[5,6]

13.3 CONCLUSION

This chapter provides a brief idea about the most abundant biopolymer "cellulose," its extraction and applications. Through various research work, the technologies developed and that helped to improve the human life but as per the current scenario the advancements in the science and technology also causing serious environmental problems. It is very important now that build up stable and ecofriendly systems. The exploitation of cellulose and nanocellulose for various applications will direct the technology expansion in an environmental friendly and sustainable manner. So, it is very important that discuss and study about sustainable materials like "cellulose."

KEYWORDS

- **cellulose**
- **biopolymer**
- **lignocellulosic biomass**
- **hemicellulose**
- **lignin**

REFERENCES

1. Abraham, E.; et al. Extraction of Nanocellulose Fibrils from Lignocellulosic Fibres: A Novel Approach. *Carbohydr. Polym.* **2011**, *86*, 1468–1475.
2. Johar, N.; et al. Extraction, Preparation and Characterization of Cellulose Fibres and Nanocrystals from Rice Husk. *Ind. Crops Prod.* **2012**, *37*, 93–99.
3. Soni, B.; et al. Chemical Isolation and Characterization of Different Cellulose Nanofibers from Cotton Stalks. *Carbohydr. Polym.* **2015**, *134*, 581–589.
4. Chirayil, C. J.; et al. Isolation and Characterization of Cellulose Nanofibrils from Helicteresisora Plant. *Ind. Crops Prod.* **2014**, *59*, 27–34.
5. Abraham, E.; et al. Environmental Friendly Method for the Extraction of Coir Fibre and Isolation of nanofibre. *Carbohydr. Polym.* **2013**, *92*, 1477–1483.
6. Nguyen, H. D.; et al. A novel Method for Preparing Microfibrillated Cellulose from bamboo fibers. *Nanosci. Nanotechnol.* **2013**, *4*, 015016.
7. Deepa, B.; et al. Structure, Morphology and Thermal Characteristics of Banana Nanofibers Obtained by Steam Explosion. *Bioresour. Technol.* **2011**, *102*, 1988–1997.

CHAPTER 14

BIOPROCESSING: A SUSTAINABLE TOOL IN THE PRETREATMENT OF LYCRA/COTTON WEFT-KNITTED FABRICS

SHANTHI RADHAKRISHNAN[*]

Department of Fashion Technology, Kumaraguru College of Technology, Coimbatore 641049, India

[*]*E-mail: shanradkri@gmail.com*

ABSTRACT

The contribution of the textile industry to the Indian economy is notable, but it has been identified as a major polluting industry. The textile wet-processing sector is highly water and chemical intensive. Some of the major aims of the textile-processing industry are to minimize the pollution of water and atmosphere during production below specified limits, to reduce the usage of harmful chemicals, and to reuse the wastewater discharge to the possible extent after treatment. Any fabric that has to be converted to the final product has to undergo certain basic processing before it is to remain white/dyed or finished. These basic treatments given to fabric to make it ready for dyeing or finishing are termed as pretreatment processes. It has been reported that the pretreatment of fabrics, whether knitted or woven, is a major contributor to pollution. Product technologies or manufacturing processes that reduce pollution or waste, energy, or material use in comparison to the technologies they replace are termed as *clean technologies.* One such technology that serves to reduce pollution and brings about good results is the use of enzymes in pretreatment of lycra/cotton weft-knitted fabrics. This study, which deals with the chemical and enzymatic pretreatment of lycra/cotton knits, reveals that the properties of enzyme pretreated cotton weft knits were equal to or better than those pretreated with chemicals. The bioprocess shows a reduction in the effluent pollution load coupled with savings in water, energy, and dye consumption. The biopretreated fabrics were ideal for further processing. Bioprocessing, a sustainable tool, is an important means of safe guarding the environment while leading to new possibilities in the production of quality textiles.

14.1 INTRODUCTION

Textile wet processing is an important sector which converts textile gray goods into finished products. It is a major industry which utilizes a large quantity of water, energy, and chemicals leading to pollution due to the unprocessed waste materials or their byproducts. A report from the United Nations Environmental Program indicates that about 400–500 million tons of hazardous and radioactive chemicals are discharged every year into the water polluting the environment and causing health problems to humans, livestock, and aquatic life. Preparatory, dyeing, printing, and finishing are the different stages in adding value to gray goods. A practical estimate reveals that 45% effluent is due to preparatory processing, 33% in dyeing, and 22% in finishing. The effluent from these processes is highly toxic and cannot be left into the environment without treatment. The textile industry utilizes 2.07% of water consumption of which a major share belongs to the wet-processing sector.[1] Water consumption depends on the type of fabrics that are to be processed. An estimate indicates that about 0.08–0.15 m^3 of water is required to process 1 kg of fabric[2-4] and about 1000–3000 m^3 of water is left out as wastewater after processing about 12–20 t of textiles per day.[5] Water as effluent results from the actual chemical processing steps or from the washing stages after the chemical processes.

Pretreatment of any fabric is an important stage in wet processing as it forms the base for all other processes like dyeing and finishing. Raw cotton contains approximately 4–12% impurities by weight which includes waxes, pectin, and protein. The goals of fabric preparation are the removal of impurities, uniform absorption of water along the length and width of the fabric, and imparting whiteness to the fabric. Scouring removes the impurities and improves absorbency while bleaching imparts whiteness to fabrics. Cellulosic fibers are bleached either by oxidation or reduction. In the case of cotton knits, hydrogen peroxide is the most commonly used bleach as chlorine-based bleaches are highly toxic to environment. Alkaline conditions and sequestering agents are required to activate the hydrogen peroxide bleach. The decomposition of hydrogen peroxide yields the hydroxide radical that act on the cellulose to give a white appearance.[6] When synthetic fibers are blended as in the case of lycra/cotton, a heat-setting process is included before the pretreatment process. Synthetic fibers tend to shrink when exposed to heat, and hence, their internal structures have to be stabilized by exposing them to a temperature higher to the processing temperature. The fabrics are held under tension in a tenter frame and heated to maintain the required dimensions.[7] The conventional chemical method uses sodium hydroxide for scouring and hydrogen peroxide for bleaching. Removal of hydrogen peroxide molecules from the bath water is essential for proper fixation of the dye and chemicals are used for their removal. Conventional scouring and bleaching result in large biological oxygen demand (BOD), chemical oxygen demand (COD), pH, oil, and grease loadings.[8]

To achieve a sustainable economy, the design, production process, transport, and consumption need to be changed with the incorporation of pollution control strategies and reuse systems, environmental care structures and modification of existing expertise, and know-how.[9] Clean technologies include a wide range of products, services, and processes that connect renewable materials and energy sources, thereby reducing the use of natural resources and curtailing emissions and waste. Clean technologies are far more superior when compared with their conventional counterparts. One such clean technology is the application of enzymes in textile processing. The wet-processing segment has greatly benefitted in terms of environmental impact and product quality by the use of enzymes. About 70–75 enzymes have been used for textile industrial processes from about 7000 enzymes that are available for use. The principal enzymes used for textiles are hydrolases and oxidoreductases.[10] Enzymes are proteins that are capable of converting a substrate into another product at a high reaction rate. Bulk of the enzymes is from microbial origin and opens opportunities for high-quantity enzyme production within a short period of time. Further, the strains can be genetically modified to improve enzyme production capacity. Enzymes aid industrial processes by using lower resources than other alternative processes.[11]

The aim of this study is to compare the effects of conventional pretreatment and biopretreatment of lycra/cotton-blended knitted fabric in terms of absorbency, weight loss, pH of fabric, bursting strength, whiteness, and yellowness index. Further, the effluent estimation after each pretreatment was evaluated with regard to pH of effluent, color, total dissolved solids (TDS), total suspended solids (TSS), COD, and BOD. Sustainability in industrial processes uses changes in technology, resources, or practices to reduce waste and uses energy and resources with increased efficiency to minimize environmental damage. This study will also highlight the sustainability concepts by estimating the effluent load in each pretreatment process and the resources utilized for the pretreatment process. This assessment will help to determine the sustainability aspects of the conventional and biopretreatment processes undertaken.

14.2 MATERIALS AND METHODS

14.2.1 MATERIALS

Cotton yarn of 1930s count was selected and weft-knitted rib fabric was manufactured in the Terrot machine with 102/84 feeders and alternate lycra arrangements. The fabric particulars are 28″ diameter, Wales per centimeter—12; courses per centimeter—20 and GSM—245. Ten kilograms of the rib fabric was subjected to chemical pretreatment in a 10-kg capacity soft flow sample machine and another 10 kg material was pretreated using enzymes—biopretreatment. The standard procedure followed by the textile-processing industries in Tirupur was followed for chemical pretreatment and the biopretreatment was undertaken using enzymes from Novozymes, Denmark.

14.2.2 METHODS

The first step in the pretreatment process is to pass the lycra/cotton gray fabric through a heat setting machine at 210°C to set the lycra for the processing treatment. The fabric is then worked for 30 min at 80°C with a demineralizing agent to remove the oils from the lycra and to make it suitable for pretreatment. The pretreatment of lycra/cotton fabric consisted of scouring (a process which removes the impurities from the cotton fabric) and bleaching (a process which imparts whiteness to the fabric). The recipes used for the chemical and biopretreatment process are given in Table 14.1.

The chemical pretreatment includes two processes scouring with sodium hydroxide and bleaching with hydrogen peroxide at 100°C for 90 min followed by a hot wash at 70°C for 10 min. The water is drained and refilled again as per the material liquor ratio (1:8) and treated with peroxide killer at 80°C for 15 min. The next process is neutralization with acid at 60°C for 30 min. The water is drained, refilled, and followed by cold wash at 35°C for 15 min. Now, the pretreated fabric is ready for dyeing or finishing. Effluent is generated from chemical pretreatment after each process, namely, scouring and bleaching, hot wash, peroxide killer treatment, acid neutralization, and cold wash.

TABLE 14.1 Recipes Used for Chemical Pretreatment and Biopretreatment of Lycra/Cotton Fabrics.

| \multicolumn{3}{c|}{Chemical pretreatment} | \multicolumn{3}{c}{Biopretreatment} |

Chemical pretreatment			Biopretreatment		
GPL	Ingredients	Quantity for 10 kg fabric	GPL	Ingredients	Quantity for 10 kg fabric
0.6	Wetting oil	0.048	1	Wetting oil	0.16
0.5	Lubricant	0.04	0.5	Lubricant	0.04
0.75	Sequesterant	0.06	0.4	Scourzyme	0.032
3	Caustic soda	0.24	0.75	Sequesterant	0.06
3	Peroxide	0.24	3	Peroxide	0.24
0.5	Stabilizer	0.04	0.5	Stabilizer	0.04
0.8	Peroxide killer	0.064	0.1	Terminox 50 L	0.016
1	Formic acid	0.16			
	MLR 1:8			MLR 1:8	

GPL, grams per liter; MLR, material liquor ratio.

The biopretreatment includes bioscouring with Scourzyme L which has pectate lyase enzyme, at 55°C for 35 min followed by bleaching with hydrogen peroxide at 95°C for 30 min. This is followed by an effluent drain and refill for removal of bleach with Terminox 50 L containing enzyme catalase at 50°C for 10 min. The catalase converts the hydrogen peroxide molecules into water, and hence, the dyeing can be carried out in the same bath; a cold wash at 35°C for 10 min can be carried out if required. Effluent is generated from biopretreatment after scouring and bleaching, and enzymatic peroxide killer treatment; cold wash is optional and hence not included in this segment.

The fabric is dried and kept in a standard testing atmosphere for 24 h before it is tested for absorbency (AATCC, Test Method 79, 2007), fabric weight (ASTM D 3776-96), pH of fabric (AATCC, Test Method 81, 2001), bursting strength (ASTM D 3786), whiteness, and yellowness index (ASTM E284). These are the basic tests which are carried out to check the efficiency of the pretreatment method. The energy and water requirement for both the pretreatments were noted. The effluent collected after each pretreatment method was estimated for parameters—pH, color, TSS, TDS, BOD, and

COD. The cost for water, energy, and effluent treatment were calculated for each type of pretreatment.

14.3 RESULTS AND DISCUSSION

The results of the fabric tests after chemical and biopretreatment of lycra/cotton weft-knitted fabrics are given in Table 14.2.

14.3.1 pH OF FABRIC

pH of fabric is important as it determines the suitability of the wet-processed textiles for subsequent dyeing or finishing operation. After pretreatment, all the samples exhibit values above 7. Values between 6.5 and 7.5 are said to be ideal as the neutral pH is 7. The chemical pretreated samples record higher values than the biopretreated counter parts. The ANOVA table revealed that the fabric and processes in main effects had significant effect at 5% level.

14.3.2 ABSORBENCY

The table highlights good absorbency values after both chemical and biopretreatment indicating that the fabrics are very suitable for the subsequent finishing treatments. The lowest absorbency time was noted in the biopretreated samples. Processes and treatment in main effects had significant effect at 5% level.

14.3.3 WEIGHT

The fabric weight was higher in the case of biopretreated samples than the chemical pretreated ones. The maximum weight recorded was after biopretreatment. It has been reported that the weight loss is directly proportional to the concentration of sodium hydroxide and the processing time.[12] The chemical formulation reveals higher concentration and time for scouring when compared to the biopretreatment procedure leading to higher weight loss in the chemical pretreated samples. The ANOVA table highlighted that processes and pretreatment in main effects showed significant effect at 1% level.

TABLE 14.2 Fabric Test Results After Chemical Pretreatment and Biopretreatment.

Sl. no.	Tests	Control	Chemical pretreated	Percentage gain or loss over control	Biopretreated	Percentage gain or loss over control	F value
1.	pH of fabric	6.8	7.4	8.82	7.3	8.70	4.662[a]
2.	Absorbency(s)	—	3	—	2	—	6.250[a]
3.	Weight (g/sq. m)	245	232	−5.30	236	−3.67	87.304[b]
4.	Bursting strength (kg/cm^2)	6.1	4.58	−24.91	4.875	−20.08	37.390[b]
5.	Whiteness index	2	59	2850	61	2950	1.595
6.	Yellowness index	33	8	−75.75	8.5	−74.24	0.087

[a]Significant at 5% level.
[b]significant at 1% level.

14.3.4 BURSTING STRENGTH

The table reveals a reduction of strength of all samples after pretreatment when compared to the original. The percentage strength loss was higher in the chemical-pretreated samples highlighting the negative effect of chemicals on the fabric. The lowest difference in strength was observed in the biopretreated sample. It has been stated that the absence of strong alkali in the liquor prevents the risk of oxycellulose formation, and hence, the strength loss is less when enzymes are used in processing.[13] From the table, it may be observed that processes in main effects showed significant effect on bursting strength at 1% level.

14.3.5 WHITENESS AND YELLOWNESS INDEX

The whiteness index has an important bearing on the final coloration of the fabric. The table shows improvement in whiteness index and reduction of yellowness index among the treated samples when compared to the originals. It may be observed that the whiteness index and yellowness index results are almost similar in both the pretreatment methods. An improvement in whiteness index and decrease in yellowness index was observed in all the pretreated samples when compared to the control. Lower whiteness index and a marginally higher yellowness index were observed in enzyme-pretreated sample when compared to the chemical counterpart. It has been reported that the whiteness index improvement achieved in the combined enzymatic/peroxide bleaching process was comparable to the whiteness increase in two consecutive peroxide bleaches with 35% H_2O_2, at boil, 1 h each.[14] The ANOVA results reveal that the processes in main effects showed no significant effect on the whiteness and yellowness index.

The results of the effluent tests from chemical and biopretreatment are given in Table 14.3. The standard norms specified by the Central Pollution Control Board for discharge of effluent water after treatment have been given in the table for reference.[15,16]

From the table, there is a decrease in effluent load after biopretreatment process when compared to the chemical pretreated process. The decrease in effluent load ranges from 23.96% in color to a maximum of 81.91% with regard to TSS. In the biopretreatment process, enzymes are used in optimized quantities and a low dose of chemicals necessary to achieve the required pH is used to activate the enzymes. In the conventional chemical pretreatment, sodium hydroxide, which is used for scouring, causes high effluent load. Bioscouring with Scourzyme L has pectate lyase enzyme, an ecofriendly alternative to sodium hydroxide.[17,18] Further, the traditional removal of hydrogen peroxide requires hot wash and inorganic salts but catalase enzyme (Terminox 50 L) converts the residual hydrogen peroxide into water which reduces the effluent load.[19, 20]

TABLE 14.3 Effluent Test Results After Chemical and Biopretreatment of Lycra/Cotton Knitted Fabric.

Sl. no.	Test parameters	Norms for water discharge by CPCB	Effluent after chemical pretreatment	Effluent after biopretreatment	Percentage decrease in biopretreatment effluent load
1.	pH	5.5–9	11.6	7.8	32.76
2.	Color	Clear	Intense 455	Hazen 346	23.96
3.	Total suspended solids	100	376	68	81.91
4.	Total dissolved solids (mg/L)	2100	4610	2605	43.49
5.	Chemical oxygen demand (mg/L)	250	3238	2455	24.18
6.	Biological oxygen demand (mg/L)	30	220	150	31.82

The resources, namely, temperature, time, energy, and water requirements for chemical and enzymatic pretreatment for 1000 kg of fabric, are given in Table 14.4. In the case of chemical pretreatment method, all the parameters showed considerable increase when compared to enzyme-treated fabrics since processes involved were greater in number. It is observed that the resources used in the bio pretreatment process are lower when compared to the chemical pretreatment process. From Table 14.4, it can be understood that all the processes in chemical pretreatment require temperature range which is either equal to or higher than those required for biopretreatment process. The table also shows that hot wash and acid neutralization steps are not required for biopretreatment. Hence, a decrease in time, energy, and water utilization is evident and the resultant effluent quantity will be lower in biopretreatment. Further, a lower use of resources in biopretreatment with regard to cycle time (50%), energy utilization (61.18%), and water consumption (40%) proves that the bioprocess is sustainable. It has been stated that industrial processes are sustainable when there is lesser use of resources and cleaner processes leading to ways of solving pollution problems and wastewater effluents for a sustainable planet.[21-23]

TABLE 14.4 Resources Used of Chemical and Biopretreatment of Lycra/Cotton-Knitted Fabrics.

Sl. no.	Process	Temperature (°C) CP	Temperature (°C) BP	Time (min) CP	Time (min) BP	Energy (electrical units) CP	Energy (electrical units) BP	Water (l) CP	Water (l) BP
1.	Scouring and bleaching	100	55	90	35	40.68	8.25	8000	8000
		100	95		30		13.22		
2.	Hot wash	70	–	10	–	2.75	–	8000	–
3.	Peroxide killing	80	50	15	10	4.64	2.39	8000	8000
4.	Neutralization	60	–	30	–	7.33	–	8000	–
5.	Cold wash	35	35	15	5	3.50	2.00	8000	8000
	Total requirement for 1000 kg of material			160	80	58.90	22.86	40,000	24,000

BP, biopretreatment; CP, chemical pretreatment.

14.4 CONCLUSION

Sustainable industrial processes are those which are friendly to environment by minimizing the use of resources and based on the fundamentals of clean technologies; the resultant product should achieve the optimum quality required for further processing or may be used as final end product. Consumers today are on the watch for ecofriendly products and do not compromise in the quality features of the product. The fabric tests reveal that biopretreatment offers good quality for further processing, while the effluent results show that the effluent load from bioprocess is lower than the chemical counterpart. The results also present savings in time, energy, and water consumption which proves that enzyme pretreatment is sustainable in comparison to the conventional chemical pretreatment method. Enzyme technology offers an abundant potential to many industries to face the challenges of depleting natural resources and rapidly increasing population. Enzymes, the protein backbone of bioprocesses, can serve as a sustainable tool in textile processing benefitting both industry and environment. Advancements in biotechnology and research will offer new and improved enzymes for future industrial applications in the field of textile processing.

KEYWORDS

- chemical pretreatment
- biopretreatment
- sustainability
- ANOVA
- pH

REFERENCES

1. Das, S. Textile Effluent Treatment—A Solution to the Environmental Pollution. *Fiber2fashion*. http://www.fibre2fashion.com/industry-article/pdffiles/Textile-Effluent-Treatment.pdf?PDFPTOKEN=709a0d279bb45f543ba777067d43d8ed8c89bf03%7C1253632944.
2. Pagga, U.; Brown, D. The Degradability of Dyestuffs: Part II: Behaviour of Dyestuffs in Aerobic Biodegradation Tests. *Chemosphere* **1986**, *15*, 479–491.
3. Kdasi, A.; Idris, A.; Saed, K.; Guan, C. Treatment of Textile Wastewater by Advanced Oxidation Processes—A Review. *Glob. Nest Int. J.* **2004**, *6*, 222–230.
4. Ghaly, A. E.; Anathashankar, R.; Alhattab, M.; Ramakrishnan, V. V. Production, Characterization and Treatment of Textile Effluents—A critical Review. *J. Chem. Eng. Process Technol.* **2014**, *5*, 1–19.
5. Jagannathan, V.; Cherurveettil, P.; Chellasamy, A.; Prempriya, M. S. Environmental Pollution Risk Analysis and Management in Textile Industry: A Preventive Mechanism. *ESJ* **2014**, *2*, 480–486.
6. Walters, A.; Santillo, E.; Johnston, P. An Overview of Textile Processing and Related Environmental concerns. *Technical Note*; University of Exeter UK, Exeter UK, June 11–15, 2005. http://www.greenpeace.to/publications/textiles_2005.pdf.
7. Hauser, P. *Textiles and Fashion: Materials, Design and Technology*, Woodhead Publishing: Sawston, Cambridge, 2015.
8. Nayak, L. *Clean Technologies—An Alternative to Reduce Wastewater Volume in Textile Industry*. http://www.academia.edu/5122743/Clean_technologies_an_alternative_to_reduce_wastewater_volume_in_textile_industry.
9. IPCC. *Climate Change 2014—Mitigation of Climate Change*. Cambridge University Press: Cambridge, MA, 2014. DOI: 10.1016/0016-3287(94)90071-X.
10. Ibrahim, D. F. Clean Trends in Textile Wet processing. *J. Text. Sci. Eng.* **2012**, *2*, 1–4.
11. Central Pollution Control Board. *General Standards for Discharge of Environmental Pollutants, Part A—Effluents*, 2008; pp 545–560. http://cpcb.nic.in/GeneralStandards.pdf.
12. Ammayappan, L.; Muthukrishnan, G.; Prabakar, C. S. A Single Stage Preparatory Process for Woven Cotton Fabric and Its Optimization. *Man-Made Text. India* **2003**, *47*, 29–35.
13. Churi, R. Y.; Khadilkar, S. M. Enzyme Based Pretreatments for Cotton Goods. *Colourage* **2005**, *53*, 88–89.

14. Tzanov, T.; Calafell, M.; Guebitz, G. M.; Cavaco-Paulo, A. Biolpreparation of Cotton Fabrics. *Enzyme Microb. Technol.* **2001,** *29,* 357–362.
15. The Environment (Protection) Rules. *Industry Specific Standards—Cotton Textile Industries (Composite and Processing),* 1986; p 412. http://www.cpcb.nic.in/Industry-Specific-Standards/Effluent/412.pdf.
16. Central Pollution Control Board. *Guidelines on Techno-Economic Feasibility of Implementation of Zero Liquid Discharge (ZLD) for Water Polluting Industries,* 2015, pp 1–22. http://www.cpcb.nic.in/Final-ZLD(Draft)1.pdf.
17. Losonczi, A. K. Bioscouring of Cotton Fabrics, *Ph.D. Thesis.* Budapest University of Technology and Economics, Hungary, 2004; pp 1–77.
18. Kirk, O.; Borchert, T. V.; Fuglsang, C. C. Industrial Enzyme Applications. *Curr. Opin. Biotechnol.* **2002,** *13,* 345–351.
19. Gavrilescu, M.; Chisti, Y. Biotechnology—A Sustainable Alternative for Chemical Industry. *Biotech. Adv.* **2005,** *23, 471–499.*
20. Jo, R. A.; Casaland, M.; Cavaco-Paulo, A. Application of Enzymes for Textile Fiber Processing. Biocatal. Biotransform. **2008,** *26,* 332–349.
21. Vigneswaran, C.; Anbumabiand, N.; Ananthasubramaniam, M. Biovision in Textile Wet Processing Industry—Technological Challenges. *JTATM* **2011,** *7,* 1–13.
22. Kumari, P.; Singh, S. S. J.; Rose, N. M. Textiles: For Sustainable Development. *IJSER* **2013,** *4,* 1379–1390.
23. Jegannathan, K. R.; Nielsen, P. H. Environmental Assessment of Enzyme Use in Industrial Production—A Literature Review. *J. Clean. Prod.* **2013,** *42,* 228–240.

CHAPTER 15

HALLOYSITE BIONANOCOMPOSITES

P. SANTHANA GOPALA KRISHNAN[1,2*], P. MANJU[1], and S. K. NAYAK[1]

[1]*Department of Plastics Technology, Central Institute of Plastics Engineering & Technology, Guindy, Chennai 600032, Tamil Nadu, India*

[2]*Department of Plastics Engineering, Central Institute of Plastics Engineering & Technology, Patia, Bhubaneswar 751024, Odisha, India*

*Corresponding author. E-mail: psgkrishnan@hotmail.com

ABSTRACT

In the last few decades, halloysite bionanocomposites have undergone significant evolution because of the interesting properties offered by them such as good mechanical strength, flame retardancy, cytocompatibility, environment friendliness, etc. The commonly used biopolymers as matrix for fabrication of the bionanocomposites are poly(lactic acid), poly(butylene succinate), starch, chitosan, and poly(vinyl alcohol). Still, numerous attempts are being made to improve or modify the properties imparted by them. The combination of different biopolymers with suitable processing techniques for the preparation of halloysite bionanocomposites can render wide range of properties for the bionanocomposites according to their field of applications.

15.1 INTRODUCTION

The growing concern in the environmental and sustainability issues has attracted the researcher's attention to replace the petroleum-based polymers with suitable alternatives derived from biomass. Thus, biopolymer gains worldwide acceptance and had a remarkable growth in different sectors in the last few decades. Researchers from various disciplines of material science are exploring the use of nanostructures, like cellulose nanostructures, carbon nanotubes (CNT), polyhedral oligomeric silsesquioxanes, and nanoclays as reinforcement to produce a new class of bionanocomposites. Because of the inherent properties of nanoparticles, they can be used as fillers or additives in polymers to serve various purposes as reinforcing agents or as specific property enhancers.[1] The types of nanoparticles used are nanoclays, CNT, metal oxides, etc. Bionanocomposites are being investigated in search of finding a solution to various drawbacks found in biopolymers, like poor mechanical strength, low thermal stability, poor barrier properties, etc. Thus, the synergy of both nanotechnology and renewable-based resources developed a wide range of applications in different fields with improved properties. The use of renewable resource-based polymers gives an added advantage of eco-friendliness to the developed bionanocomposites.

Nowadays, halloysite nanotubes (HNTs) have become the extensively chosen nanostructure in developing the bionanocomposites. This is because of the advantages offered by this nanomaterial, like its tubular structure, nanoscale lumen, high aspect ratio, hydrophilic nature, low cost, abundant availability, easy to disperse in polymer matrix by applying shear force, environment friendliness and cytocompatibility, high mechanical strength, and low hydroxyl group density on their surface. Thus, it enhances the thermal, mechanical, flame, and barrier properties, resulting in the development of hybrid class of bionanocomposites.[1,2] The application of these materials spreads in different areas including biomedical, tissue engineering scaffolds, drug delivery, cancer cell isolation, bone implant, and cosmetics. Various HNT bionanocomposites derived from poly(lactic acid) (PLA), poly(butylene succinate) (PBS), starch, chitosan, and poly(vinyl alcohol) (PVOH) are discussed in this chapter.

15.2 CLASSIFICATION OF CLAYS

Clay is a soft loose earthy material composed of clay minerals, quartz, feldspar, carbonates, ferruginous material, and other nonclay materials. Clay minerals are essentially hydrous aluminum silicates with a sheet-like structure (phyllosilicates), in which magnesium or iron may substitute wholly or partly for aluminum and with alkalis or alkaline earths as essential constituents resulting in variable chemical compositions.[1] The most common clay minerals are kaolinite, halloysite, smectite, allophane, chlorite, and illite.

Among the layered silicate types of clay, phyllosilicates are extensively used in preparing clay-based nanocomposites.[1,2]

Clay minerals can be divided into four major groups: (1) kaolinite group, (2) montmorillonite/smectite group, (3) illite group, and (4) chlorite group based on the variation in their layered structure (Table 15.1).[1–5]

TABLE 15.1 Types of Clay.[1–5]

Clay group	General formula	Members	Structural characteristics	Uses
Kaolinite	$Al_2Si_2O_5(OH)_4$	Kaolinite, dickite, halloysite, and nacrite	Each member is composed of silicate sheets (Si_2O_5) bonded to aluminum oxide/hydroxide layers ($Al_2(OH)_4$); the two types of layers are tightly bonded	Fillers in ceramics, paint, rubber, paper, and plastics
Smectite	$(Ca, Na, H)(Al, Mg, Fe, Zn)_2(Si, Al)_4O_{10}(OH)_2 \cdot XH_2O$	Montmorillonite, talc, pyrophyllite, saponite, and nontronite	The layer structure contains silicate layers, sandwiching an aluminum oxide/hydroxide layer ($Al_2(OH)_4$)	Fillers in paints, rubbers, as plasticizer in molding sands, in drilling muds, and as electrical, heat, and acid-resistant porcelain, talc in facial powder
Illite	$(K,H)Al_2(Si,Al)_4O_{10}(OH)_2 \cdot XH_2O$	Illite	The structure of this group is similar to the montmorillonite group with silicate layers sandwiching an aluminum oxide/hydroxide layer in the same stacking sequence	Filler in some drilling mud, and it is a common constituent in shales
Chlorite	These clays do not possess a general formula	Amesite, chamosite, cookeite, daphnite	Each member is having different formula and different structures	Anti-adhesive agent and filler

15.3 HALLOYSITE NANOTUBES

HNT is an important clay mineral of the kaolin group which was named after Baron Omalius d'Halloy (1707–1789), a Belgian geologist who first noted the mineral. HNT is a two-layered aluminosilicate, with the chemical formula $Al_2Si_2O_5(OH)_4 \cdot nH_2O$.[5] It is chemically similar to kaolinite, dickite, or nacrite. However, it differs mainly in the morphology of crystals of unit layers that are separated by a monolayer of water molecule.[4] It occurs widely in both weathered rocks and soils. Raw HNT is mined from natural deposits and usually is white in color and can be easily ground into powder. HNT is being mined from different countries like China, New Zealand, America, South Africa, Brazil, and France.[6] HNT can have different morphologies like spheroidal, short tubular, platy particle. But the most common structure is the hollow tubular structure, with diameter less than 100 nm, where the aluminosilicate sheets are rolled up like a scroll (see Fig. 15.1). HNT length typically varies from 0.2 to 1.2 μm, while the inner and outer diameter of tube is 10–30 and 40–70 nm, respectively.[7]

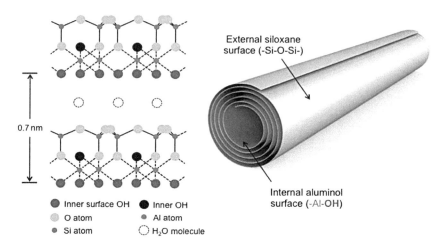

FIGURE 15.1 Crystalline structure of halloysite. Adapted from http://phantomplastics.com/functional-fillers/halloysite/.

The structural unit of HNT is formed by the arrangement of two sheets: Si-tetrahedra and Al-octahedra. It contains two types of hydroxyl groups, inner aluminol (Al–OH) and outer siloxane (Si–O–Si) groups, which are situated between layers and on the surface of the nanotubes, respectively.[1,7] These hydroxyl groups can act as anchoring sites for the surface modification of HNTs with different functional groups to enrich the surface chemistry of HNTs. The HNT mineral contains a layer of water in its interlayer space which results in an increase in layer thickness up to 1.1 nm. As these water molecules are attached through weak interactions, they can be easily eliminated by heating it.[8] This results in formation of HNT, with a reduction in interlayer spacing, ranging from 0.7 to 1.1 nm. Thus, it is considered that hydrated HNTs have 1.1 nm and dehydrated HNTs have 0.7 nm interlayer spacing.[9]

The surfaces of layered silicates are generally associated with a charge, which is not a constant value and varies with layers. This is known as cation exchange capacity (CEC). CEC is the number of exchangeable cations per dry weight that a soil is capable of holding, at a given pH value, and available for exchange with the soil water solution. The SI unit is cmol/kg.[10,11] The different CEC values of HNTs according to their origin are given in Table 15.2. The CEC value varies in the range of 2–60 cmol/kg.[12–17]

TABLE 15.2 CEC Values of HNT Clay.[13–17]

Samples	Cation exchange capacity (cmol/kg)
Yoake (Japan)	40–50
Katy Soil (Texas)	58.1
Komaki (Japan)	60
Cameroon (Central Africa)	30–51
Australia	12
Matauri Bay (New Zealand)	2
Indonesia	6–10

15.4 BIOPOLYMERS

Biopolymers are the important component of bionanocomposites which attract researcher's attention. They are the perfect choice as the matrix of bionanocomposites and offer immense advantages which make bionanocomposites widely accepted. The important biopolymers of choice are PLA, PBS, starch, chitosan, PVOH, etc. (*see* Table 15.3). The use of biopolymers is a perfect alternative because of its renewable nature and in turn decreases the carbon foot print and solves the degradability issues as most of them are biodegradable in nature.[18]

TABLE 15.3 Biopolymers Studied.

Biopolymer	Structure	References
PLA		[40–47]
PBS		[48–50]
Starch		[51–54]
Chitosan		[55,56]
PVOH		[57]

15.5 PROCESSING TECHNIQUES

15.5.1 SOLUTION MIXING

Solution mixing is the most commonly used method for fabricating HNT nanocomposites (*see* Fig. 15.2). The advantage of this method is that it can be easily performed and is suitable for small sample sizes. The disadvantage is, it cannot be used for insoluble polymers and also solvent removal is a tedious and time-consuming process.

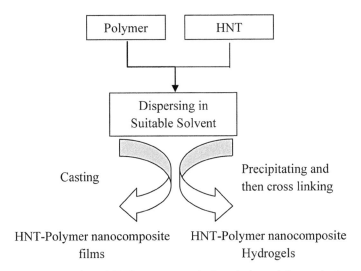

FIGURE 15.2 Preparation of HNT nanocomposite by solution-mixing method.

Stirring enables easy dispersion of HNT in the solvents by overcoming the relatively weak tube–tube interactions. Ultrasonication can also be employed to improve their dispersion in the solvents. HNT-bionanocomposite films of PVOH, chitosan, and potato starch have been reported. Cavallaro et al. reported HNT–alginate hydrogels by mixing HNT in water followed by cross-linking using calcium ions.[19]

15.5.2 MELT PROCESSING

Melt processing is an appropriate method for production of HNT nanocomposites industrially. As HNTs can be easily dispersed in polymer melt by shear force, this method is suitable for processing them. Almost all thermoplastics such as polypropylene,[20,21] linear low-density polyethylene,[22] nylon-6,[23] and biopolymers like soy protein,[24] poly(ε-caprolactone),[25] wheat starch,[26] and PLA[27] can be mixed with HNTs to produce the nanocomposites. Generally, melt processing involves mixing HNTs with a molten polymer via shear force using an extruder. Afterward, HNT nanocomposites were processed via compression molding, injection molding, or extrusion. The disadvantage of melt processing is the unexpected polymer degradation and oxidation that occur at high temperatures, as well as the strong shear force, leading to the decreased polymer properties.

15.5.3 IN SITU POLYMERIZATION

In this method, HNTs are dispersed in a liquid monomer or solvent in which monomer is dissolved. This system is then polymerized to prepare HNT nanocomposites. The advantage of this method is that in situ polymerization can improve the initial dispersion of the nanotubes in the monomer.[9,28,29]

15.5.4 ELECTROSPINNING

Electrospinning is a simple and versatile fiber synthesis technique in which a high-voltage electric field is applied to the polymer solution (Fig. 15.3). This method forms continuous micro/nanofibers by which they can be used to form a network having high porosity with very small pore size and a very large surface-to-volume ratio. Hence, these materials found application in biomedical field, such as drug delivery, artificial organs, wound dressing, and medical prostheses. The polymers used for electrospinning with HNTs range from PLA (or its copolymer or blend) to PVOH.[9] Biodegradable PLA and PVOH have good electrospinning properties, gaining them applications in many areas. Dichloromethane or chloroform was employed to dissolve the PLA with dimethyl formamide to enhance the electric conductivity. Water was used as a solvent for PVOH.[30–33]

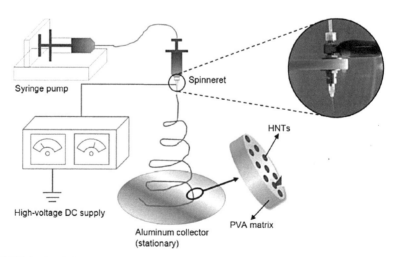

FIGURE 15.3 Schematic representation of electrospinning of PVOH (modified from Lee, I. W.; Li, J.; Chen, X.; Park, H. J. Electrospun Poly(Vinyl Alcohol) Composite Nanofibers with Halloysite Nanotubes for the Sustained Release of Sodium D-Pantothenate. *J. Appl. Polym. Sci.* **2016**, *133*, 42903. With permission from John Wiley & Sons).

15.5.5 LAYER-BY-LAYER DEPOSITION METHOD (LbL METHOD)

This method is employed for fabrication of thin films by depositing alternating layers of oppositely charged materials with washing steps in between (Fig. 15.4). HNTs are negatively charged; therefore, they can assemble via the LbL method with polycations, such as poly(allyl amine hydrochloride) chitosan, etc. LbL is an easy and inexpensive process for nano-architectures and allows a variety of materials to be incorporated within the films. The drug or co-enzyme-loaded HNTs can be incorporated within organic–inorganic composite films offering a novel method for fabricating core–shell nanomaterials with potential applications in drug-release systems.[35]

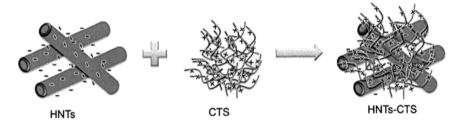

FIGURE 15.4 Schematic representation of layer-by-layer deposition method of chitosan/halloysite (CT/HNT) (modified from Yu, L.; Wang, H.; Zhang, Y.; Zhang, B.; Liu, J. Recent Advances in Halloysite Nanotube-Derived Composites for Water Treatment. *Environ. Sci. Nano* **2016**, *3*, 42. With permission from Royal Society of Chemicals).

15.5.6 ELECTROPHORETIC DEPOSITION

This method can also be adopted for making polymer nanocomposite films. The electrophoretic motion of colloidal particles or polymer macromolecules under an electric field forms a deposit at the electrode surface, thereby enables the formation of uniform films with a controlled thickness (Fig. 15.5). Nanocomposite films containing HNTs and a polyelectrolyte (chitosan or hyaluronic acid) can be obtained using the cathodic and anionic deposition methods. The prepared nanocomposites have promising applications in biomedical implants with improved bioactivity, biocompatibility, and corrosion protection for metallic implants.[37,38]

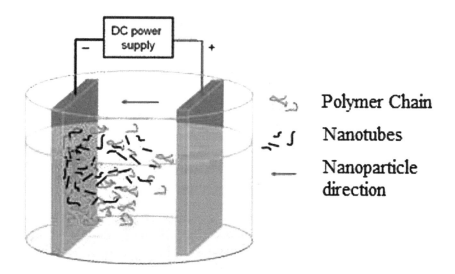

FIGURE 15.5 (See color insert.) Electrophoretic deposition (modified from Federica M.; De, R.; Virginia, M. New Method to Obtain Hybrid Conducting Nanocomposites Based on Polyaniline and Carbon Nanotubes. *ENI* **2011**, *6*, 86).

15.6 BIONANOCOMPOSITES

The important attractiveness of bionanocomposites is the bio-based polymer matrices because of the advantages offered by them like degradability, renewability, biocompatibility, eco-friendliness, etc. Hence, they are widely being commercialized. The HNT bionanocomposites discussed here contain PLA, PBS, starch, PVOH, and chitosan as matrix.

15.6.1 PLA–HNT BIONANOCOMPOSITES

Liu et al. reported the PLA–HNT bionanocomposite having substantially higher modulus, strength, and toughness than that of PLA. Both tensile and flexural properties of PLA increased proportionally with HNT loading. The tensile strength, flexural strength, and flexural modulus of HNT–PLA nanocomposites with 30 phr HNT were 74.1, 108.3, and 6.56 GPa, which were 34%, 25%, and 116% higher than those of the neat PLA, respectively. In addition, toughness and impact strength of PLA–HNT bionanocomposite also improved by the loading of HNT. Impact strength of HNT nanocomposites containing 10 wt% HNT was 70% higher than that of virgin PLA. But as the loading of HNT increased beyond 20 phr, the properties found to decrease, and this may be attributed due to the agglomeration of the nanoparticles leading to weaker interaction between the matrix and reinforcement.[40]

Yeniova Erpek et al. prepared plasticized PLA–HNT bionanocomposite using PEG as plasticizer. The mechanical property studies showed a marginal improvement in elongation-at-break values of plasticized PLA when compared with neat PLA. The composite containing 5 wt% of HNT yields 25% improvement in the elongation-at-break value with respect to neat PLA. Also with increasing HNT content from 5 to 10 wt%, there is a slight enhancement in the strength of the nanocomposites but decreases the elongation-at-break values of the plasticized PLA nanocomposites.[41]

PLA–HNT bionanocomposites were prepared by solution-mixing technique and were casted into films by solution-casting method. Tensile strength of the films was increased by 40% with the increase of HNT concentration from 0 to 5 wt% and then decreased thereafter. Also, the elastic modulus value of films with 5% HNT was 58% higher than that of neat PLA.[42]

PLA–HNT bionanocomposites were formed by melt compounding and the foams were prepared by using supercritical carbon dioxide as physical blowing agent. The flexural and tensile modulus of PLA–HNT nanocomposites improved significantly with the incorporation of HNT. Comparing with the neat PLA, the incorporation of HNT in the PLA matrix results in a significant enhancement in modulus from 3.38 GPa up to 4.02 GPa in the flexural modulus, 1.72 GPa up to 2.05 GPa in tensile modulus, respectively. PLA–HNT nanocomposite foams showed much higher cell density and smaller mean cell size due to HNT serving as the heterogeneous nucleating agents. Presence of the HNT can act as nucleating agents on the crystallization of PLA.[43]

Teo and Chow investigated the effect of HNT added to PLA–PMMA blend, prepared using melt compounding. The impact strength of PLA–PMMA blend was further improved by the addition of HNT. Addition of 5 phr of HNT increased the impact strength of the blend by 9.6%. The enhancement of impact strength may be because of HNTs playing a role in hindering the crack patch caused by impact. Addition of HNT into PLA–PMMA blend maintained the melting temperature at about 151°C. The melting temperature of PLA–PMMA–HNT composite is also slightly lower than that of neat PLA. HNT in the nanocomposites worked as a nucleating agent of PLA which cause the composites to crystallize more quickly than neat PLA.[44]

The effect of HNT on the crystallization behavior of PLA was investigated by Kaygusuz and Kainayak. The degree of isothermal crystallinity of PLA increased with the incorporation of HNTs.

For instance, 3 wt% HNT increased crystallinity of PLA from 24% to 37%, that is, an increase of 40%.[45]

De Silva et al. compared the properties of the PLA nanocomposites reinforced with ZnO and ZnO–HNT. The results showed that incorporation of ZnO into PLA decreased the tensile strength by approximately 4% and 10.4% with the addition of 5 and 10 wt% of ZnO, compared to the pure PLA films. But on addition of HNT with ZnO, even though ZnO led to poor tensile properties, ZnO–HNT exhibited enhancement in the properties. The addition of ZnO–HNT increased the tensile strength by 28.2% and 30.8% by incorporating 2.5 and 5 wt% of ZnO–HNT, respectively. Besides, modulus also increased by 30% and 65% with the addition of 5 wt%, respectively. Antimicrobial tests revealed that ZnO–HNT can act as a promising antimicrobial agent against bacteria such as *Escherichia coli* and *Staphylococcus aureus*, where the bacteria count reduced by more than 99%.[46]

Xu et al. proposed a novel method for preparation of PLA–HNT nanocomposites after a set of surface modifications performed on HNT. Surface modification of HNTs with L-lactic acid was performed via direct condensation polymerization of L-lactic acid with HNT and also through treating L-lactic with HNT. The hydroxyl groups present in HNTs enable easy grafting of L-lactic acid and PLA over the surface of HNT. Two modified HNTs were obtained: grafting HNT with lactic acid (L-HNT) and PLA (p-HNT). Then, a series of HNT–PLLA, L-HNT–PLLA, and p-HNT–PLLA composite membranes was prepared by solution-casting method and was characterized. The results showed that the L-HNT–PLLA and *p*-HNT–PLLA composites had better tensile properties than that of the HNT–PLLA composites. With the filler content increased, the tensile strength and modulus increased first and then decreased, as the filler content was 10 wt%, the highest tensile strengths of 24.82, 29.74, and 37.2 MPa, and moduli of 1.15, 1.48, and 1.64 GPa were obtained for HNT–PLLA, L-HNT–PLLA, and p-HNT–PLLA composite films, respectively. Also, polarization optical microscopy results show that t the crystallization rate of HNT–PLLA faster than that of PLLA, and the more and smaller spherulites were formed in the HNT–PLLA composite due to the nucleation effect of HNTs. The composites prepared with surface-modified HNTs showed much higher crystallization rate than those of the HNT–PLLA composite and pure PLLA.[47]

15.6.2 PBS–HNT BIONANOCOMPOSITES

Wu et al. studied the effect of the addition of HNT to the polymer matrix after forming nanocomposite foams using super critical carbon dioxide as the physical blowing agent. The PBS–HNT nanocomposites show higher values of storage modulus and loss modulus than those of pure PBS. Moreover, the increase of storage modulus and loss modulus is the highest when the content of HNT is 7 wt%, which is caused by the strong interaction of HNT with PBS and the homogeneous dispersion of HNT within the PBS matrix. Enthalpy of fusion of PBS increases gradually with an increase of HNT content due to the nucleating effect of HNT. Besides, cell density and volume expansion ratio of foams increase with the increment of HNT content. By increasing the HNT content to 5 wt%, the cell density increases to 2.17×10^8 cells/cm^3 and the volume expansion ratio increases to 5.70.[48]

Wu et al. studied the morphology, thermal, and mechanical properties of PBS–HNT nanocomposites fabricated using melt compounding. The crystallization studies suggest that the crystallization temperature and crystallization enthalpy of PBS–HNT have increased slightly with increasing HNT content, with that of neat PBS. It indicates that the presence of the HNT can act as heterogeneous nucleating agents. The mechanical properties of neat PBS and PBS–HNT nanocomposites were compared and showed that the flexural modulus and strength of PBS gradually increase with increasing of HNT content. The HNT can enhance the flexural modulus of PBS nanocomposite from 825.3 to 936.2 MPa as the HNT loading is increased up to 7 wt%. In addition, the crystallinity of the nanocomposites is higher than neat PBS, which may also contribute to the improvement of the flexural modulus and strength.[49]

Gradzik et al. prepared HNT nanocomposites with matrix of PBS-ethylene glycol multiblock copolymer (PBS-EG) using solvent casting. The neat PBS-EG copolymer shows a water contact angle of 33.4°, while the nanocomposites, both containing unmodified and functionalized HNTs, show a reduced contact angle of 25° and 26° for unmodified and modified HNTs, respectively. The reduction in water contact angle can be related to the hydrophilicity of HNT to the HNT enrichment at the surface.[50]

15.6.3 STARCH–HNT BIONANOCOMPOSITES

Sadagh and Nafchi prepared potato starch-based bionanocomposite films incorporated with HNT by solvent-casting method. Tensile strength increased from 7.33 to 9.82 MPa and heat seal strength increased from 375 to 580 N/m. Moreover, the incorporated clay nanoparticles acted as a barrier and decreased permeability of the gaseous molecules. The addition of HNT improves the barrier and mechanical properties of potato starch films and these bionanocomposites have high potential to be used for food packaging purposes.[51]

Porous bionanocomposites based on HNT as nanofillers and plasticized starch as polymeric matrix were successfully prepared by melt extrusion. Here, HNTs act as filler as well as nucleating agent that increases the porosity of the composites. This increase in cell size or porosity is useful for the biomedical application as scaffold because it allows the formation of cellular and extracellular components of bone and blood vessels.[52]

Schmitt et al. prepared starch-based HNT bionanocomposites, glycerol as plasticizer, using melt extrusion. Addition of HNT increased storage modulus and young's modulus without loss of ductility. DMA analysis shows that the storage modulus of all nanocomposites is higher than that of neat starch–glycerol. Storage modulus of nanocomposite increased with increase in HNT content which is caused by the restrictions of the segmented motion of the starch chains. Young's modulus significantly increases by +84% and the yield stress by +75.5% compared to plasticized starch for nanocomposite containing 4 wt% of HNT.[53]

Addition of HNT slightly enhances the thermal stability of starch. The tensile properties of starch are significantly improved up to +144% for Young's modulus and up to +29% for strength upon addition of both modified and unmodified HNT, interestingly without loss of ductility. Modified HNTs lead to significantly higher Young's modulus than unmodified HNT.[54]

15.6.4 CHITOSAN–HNT BIONANOCOMPOSITES

Bionanocomposite film of chitosan–polyvinyl alcohol blend (CS–PVOH) incorporated with HNT was prepared by solution-casting method. Normally inorganic fillers are said to increase the thermal stability of the composites. The thermal analyses of the composites were in well agreement with this fact as the decomposition temperature of HNT-incorporated CS–PVOH film was more than that of CS–PVOH films. Likewise, the tensile strength and elongation-at-break values were also enhanced by the addition of HNT. The tensile strength of CS–PVOH blend film having 3 wt% of HNT increased by 39.72%, and the elongation at break of CS–PVOH blend film having 2 wt% of HNT increased by 26.14% compared with the neat CS–PVOH film having 36.01 MPa and 99.92% tensile strength and elongation-at-break values, respectively.[55]

Liu et al. developed novel chitosan–HNT nanocomposite scaffolds by combining solution-casting and freeze-drying technique. The mechanical and thermal property values indicated that the scaffolds exhibited significant enhancement in compressive strength, compressive modulus, and thermal stability compared with the neat chitosan scaffold. The stress at 80% strain for the newly developed scaffold is 0.55 MPa, which is about 17 times higher than that of the pure chitosan scaffold. Also, the compression modulus of NC scaffolds is 0.45 MPa with 80 wt% HNTs significantly higher than that of pure chitosan. The thermal analysis of the nanocomposites exhibits higher degradation temperatures when compared with pure chitosan scaffolds. The mechanical property values of scaffolds can be correlated with the density of the scaffold and Young's modulus of the materials from which the scaffold is fabricated. Increasing the HNT content of the scaffold increases the relative density of scaffold. Hence, the interaction between the HNT and chitosan can be improved and the dispersion of HNT in chitosan can be enhanced. As a consequence, the compression property of chitosan nanocomposites scaffolds significantly increases compared with that of pure chitosan scaffold.[56]

15.6.5 PVOH–HNT BIONANOCOMPOSITE

Zhou et al. prepared PVOH–HNT nanocomposite films and compared the properties with that of neat PVOH films. The tensile strength and elongation-at-break values were more for PVOH–HNT films than the PVOH films. The tensile strength and elongation-at-break value for films with 5% HNT were 56 MPa and 320%, respectively, which is higher than that of neat PVOH films. But on increasing the HNT concentration beyond 7%, the mechanical properties found to decrease and may be due to the agglomeration of the HNT particles.[57]

15.7 APPLICATIONS OF BIONANOCOMPOSITES

Bionanocomposites have been used in various applications since last decades. The evolution of nanotechnology is offering new fields of application according to the properties of the developed composites.[58] CS/HNT bionanocomposite films can be used as an excellent packaging material since they have good barrier and mechanical properties.[55] HNT bionanocomposites films made from potato starch can also be used for packaging application.[51] The biocompatible and biodegradable nature of almost all bio-based polymers makes it a suitable choice in the biomedical field.[57] Apart from the bio-based polymer matrixes, HNTs are also suitable for medical applications due to cytocompatibility. Many nanoparticles such as fullerene, carbon nanotubes, metal oxide nanoparticles, etc. cannot be directly utilized in biomedical field. Recently, the bionanocomposite scaffolds are the ones which are attracting the researchers attention.[52] Apart from scaffolds, HNT nanocomposites are applicable in tissue engineering scaffolds, drug delivery, cancer cell isolation, bone implant, and cosmetics.[59,60] The other common uses of HNT nanocomposites are in bioreactors, time-release capsules, catalyzing polymer degradation, templates, high tech ceramic applications, as heterogeneous nucleating agents in polymer matrix, corrosion inhibitor, etc.

15.8 CONCLUSIONS

Bionanocomposites have drawn wide attention because of their unique properties offered by them which can be utilized in different applications. The development of nanotechnology and commercialization of bio-based polymers have paved the way for it. The use of bio-based polymers gives the added advantage of sustainability and eco-friendliness to the bionanocomposites. The inherent disadvantages in bio-based polymers such as low thermal stability, toughness, hydrophilicity, and crystallinity of the bio-based polymers can be overcome by fabricating novel bionanocomposites.

KEYWORDS

- **biopolymers**
- **halloysite nanotubes**
- **bionanocomposites**
- **processing techniques**
- **applications**

REFERENCES

1. Uddin, F. Clays, Nanoclays, and Montmorillonite Minerals. *Metall. Mater. Trans. A* **2008**, *39*, 2804–2814.
2. De, S. S. P.; De, S. S. H.; Brindley, G. W. Mineralogical Studies of Kaolinite–Halloysite Clays' Part a Platy Mineral with Structural Swelling and Shrinking Characteristics. *Am. Miner.* **1900**, *51*, 5–6.
3. Mackenzie, R. C. The Classification and Nomenclature of Clay Minerals. *Clay Min. Bull.* **1959**, *4*, 52–66.
4. Brindley, G. W. Structural Mineralogy of Clays. *Clays Clay. Miner.* **1952**, *1*, 33–43.
5. Sakiewicz, P.; Nowosielski, R.; Pilarczyk, W.; Gołombek, K.; Lutyński, M. Selected Properties of the Halloysite as a Component of Geosynthetic Clay Liners (GCL). *JAMME* **2011**, *48*, 177–191.
6. Kamble, R.; Ghag, M.; Gaikawad, S.; Panda, B. K. Halloysite Nanotubes and Applications: A Review. *JASR* **2012**, *3*, 25–29.
7. Zhang, Y.; Tang, A.; Yang, H.; Ouyang, J. Applications and Interfaces of Halloysite Nanocomposites. *Appl. Clay Sci.* **2016**, *119*, 8–17.
8. Joussein, E.; Petit, S.; Churchman, J.; Theng, B.; Righi, D.; Delvaux, B. Halloysite Clay Minerals—A Review. *Clay Miner.* **2005**, *40*, 383–426.
9. Liu, M.; Jia, Z.; Jia, D.; Zhou, C. Recent Advance in Research on Halloysite Nanotubes–Polymer Nanocomposite. *Prog. Polym. Sci.* **2014**, *39*, 1498–1525.
10. Rawtani, D.; Agrawal, Y. K. Multifarious Applications of Halloysite Nanotubes: A Review. *Rev. Adv. Mater. Sci.* **2012**, *30*, 282–295.
11. Christie, A. B.; Thompson, B.; Brathwaite, B. Mineral Commodity Report 20—Clays. *New Zeal. Min.* **2000**, *27*, 26–43.
12. Ng, K. M.; Lau, Y. T. R.; Chan, C. M.; Weng, L. T.; Wu, J. Surface Studies of Halloysite Nanotubes by XPS and ToF-SIMS. *Surf. Interface Anal.* **2011**, *43*, 795–802.
13. Stul, M. S.; Van Leemput, L. Particle-Size Distribution, Cation Exchange Capacity and Charge Density of Deferrated Montmorillonites. *Clay. Miner.* **1982**, *17*, 209–215.
14. Garrett, W. G.; Walker, G. F. The Cation-Exchange Capacity of Hydrated Halloysite and the Formation of Halloysite–Salt Complexes. *Clay Miner.* **1959**, *4*, 75–80.
15. Ma, C.; Eggleton, R. A. Cation Exchange Capacity of Kaolinite. *Clay Clay Miner.* **1999**, *47*, 174–180.
16. Chaikum, N.; Sooppipatt, N.; Carr, R. M. The Cation Exchange Capacity of Some Kaolin Minerals. *J. Sci. Soc. Thailand* **1981**, *7*, 100–109.
17. Yuan, P.; Tan, D.; Annabi-Bergaya, F. Properties and Applications of Halloysite Nanotubes: Recent Research Advances and Future Prospects. *Appl. Clay. Sci.* **2015**, *112*, 75–93.
18. Bhat, A. H.; Khan, I.; Usmani, M. A.; Rather, J. A. Bioplastics and Bionanocomposites Based on Nanoclays and Other Nanofillers. In *Nanoclay Reinforced Polymer Composites*. Springer: Singapore, 2016; pp 115–139.
19. Cavallaro, G.; Gianguzza, A.; Lazzara, G.; Milioto, S.; Piazzese, D. Alginate Gel Beads Filled with Halloysite Nanotubes. *Appl. Clay Sci.* **2013**, *72*, 132–137.
20. Du, M.; Guo, B.; Wan, J.; Zou, Q.; Jia, D. Effects of Halloysite Nanotubes on Kinetics and Activation Energy of Non-Isothermal Crystallization of Polypropylene. *J. Polym. Res.* **2010**, *17*, 109–118.
21. Du, M. L.; Guo, B. C.; Jia, D. M. Thermal Stability and Flame Retardant Effects of Halloysite Nanotubes on Poly(Propylene). *Eur. Polym. J.* **2006**, *42*, 1362–1369.

22. Jia, Z.; Luo, Y.; Guo, B.; Yang, B.; Du, M.; Jia, D. Reinforcing and Flame Retardant Effects of Halloysite Nanotubes on LLDPE. *Polym. Plast. Technol. Eng.* **2009**, *48*, 607–613.
23. Handge, U. A.; Hedicke-Hochstotter, K.; Altstadt, V. Composites of Polyamide 6 and Silicate Nanotubes of the Mineral Halloysite: Influence of Molecular Weight on Thermal, Mechanical and Rheological Properties. *Polym. J.* **2010**, *51*, 2690–2699.
24. Nakamura, R.; Netravali, A. N.; Morgan, A. B.; Nyden, M. R.; Gilman, J. W. Effect of Halloysite Nanotubes on Mechanical Properties and Flammability of Soy Protein Based Green Composites. *Fire Mater.* **2013**, *37*, 75–90.
25. Lee, K. S.; Chang, Y. W. Thermal, Mechanical, and Rheological Properties of Poly(ε-Caprolactone)/Halloysite Nanotube Nanocomposites. *J. Appl. Polym. Sci.* **2013**, *128*, 2807–2816.
26. Schmit, H.; Prashantha, K.; Soulestin, J.; Lacrampe, M. F.; Krawczak, P. Preparation and Properties of Novel Melt-Blended Halloysite Nanotubes/Wheat Starch Nanocomposites. *Carbohydr. Polym.* **2012**, *89*, 920–927.
27. Liu, M.; Zhang, Y.; Zhou, C. Nanocomposites of Halloysite and Polylactide. *Appl. Clay Sci.* **2013**, 52–59.
28. Ye, Y.; Chen, H.; Wu, J.; Ye, L. High Impact Strength Epoxy Nanocomposites with Natural Nanotubes. *Polym. J.* **2007**, *48*, 6426–6433.
29. Huttunen-Saarivirta, E.; Vaganov, G. V.; Yudin, V. E.; Vuorinen. J. Characterization and Corrosion Protection Properties of Epoxy Powder Coatings Containing Nanoclays. *Prog. Org. Coat.* **2013**, *76*, 757–767.
30. Dong, Y.; Chaudhary, D.; Haroosh, H.; Bickford, T. Development and Characterisation of Novel Electrospun Polylactic Acid/Tubular Clay Nanocomposites. *J. Mater. Sci.* **2011**, *46*, 6148–6153.
31. Zhao, Y.; Wang, S.; Guo, Q.; Shen, M.; Shi, X. Hemocompatibility of Electrospun Halloysite Nanotube- and Carbon Nanotube-Doped Composite Poly(Lactic-*co*-Glycolic Acid) Nanofibers. *J. Appl. Polym. Sci.* **2013**, *127*, 4825–4832.
32. Dong, Y.; Bickford, T.; Haroosh, H.; Lau, K. T.; Takag, I. H. Multi-Response Analysis in the Material Characterisation of Electrospun Poly(Lactic Acid)/Halloysite Nanotube Composite Fibres Based on Taguchi Design of Experiments: Fibre Diameter, Non-Intercalation and Nucleation Effects. *Appl. Phys. A* **2013**, *112*, 747–757.
33. Qi, R.; Cao, X.; Shen, M.; Guo, R.; Yu, J.; Shi, X. Biocompatibility of Electrospun Halloysite Nanotube-Doped Poly(Lactic-*co*-Glycolic Acid) Composite Nanofibers. *J. Biomater. Sci. Polym. Ed.* **2012**, *23*, 299–313.
34. Lee, I. W.; Li, J.; Chen, X.; Park, H. J. Electrospun Poly(Vinyl Alcohol) Composite Nanofibers with Halloysite Nanotubes for the Sustained Release of Sodium D-Pantothenate. *J. Appl. Polym. Sci.* **2016**, *133*, 42901–42911.
35. Lvov, Y.; Price, R.; Gaber, B.; Ichinose, I. Thin Film Nanofabrication via Layer-by-Layer Adsorption of Tubule Halloysite, Spherical Silica, Proteins and Polycations. *Colloids. Surf. A* **2002**, *198–200*, 375–382.
36. Yu, L.; Wang, H.; Zhang, Y.; Zhang, B.; Liu, J. Recent Advances in Halloysite Nanotube Derived Composites for Water Treatment. *Environ. Sci. Nano* **2016**, *3*, 28–44.
37. Deen, I.; Pang, X.; Zhitomirsky, I. Electrophoretic Deposition of Composite Chitosan–Halloysite Nanotube-Hydroxyapatite Films. *Colloids. Surf. A* **2012**, *410*, 38–44.
38. Deen, I.; Zhitomirsky, I. Electrophoretic Deposition of Composite Halloysite Nanotube–Hydroxyapatite–Hyaluronic Acid Films. *J. Alloys Compd.* **2013**, *586*, 531.
39. Federica, M.; De, R.; Virginia, M. New Method to Obtain Hybrid Conducting Nanocomposites Based on Polyaniline and Carbon Nanotubes. *ENI* **2011**, *6*, 86.

40. Liu, M.; Zhang, Y.; Zhou, C. Nanocomposites of Halloysite and Polylactide. *Appl. Clay Sci.* **2013**, *75*, 52–59.
41. Yeniova Erpek, C. E.; Ozkoc, G.; Yilmazer, U. Effects of Halloysite Nanotubes on the Performance of Plasticized Poly(Lactic Acid)-Based Composites. *Polym. Compos.* **2016**, *11*, 3134–3148.
42. De, Silva, R. T.; Pasbakhsh, P.; Goh, K. L.; Chai, S. P.; Chen, J. Synthesis and Characterisation of Poly(Lactic Acid)/Halloysite Bionanocomposite Films. *J. Compos. Mater.* **2014**, *48*, 3705–3717.
43. Wu, W.; Cao, X.; Zhang, Y.; He, G. Polylactide/Halloysite Nanotube Nanocomposites: Thermal, Mechanical Properties, and Foam Processing. *J. Appl. Polym. Sci.* **2013**, *130*, 443–452.
44. Teo, Z. X.; Chow, W. S. Impact, Thermal and Morphological Properties of Poly(Lactic Acid)/Poly(Methyl Methacrylate)/Halloysite Nanotube Nanocomposites. *Polym. Plast. Technol. Eng.* **2016**, *14*, 1474–1480.
45. Kaygusuz, I.; Kaynak, C. Influences of Halloysite Nanotubes on Crystallisation Behaviour of Polylactide. *Plast. Rubber Compos.* **2015**, *44*, 41–49.
46. De, Silva, R. T.; Pasbakhsh, P.; Lee, S. M.; Kit, A. Y. ZnO Deposited/Encapsulated Halloysite–Poly(Lactic Acid) (PLA) Nanocomposites for High Performance Packaging Films with Improved Mechanical and Antimicrobial Properties. *Appl. Clay Sci.* **2015**, *111*, 10–20.
47. Xu, W.; Luo, B.; Wen, W.; Xie, W.; Wang, X.; Liu, M.; Zhou, C. Surface Modification of Halloysite Nanotubes with L-Lactic Acid: An Effective Route to High-Performance Poly(L-Lactide) Composites. *J. Appl. Polym. Sci.* **2015**, *132*, 41451–41459.
48. Wu, W.; Cao, X.; Lin, H.; He, G.; Wang, M. Preparation of Biodegradable Poly(Butylene Succinate)/Halloysite Nanotube Nanocomposite Foams Using Supercritical CO_2 as Blowing Agent. *J. Polym. Res.* **2015**, *22*, 1–11.
49. Wu, W.; Cao, X.; Luo, J.; He, G.; Zhang, Y. Morphology, Thermal, and Mechanical Properties of Poly(Butylene Succinate) Reinforced with Halloysite Nanotube. *Polym. Compos.* **2014**, *35*, 847–855.
50. Gradzik, B.; Stenzel, A.; Boccaccini, A. R.; El, Fray, M. Influence of Functionalized Halloysite Clays (HNT) on Selected Properties of Multiblock (e) PBS–EG Copolymer Obtained by Enzymatic Catalysis. *Des. Monomers Polym.* **2015**, *18*, 501–511.
51. Sadegh-Hassani, F.; Nafchi, A. M. Preparation and Characterization of Bionanocomposite Films Based on Potato Starch/Halloysite Nanoclay. *Int. J. Biol. Macromol.* **2014**, *67*, 458–462.
52. Schmitt, H.; Creton, N.; Prashantha, K.; Soulestin, J.; Lacrampe, M. F.; Krawczak, P. Preparation and Characterization of Plasticized Starch/Halloysite Porous Nanocomposites Possibly Suitable for Biomedical Applications. *J. Appl. Polym. Sci.* **2015**, *132*, 41341–41350.
53. Schmitt, H.; Prashantha, K.; Soulestin, J.; Lacrampe, M. F.; Krawczak, P.; Raquez, J. M. Processing and Mechanical Behaviour of Halloysite Filled Starch Based Nanocomposites. In *Advanced Materials Research*; Trans Tech Publications: Switzerland, 2012; Vol. 584, p 445.
54. Schmitt, H.; Prashantha, K.; Soulestin, J.; Lacrampe, M. F.; Krawczak, P. Preparation and Properties of Novel Melt-Blended Halloysite Nanotubes/Wheat Starch Nanocomposites. *Carbohydr. Polym.* **2012**, *89*, 920–927.
55. Huang, D.; Wang, W.; Kang, Y.; Wang, A. A. Chitosan/Poly(Vinyl Alcohol) Nanocomposite Film Reinforced with Natural Halloysite Nanotubes. *Polym. Compos.* **2012**, *33*, 1693–1699.

56. Liu, M.; Wu, C.; Jiao, Y.; Xiong, S.; Zhou, C. Chitosan–Halloysite Nanotubes Nanocomposite Scaffolds for Tissue Engineering. *J. Mater. Chem. B* **2013**, *1*, 2078–2089.
57. Zhou, W. Y.; Guo, B.; Liu, M.; Liao, R.; Rabie, A. B. M.; Jia, D. Poly(Vinyl Alcohol)/Halloysite Nanotubes Bionanocomposite Films: Properties and In Vitro Osteoblasts and Fibroblasts Response. *J. Biomed. Mater. Res. A* **2010**, *93*, 1574–1587.
58. Barton, J.; Niemczyk, A.; Czaja, K.; Sacher-Majewska, B. Polymer Composites, Biocomposites and Nanocomposites. Production, Composition, Properties and Application Fields. *CHEMIK* **2014**, *1*, 280–287.
59. Gaaz, T. S.; Sulong, A. B.; Akhtar, M. N.; Kadhum, A. A. H.; Mohamad, A. B.; Al Amiery, A. A. Properties and Applications of Polyvinyl Alcohol, Halloysite Nanotubes and Their Nanocomposites. *Molecules* **2015**, *20*, 22833–22847.
60. Fakhrullin, R. F.; Lvov, Y. M. Halloysite Clay Nanotubes for Tissue Engineering. *Nanomedicine (Lond.)* **2016**, *11*, 2243–2246.
61. Phantom Plastics. *Halloysite Nanotubes*. http://phantomplastics.com/functional-fillers/halloysite/ (accessed August 24, 2016).

CHAPTER 16

FLAME SPRAYING OF POLYMERS: DISTINCTIVE FEATURES OF THE EQUIPMENT AND COATING APPLICATIONS

YURY KOROBOV[1] and MARAT BELOTSERKOVSKIY[2,*]

[1]*UrFU, Mira St. 19, Ekaterinburg 620002, Russia*

[2]*JIME NSA Belarus, Academic Street 12, Minsk 220072, Republic of Belarus*

*Corresponding author. E-mail: yukorobov@gmail.com

ABSTRACT

Flame spraying of polymers provides functional coatings to protect against wear and corrosion. The specific features of polymer flame-spraying guns are described. A range of initial polymer powders is represented. Purposefully, change of the properties of the resulting coatings by adding various fillers to the initial polymer powder is shown. The coating reliability is strongly influenced by its adhesion strength. The influence of the characteristics of initial polymers, the share of inorganic fillers, and technological spraying parameters on the adhesion strength were analyzed. Typical application cases are represented.

16.1 OVERVIEW OF POLYMER COATINGS

The most widely used methods for powder polymer deposition include the following: coating from the fluidized bed, electrostatic coating, and thermal spraying.

Only special chambers, baths, and ovens can realize the first two methods. Moreover, in case of large parts, their use is limited by excessive energy consumption. Therefore, thermal spraying is one of the most economical and easiest methods to implement polymer coating. It allows melting and forming a layer in a single operation.

Polymer processing should meet the following requirements:

- low heating temperature to escape a thermal destruction;
- reliability;
- low costs; and
- processability.

Plasma spraying and high-velocity oxygen-fuel spraying (HVOF) allow producing polymer coatings of good quality in some regime intervals.[1-6] However, their cost is minimum 3–5 times as high comparing to flame spraying.[7] So, flame spraying is of interest to develop for the deposition of polymers because of its simplicity and low cost.

Thermal spraying of polymers can be traced back to the 1940s, when polyethylene (PE) was first produced. Early attempts were unsuccessful because flame-spraying guns, designed for spraying metals, were used. They produced a flame that was both too hot and too short to melt the PE without degradation.[8]

16.2 DESIGN OF THE POLYMER FLAME-SPRAYING GUN

Since the 1970s of the last century, guns were developed for flame spraying of polymer materials.[1,9-11] Common features of gun design are seen in Figure 16.1. Combustion mixtures include oxidizer (oxygen or air) and fuel gas (acetylene or propane). A powder-feed nozzle is located at the center of the flame-spray gun. Two ring-shaped nozzle outlets, the inner ring being for air and the outer ring for the flame, surround this. As a result, the polymer powder is soft heated by air blanket excluding a direct contact with the flame. Such design provides longer residence time at a lower temperature for the sprayed powder.

Heating of the powder is determined by the ratio of velocities of air blanket and the surrounding flame. Control of the ratio leads to a visible change in the shape of the torch.

Fuel mixture air–propane seems to be more suitable to spray polymers. First, combustion temperature is about 800–1000°C lower comparing to oxygen–fuel mixtures. Use of compressed air instead of pure oxygen decreases flame temperature due to heating of nitrogen. In addition, the actual flame temperature can be changed additionally by adjusting the oxidizer/fuel flow rate ratio (Fig. 16.2).

Flame Spraying of Polymers: Distinctive Features

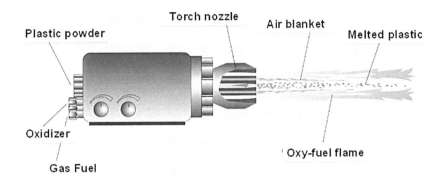

FIGURE 16.1 Scheme of polymer-spraying gun.

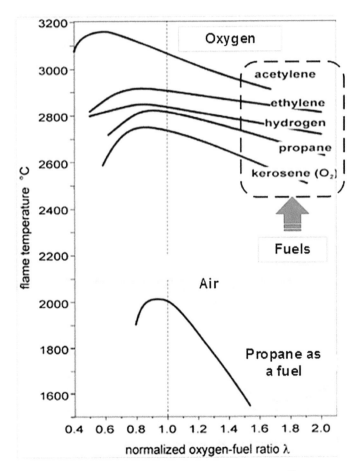

FIGURE 16.2 Flame temperature of various combustion mixtures.[12]

In recent years, the most widespread guns operate using air as the oxidizer. Heat distribution modeling showed that specific heat input of the torch at spraying should be within a range of $(1-3) \cdot 10^6$ W/m^2 to escape its thermal destruction of polymer coatings.[13,14] In this case, short-term overheating of the material is not more than 1.5 times higher above the range of polymer melting temperatures, 360–670 K. It provides dense polymer coatings with a minimum content of low-polymeric thermal degraded substance.

Typical polymer flame-spraying gun Terco-P was taken to show basic coating features (Fig. 16.3).

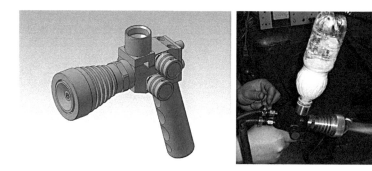

FIGURE 16.3 Polymer flame-spraying gun Terco-P.[11]

Specific features of gun Terco-P are the following:

- polymer melting point—360–650 K;
- output—2.8 kg/h;
- maximum gas pressure (MPa): air—0.5, propane—0.2;
- initial polymer form is spherical powder; and powder diameter is in the range of 50–300 μm.

Coating properties are the following:

- adhesion strength—7.5–10.0 MPa;
- Tensile strength—70 MPa;
- Deposition efficiency—0.92;
- Coating thickness—1–4 mm;
- Thermal destruction—negligible.
- Substrates: metals, ceramics, glass, and concrete.

Flame Spraying of Polymers: Distinctive Features

It allows adjustment of the heat exchange in the system "torch—polymer particle" by changing ratio between velocities of the air blanket and the concurrent combustible mixture. Results are visible as a flame shape change (Fig. 16.4).

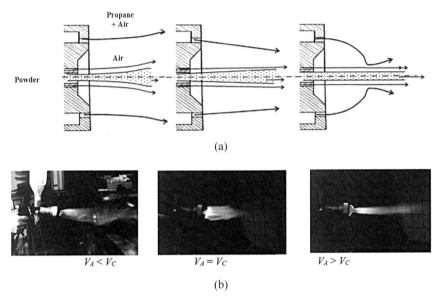

FIGURE 16.4 Torch shape depending on the ratio between velocities of the air blanket (V_A) and the combustible mixture (V_C). (a) Scheme of gas inlet and (b) view of flame.

16.3 FEEDSTOCK POLYMER MATERIALS

Many plastics can now be completely melted in-flight and allow heat-sensitive components to be coated. Polymers have been sprayed by common thermal spray processes, such as flame spraying, plasma spraying, and HVOF (Table 16.1).[1,2]

TABLE 16.1 Thermally Sprayed Polymers and Polymer Composites.

Polymeric material	Flame	HVOF	Plasma
Bismaleimide-phenolic resin			+
Cyanate ester thermosets			+
EMAA copolymer/Al_2O_3	+		
Epoxy or epoxy–nylon blend filled with Cu or Cu/Ni	+		
Epoxy enamels			+

TABLE 16.1 *(Continued)*

Polymeric material	Flame	HVOF	Plasma
Epoxy filled with TiO_2			+
Epoxy thermosets			+
EVA			
EAA copolymer	+		+
EMAA copolymer	+		
ETFE	+		
LCP	+		
Nylon-11/glass or Al_2O_3 or SiO_2 or carbon black	+	+	+
PE	+		+
PE modified with methacrylic acid	+		
PE (ultrahigh molecular weight)/WC–Co or $MgZrO_3$			+
PE/Al_2O_3			+
Phenolic thermosets			+
PA—Nylons	+		+
PAS		+	
PAEK			+
PC and PC/SiO_2		+	+
PES	+		+
PES or polyurethane filled with TiO_2, SiO_2, and Al_2O_3	+		
Polyester–epoxy resin	+		+
Polyether block amide copolymer	+		
PEA copolymer			+
PEEK	+	+	+
$PEEK/Al_2O_3$		+	
PET	+	+	+
Polyethylene–polypropylene copolymer	+		
PMMA			+
PPS	+	+	+
PPS/Al_2O_3		+	
PP			+
Polysulphone			+
PTFE and its copolymers	+		+
PVDF			+
PCCP	+		
PVDF–HFP copolymer			+
PVDF/WC-Co or $MgZrO_3$			+
Polyimide (PMR)/WC-Co		+	
Urethane	+		

EAA, ethylene-acrylic acid copolymer; EMAA, Ethylene-methacrylic; ETFE, ethyltetrafluoroethylene; EVA, ethylene vinylene acetate; HFP, hexafluoropropylene; HVOF, high-velocity oxygen-fuel spraying; LCP, liquid-crystalline polymers; PA, polyamides; PAS, polyarylene sulfide; PAEK, polyaryletherketone; PC, polycarbonate; PCCP, postconsumer commingled polymers; PE, polyethylene; PEA, polyether-amide; PEEK, polyetheretherketone; PES, polyester; PET, polyethylene terephthalate; PMMA, polymethylmethacrylate; PMR, polyimide; PP, polypropylene; PPS, polyphenylene sulfide; PTFE, polytetrafluoroethylene; PVDF, polyvinylidene fluoride.

As seen from Table 16.1, polymer coatings with added of inorganic fillers are also produced. It allows improving wear, corrosion, and radiation resistance. Such composite polymer coatings are intensively explored applying to various sectors of industry.

16.4 ADHESION STRENGTH OF THE COATING

Adhesion strength of the coating to substrate is one of the main factors determining the performance of polymer coatings. An influence of air–fuel ratio of the torch, type of polymer, particle size of the polymer powder, and a share of inorganic fillers in the blend on the adhesion strength of the coatings is described below.

The efficiency of flame polymer coatings during their exploitation under the influence of external factors is determined, above all, by the strength of the coating adhesion to the substrate. The efficiency of the application process, the adhesive strength and properties of the resulting flame composite polymer coatings obtained with flame devices depends on the composition of the material being deposited, and the technological features of spraying. The value of the adhesive strength of the polymer coating composition onto the prepared substrate surface depends on the structure of sprayed powder particles, and the type and percentage of fillers.

To improve the adhesion of flame-sprayed coatings, volume modification of applied polymer with inorganic fillers was used. It was successfully applied in different methods of formation of polymer coatings.[1,10] Fillers are used to modify the mechanical, structural, chemical, and other properties of polymers, including strength of adhesive polymer–metal compounds. Depending on the type and nature of fillers, they can increase or decrease the adhesion strength of polymer coatings to the substrate.

Therefore, the influence of the type and the percentage of filler of the polymer on adhesion strength of the polymer coating was investigated.[15] The following polymer powders were tested: polyamide (PA), high-density polyethylene (HDPE) and polyethylene terephthalate (PET). The following components were added to polymers: 200–300 μm glass-ceramic (GC) and 25–50 μm aluminum powder (Al). The thickness of the coatings was 0.5–1 mm.

Adhesion strength of the coatings was evaluated by the method of separation of a conical pin.[16] According to a schematic view of the test (Fig. 16.5), the washer 1 serves as the basis, pin 2 is inserted into its hole so that its end face surface is flush with the external plane of the washer. The total surface of the pin and the washer after preparation is coated with the coating 3. The pin and the washer were made from low carbon steel like C1020 ASTM.

The test consists of pulling the pin by applying a force (P). The adhesion strength (σ) is calculated by an equation:

$$\sigma = P/S,$$

where S is the square of the end-face surface. The results were averaged for five samples.

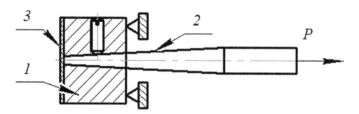

FIGURE 16.5 Scheme of conical pin separation.

The adhesion strength of the coating HDPE + GC/Al changes due to filler share (5–30 vol%). Peak value is indicated at 15–20% of fillers (Fig. 16.6).

The adhesion strength of the coating PA/HDPE/PET + 10 vol% Al changes due to polymer particle size. Min/max ratio of diameters is 1.5. Peak value is indicated at mean particle diameter 200–250 μm (Fig. 16.7).

These results can be explained by the following:

- Particle less than 50 μm is subjected to burning when spraying. Such burning products may be placed on a surface to be coated which leads to reduction of the adhesion strength.

- Particle has no time to melt by the heat of the torch if its size exceeds 300 μm. Presence of incompletely melted particles in coating–substrate interface also leads to reduction of the adhesion strength.

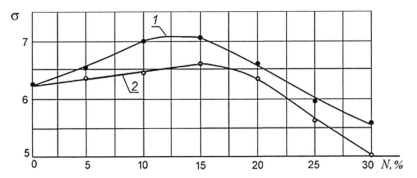

FIGURE 16.6 The change of the adhesion strength σ of polymer coating. (a) Due to the share of filler content (N, vol%), 1—glass-ceramic; 2—aluminum powder.

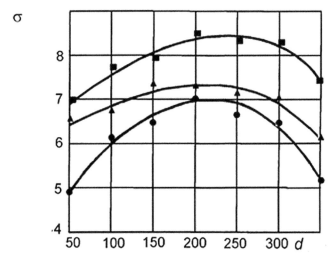

FIGURE 16.7 The change of the adhesion strength σ of polymer coating due to particle size of polymers. PET (■), HDPE (▲), and PA (●).

The results of experimental studies showed that the maximum adhesive strength is observed when ratio (max particle diameter, d_{max})/(min particle diameter, d_{min}) is less than 1.8–2.0 (Fig. 16.8). Polymer powders sizes range from 100 to 300 μm. So, using powders of uniform particle size leads to improving adhesion strength.

In addition, studies have shown (Figs. 16.7 and 16.8) that coating based on PET has the highest adhesion strength comparing to PA and HDPE. As noted, there is a certain correlation with the adhesion energy of the molecules bond in the polymer (cohesive energy).[8] The higher the energy of cohesion of the functional groups of the polymer, the higher is the adhesiveness. PET, due to presence of high-energy functional groups, has the highest adhesion.

As known, the adhesion strength of polymer coatings with surface of mild steel is drastically increased when oxygen-containing groups appear in polymers (–OH, –COOH, etc.).[8,17]

Flame, depending on the share of the combustible gas in the gas mixture, obtains "oxidizing," "normal," or "reducing" mode.[18] Normal flame is formed by combustion of the stoichiometric mixture of air–fuel, when all of the hydrocarbon molecules are reacted with oxygen molecules. Oxidizing flame is formed by combustion with excess oxygen in the mixture. An excess of fuel gas forms reducing flame. Oxidizing flame has a limit of oxidant concentration above which the combustion process is terminated.

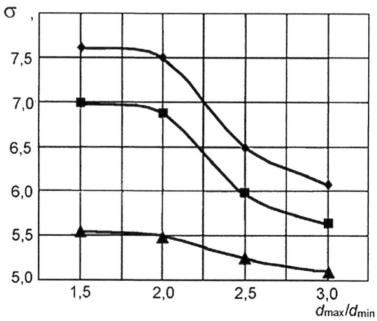

FIGURE 16.8 The change of the adhesion strength σ of polymer coating due to the d_{max}/d_{min} ratio (filler content—10 vol% Al). (♦) PET, (■) HDPE, and (▲) PA.

For flame-spray guns, the lower limit of the reducing propane–air flame is 16 volumes of air per 1 volume of propane (oxidizer-to-fuel ratio $\beta = 16$). With a further decrease in air content, there is a large amount of unreacted carbon. The upper limit of the oxidizer-to-fuel ratio is ($\beta = 32$). The subsequent increase of oxidant leads to the disruption of the flame.

The adhesion strength of the coating changes due to oxidizer-to-fuel ratio and particle size. The tests showed that its highest values are obtained at oxidizing flame ($\beta = 22–32$) and with particles of smaller size (Fig. 16.9). It results from the largest specific surface area of particles–air interaction, which promotes the formation of a significant number of oxygen-containing groups.

So, various stages of the process require various modes. Spraying the underlayer should be performed with the powder of particle size less than 60 µm, $\beta = 24–32$. Spraying the base coating should be performed with the powder of particle size 100–300 µm, $\beta = 20–24$. In this case, a significant oxidation of the coating material, which may cause some reduction in polymer properties and adversely affect the adhesion, is eliminated. Melting the coating should be carried out at $\beta = 16–20$ to decrease oxidation that can be realized by reducing flame treatment.

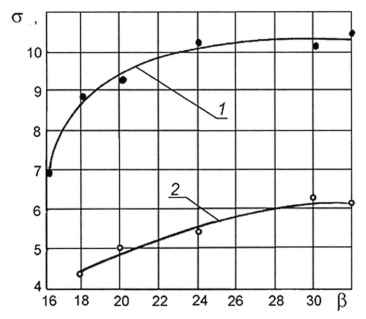

FIGURE 16.9 The change of the adhesion strength of polymer coating (σ) due to the oxidizer-to-fuel ratio (β). Polymer powder was PA–HDPE 50/50, particle size 50–60 µm (1) and 100–200 µm (2).

16.5 APPLICATION CASES

Polymer coatings are sprayed on metals, ceramics, glass, wood, and construction materials (concrete, brick, and slate). With regard to these materials, the polymer coatings are applied to a variety of purposes:

- protection of structural elements from aggressive environments;
- wear protection, application of antifriction layer;
- electrical connections of electrical power fittings;
- isolation of the contact of dissimilar metals to avoid electrochemical processes;
- restoration of defective polymer coating onsite; and
- the addition of fillers to the starting polymer, see Figure 16.10, changes the properties of the coatings as follows: rubber crumb improves antifriction properties; alumina, metals improve durability; and elements for thermal neutron absorption (boron and its compounds) increase the radiation resistance.

Some examples are shown below.

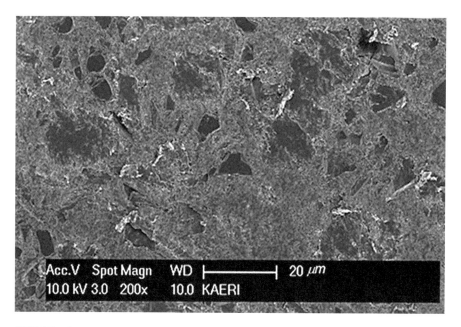

FIGURE 16.10 Structure of polymer coating with filler: UHMW-PE (ultra-high molecular weight polyethylene) + 20% B4C.

Flame Spraying of Polymers: Distinctive Features   247

This process has been successfully used to protect concrete structures (foundations, walls), which can be destroyed by the corrosion of steel reinforcement (Fig. 16.11).

FIGURE 16.11 Polymer coating is sprayed on concrete wall of reservoir.

Polymeric coatings are characterized by high performance in a dry friction conditions. Bench tests have shown that coefficient of dry sliding friction is 0.15 at specific load of 5 MPa and 0.08 at specific load of 10 MPa. It allows the decrease of cranking shall effort, contact temperature, and wear of mating parts, (Table 16.2).

TABLE 16.2 Results of Bench Tests.

Mated parts	Section moment to the cranking shall (N m)	Temperature in the friction zone (°C)	Linear wear (µm)
Steel–steel	110–130	55–65	60–85
Steel–polymer coating	60–80	30–35	10–20

Such coatings have been used on the surface of a cereal overload sleeve of combine harvester (Fig. 16.12). Spraying process is shown in Figure 16.13.

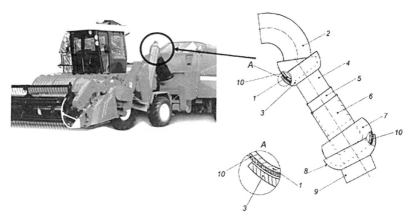

FIGURE 16.12 Polymer coating on a cereal overload sleeve of combine harvester. 1, 7—truncated spherical nozzles; 2—feed conveyor; 3—hollow truncated spherical support; 4, 5, 6—upper, middle, and lower tube adapters of telescopic extension; 8—truncated spherical tank; 9—box-shaped receptacle; 10—friction reducing polymer coating.

During the run tests, the sleeve with a polymer coating showed minimum grain loss and excellent durability of spherical joints at overload of the cereal. In truncated spherical nozzles, without coating failure was observed after 80 h, the damaged span was 45–70% of the contact area. Coated joints worked without failure 220 h in the absence of wear.

FIGURE 16.13 Polymer spraying of the truncated spherical support.

Flame Spraying of Polymers: Distinctive Features   249

Polymer coating is applied at processing galvanic baths that operate at aggressive chemical environment (Fig. 16.14).

FIGURE 16.14 Polymer coating is sprayed on a galvanic bath.

Polymer coating is used as road markings and concrete block paving (Fig. 16.15). Concrete block of which the surface is coated by powder polymer satisfied all the standards of the traffic paint in Japan in terms of whiteness, reflectance ratio, weathering resistance, and abrasion resistance.[19]

FIGURE 16.15 Polymer coating at road markings and concrete block paving.

16.6 CONCLUSIONS

1. Flame-sprayed coatings can be produced from various polymers of 360–670 K without destruction.
2. Applying to various polymeric coatings maximum adhesion strength is achieved by spraying the powders with a particle size 150–300 μm; the ratio between the minimum and maximum diameter of the particles should be less than 2.0.
3. Inorganic fillers in the polymer powder allow to increase the adhesion strength by 15–20%, and peak value of the filler content is 15 vol%.
4. To get the highest coating properties, the process should be divided to stages of spraying intermediate layer, base layer, and melting which differs by size of used powder and oxidizer-to-fuel ratio.
5. Polymer coatings are applied to protect structural elements against aggressive environments, wear, and to provide antifriction properties and electro insulation.

KEYWORDS

- **polymer powder**
- **polymer flame spraying**
- **adhesion strength**
- **polymer-spraying gun**

REFERENCES

1. Petrovicova, E.; Schadler, L. S. Thermal Spraying of Polymers. *Int. Mater. Rev.* **2002**, *47* (4), 169–190.
2. Gupta, V. Thermal Spraying of Polymer-Ceramic Composite Coatings with Multiple Size Scales of Reinforcements. *Master Thesis*, Drexel University, Germany, 2006.
3. Rodchenko, D. A., Kovalkov, A. N., Barkan, A. I. Heating of the Polymer Particles by Spraying a Jet of Plasma. *Proc. Univ. Mech. Eng.* **1985**, *9*, 108–113 (in Russian).
4. Bao, Y. Plasma Spray Deposition of Polymer Coatings. *PhD Thesis*, Brunel University, Uxbridge, West London, 1995.
5. Gupta, V.; Niezgoda, S.; Knight, R. HVOF Sprayed Multi-Scale Polymer/Ceramic Composite Coatings. In *Thermal Spray 2006: Science, Innovation, and Application,*

Proceedings of International Thermal Spray Conference, Materials Park, OH; Marple, B. R., Moreau, C., Eds.; ASM International: Materials Park, OH, 2006.
6. Vuoristo, P. M. J. Functional and Protective Coatings by Novel High-Kinetic Spray Processes, In *Kokkola Material Week-2014, Proceedings of International Conference*, Kokkola, Finland, 2014. TUT, Tampere, 2014.
7. Brogan, J. A. Thermal-Spraying of Polymers and Polymer Blends. *MRS Bull.* **2000,** *25* (7), 48–53.
8. Wicks, Z. W.; Jones, F. N.; Pappas, S. P. *Organic Coatings. Science and Technology*; John Wiley & Sons Ltd.: Chichester, 1994; Vol 2, p 438.
9. ASM International. *Thermal Spray Coatings: ASM Handbook*; ASM International: Welding, Brazing, and Soldering, 1993; Vol 6, pp 1004–1009.
10. Chen, H., Qu, J., Shao, H. Erosion-Corrosion of Thermal-Sprayed Nylon Coatings. *Acta Mater.* **1999,** *57*, 980–992.
11. Belotserkovsky, M. A., Chekylaev, A. V., Korobov, Yu. S. In *Flame Spraying of Polymer Coatings*. Technologies of Repair, Restoration and Strengthening of Machine Parts, Machinery, Equipment, Tools and Tooling, Proceedings of 9th International Conference. *In 2 parts*, St. Petersburg, Russia, Polytechnic. University Publishing House: St. Petersburg, 2007; Part 1, pp 26–33 [in Russian].
12. Kreye, H., Gartner, F., Kirsten, A., Schwertzke, R. High Velocity Oxy-Fuel Flame Spraying. State of Art, Prospects and Alternatives. In *5th Colloquium "High Velocity Oxy-Fuel Spraying"*. GTS e.V.: Erding, Germany, 2000.
13. Belotserkovsky, M. Development of Up-to-Date Coating and Hardening Technologies in Belarus. In *Reports of the Korea*. Eurasia Technology Cooperation Workshop: Seoul, Korea, 2007.
14. Rekhlitsky, O., Solovei, N., Belotserkovsky, M., Korotkevitch, S. In *Anti-Friction and Mechanical Properties of Flame Sprayed Polymer Coatings*. Proceedings of "BaltTrib – 2013", Lithuania, Kaunas, 2013.
15. Korobov, Yu. S., Belotserkovsky, M. A., Timofeev, K. M., Thomas, S. Adhesive Strength of Flame-Sprayed Polymer Coatings. In *AIP Conference Proceedings*, 16–20 May 2016, 1785, 030011, 2016, Ekaterinburg, Russia. http://dx.doi.org/10.1063/ 1.4967032.
16. Tyshinski, L. I. Plokhov, A. V., Tokarev, O. A., Sindeev, V. I. *Methods of Study the Materials*. Mir: Moscow, 2004; p 384 [in Russian].
17. Rama, K., Layek, A., Nandi, K. A Review on Synthesis and Properties of Polymer Functionalized Grapheme. *Polymer* **2013,** *54* (19), 5087–5103.
18. *Welder's Handbook for Gas Shielding Arc Welding, Oxy Fuel Cutting & Plasma Cutting*, 3rd ed.; Air Products PLC: New York, 1999; p 274.
19. Sano, Y., Fukushima, N., Niki, T. Consideration on Powder Flame Sprayed Concrete Blocks. https://docs.google.com/viewerng/viewer?url=http://www.sept.org/techpapers/ 133.pdf [online].

PART III
Research on Electrochemistry

CHAPTER 17

GRAPHENE-BASED HYBRIDS FOR ENERGY STORAGE AND ENERGY CONVERSION APPLICATIONS

ANJU K. NAIR[1,2*], SABU THOMAS[1,3], KALA M. S.[2], and NANDAKUMAR KALARIKKAL[1,4*]

[1]*International and Inter University Centre for Nanoscience and Nanotechnology, Mahatma Gandhi University, Kottayam 686560, Kerala, India*

[2]*Department of Physics, St. Teresa's College, Ernakulam 682011, Kerala, India*

[3]*School of Chemical Sciences, Mahatma Gandhi University, Kottayam 686560, Kerala, India*

[4]*School of Pure and Applied Physics, Mahatma Gandhi University, Kottayam 686560, Kerala, India*

*Corresponding authors. E-mail: nkkalarikkal@mgu.ac.in

ABSTRACT

Graphene, a two-dimensional, atomically thin layer of sp^2-hybridized carbon, possesses remarkable attention and intensive research interest due to its spectacular properties such as very high electrical conductivity, better thermal properties, and exceptional mechanical strength, which makes as an exciting material for many fields. This chapter discusses the recent advancement in the development of graphene-based hybrids. We focus on the advantages of graphene-based composites for the applications of energy storage and energy conversion applications, especially in the fields of lithium ion batteries, electrochemical supercapacitors, and fuel cells.

17.1 INTRODUCTION

The importance of developing new types of energy sources is evident from the fact that the global energy consumption due to the depletion of fossil fuels, global warming, and environmental pollution has been accelerating at an alarming bell to human society due to the rapid economic expansion worldwide, increase in world population, and ever increasing human reliance on energy-based appliances. Therefore, renewable energy storage and conversion materials, as well as their devices, are highly essential. Nanotechnology has opened up new frontiers in materials science and engineering to meet this challenge.[1-3] In particular, carbon nanomaterials and nanotechnologies have been demonstrated to be an enabling technology for creating high-performance energy conversion and storage devices.[4-6]

Comparing to conventional energy materials, carbonaceous nanomaterials possess some unusual morphological, electrical, optical, and mechanical properties useful in enhancing energy conversion and storage performance.[7-10] Graphene—a material that appears in almost every research area nowadays—is a shining star on account of its intrinsic outstanding physicochemical properties. The two-dimensional (2D) sp^2 carbon network in the graphene structure makes it the thinnest but strongest material in the universe.[11,12] Graphene exhibits extraordinary mechanical, thermal, and electronic properties such as high specific surface area per unit volume, high chemical stability, excellent thermal conductivity (up to 5000 W m^{-1} K^{-1}), and a unique band structure with band-tuning ability and extremely high carrier mobility (in excess of 100,000 cm^2 $V^{-1} \cdot s^{-1}$).[13,15] Thus, it is envisioned as a superior alternative to silicon and is widely used in applications such as solar cells, field-effect transistors, batteries, and super capacitors.[16,19]

17.2 STRUCTURE OF GRAPHENE

Graphene is a 2D single layer of sp^2-hybridized carbon network with hexagonal structure. The electronic properties of graphene mainly depend on the arrangement and number of graphene layers. Few-layer graphene exhibits different electronic structure than that of bulk graphite.

17.3 PROPERTIES OF GRAPHENE

17.3.1 ELECTRONIC PROPERTIES

The graphene acts as massless particles or termed as Dirac fermions.[20] It should be noted that graphene exhibits as zero band gap material with a small overlap between valence band and conduction band. It also exhibits room temperature nobilities of ~10,000 cm^2 V^{-1} s^{-1} and the suspended graphene reveals low temperature mobility of ~20,000 cm^2 V^{-1} s^{-1} for carrier density below 5 × 10^9 cm^{-2} which is not seen in semiconductors or nonsuspended graphene sheets. Moreover, it exhibits a remarkable half-integer quantum Hall effect for both hole and electron carriers by adjusting the chemical potential with the application of electric field.[21]

17.3.2 OPTICAL PROPERTIES

The single-layer graphene has the ability to absorb 2.3% white light with a negligible reflectance of (<0.1%), and this absorbance value increases linearly with the increase of number of layers.[22] Due to this unique electronic properties of graphene materials, the electrons behave like Dirac massless electrons with very high mobility. It proves that the dynamic conductivity of graphene G depends on the equation, $G = \pi e^2/2h$ over the visible frequency region, where c corresponds to speed of light, h is Planck's constant. Further, the transparency property of graphene only depends on the fine structure constant $a = 2\pi e^2/hc$, which suggests the strong coupling between the light and relativistic electrons.[23] It has been noted that once optical intensity attains a threshold value, saturable absorption takes place (light causes an absorption reduction at very high intensity light). This is an important parameter that corresponds to the mode-locking of fiber lasers. The optical transitions and interband transitions of single layer and monolayer graphene are modulated by electrical gating with the aid of infrared spectroscopy, which opens up promising applications in the field of infrared optics and optoelectronics devices.[24]

17.3.3 THERMAL PROPERTIES

The in-plane thermal conductivity of single-layer freely suspended graphene at room temperature has been calculated as 3000–5000 W m^{-1} K^{-1} is among the highest of any known material.[25] Seol et al. reported that the thermal conductivity of graphene is ~600 W m^{-1} K^{-1} when it is making hybrid with amorphous silica. This decrease in thermal conductivity is mainly ascribed to the phonons leakage across the graphene–silica interface and very high interface-scattering.[26] However, this value is still many times higher than copper and silicon, which are mainly used in the electronics industry today.

17.4 SYNTHESIS TECHNIQUES

Till date, remarkable efforts have been made to widen synthesis modes for graphene and its graphene derivatives to attain high yield of production. Various synthesis techniques can be adopted for the preparation of graphene/nanoparticle hybrids.

17.4.1 EX SITU SYNTHESIS OF NANOPARTICLES ON GRAPHENE NETWORK

Researchers have put forth many scientific efforts to synthesize inorganic nanostructures of different shape and size over the graphene surface to further enhance their properties. These hybrids are commonly applied in fields like electrochemical energy conversion and storage, photovoltaic, sensors, electronics, and optics. The inorganic nanostructures like metals and metal oxides have been made composite with graphene which include Ag,[27] Au,[28] Pd,[29] Pt,[30] Ni,[31] Cu,[32] and oxides like TiO$_2$,[33] SnO$_2$,[34] MnO$_2$,[35] ZnO,[36] SiO$_2$,[37] NiO,[38] Fe$_3$O$_4$,[39] etc. The synthesis procedures are mainly classified as ex situ synthesis and in situ synthesis.

The ex situ synthesis occupies the external mixing of graphene nanosheets (GNSs) and preprepared or commercially obtainable nanoparticles in solutions. The surface modification of the nanoparticles and/or graphene is regularly carried out so that they can attach easily through the binding of either the chemical bonding or the noncovalent interactions. For instance, benzyl-mercaptan-modified CdS nanoparticles[40] or 2-mercaptopyridine capped Au NPs[41] have been effectively binded to GO or rGO surfaces through π–π stacking. In another work, bovine serum albumin protein has been used to alter the rGO surface through π–π interaction to absorb metal nanoparticles of Au, Ag, Pt, and Pd.[42]

17.4.2 IN SITU SYNTHESIS OF NANOPARTICLES ON GRAPHENE NETWORK

Although the ex situ synthesis undergoes low density and nonuniform coverage of the nanomaterials on the graphene surfaces.[41] On the contrary, the in situ synthesis provides homogeneous surface coverage of nanoparticles by controlling the nucleation sites on graphene structure through surface functionalization. Accordingly, monodisperse distribution of NPs on graphene surfaces can be achieved. Chemical reduction is the mainly followed popular scheme for synthesis of metal nanostructures. The graphene–Au NP composites can be achieved by the reduction of metal precursors $HAuCl_4$ with $NaBH_4$ as reducing agent in a rGO–octadecylamine solution.[43] Nevertheless, while rGO has less oxygen-containing functional groups, it limits aqueous solution-based procedures. To tackle this problem, 3,4,9,10-perylene tetracarboxylic acid (PTCA)-functionalized rGO sheets have been employed as support matrix for the in situ reduction of $HAuCl_4$. Due to the presence of added carboxylic groups from PTCA, the getting rGO–Au NP composites are more water soluble and provide an intense coverage of NPs.[44]

In another work, $CoMoO_4$ nanoparticles prepared on reduced graphene oxide ($CoMoO_4NP/rGO$) were carried out by a simultaneous hydrothermal synthesis.[45] This synthesis procedure helps to create multilevel porous structure along with high surface area. After the hydrothermal treatment, the nanoparticles exhibit strong interactions or make covalent bonds with graphene and at the same time, GO sheets reduced to RGO. The $CoMoO_4$ remains after oxidation in air in thermogravimetric analysis (TGA) are found to be 65%, 74%, and 88% for the $CoMoO_4NP/rGO$ nanocomposites synthesized with 1, 3, and 6 mL $CoMoO_4$ precursor mixed with 10 mL of GO suspension, respectively (Fig. 17.1). The XRD pattern of the composites display the diffraction peaks (0 0 1), (2 0 0), (2 2 0), (2 2 2), (1 5 2), and (2 6 0) correspond to the monoclinic $CoMoO_4$. Moreover, in the XRD pattern, the (0 0 2) peak of RGO peak is invisible, indicating that rGO flakes are well-separated by $CoMoO_4$ nanoparticles. SEM images reveal the porous structure of graphene sheets indicates that the scaffold is built of stacked graphene sheets and the nanoparticles are not visible with the resolution of the SEM. TEM images show that tiny $CoMoO_4$ nanoparticles of 3–5 nm size are thickly assembled on the surface of graphene sheets without any sort of aggregation. Herein, GNSs play an important role in attaining good dispersion.

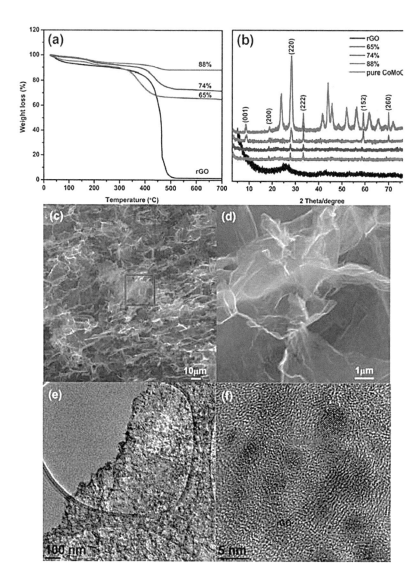

FIGURE 17.1 (See color insert.) Thermogravimetric analysis (TGA) of CoMoO$_4$NP/ rGO with different (a) CoMoO$_4$ contents and (b) XRD patterns. (c–f) Characterization of the morphology and elemental analysis of CoMoO$_4$NP/rGO (74%) nanocomposite. (c,d) Typical SEM images of the nanocomposite with different magnifications, revealing that the nanoparticles changed the structure of the graphene sheets. (e,f) TEM images of the nanocomposite, showing that the CoMoO$_4$ nanoparticles are confined in the matrix of the GNSs. Copyright reserved to the American Chemical Society.[45]

17.5 GRAPHENE–NANOPARTICLE HYBRIDS FOR ENERGY STORAGE AND ENERGY CONVERSION DEVICES

Graphene/nanoparticle-based hybrid structures have been studied for a wide range of applications in energy conversion systems, such as solar cells and fuel cells. Here, we present the broad scope of graphene/nanoparticles hybrids for cutting-age research and development in the field of energy conversion and energy storage in-depth.

17.5.1 LITHIUM-ION BATTERIES

Lithium-ion batteries have received great attention in portable electronics market, with its high gravimetric and volumetric capacities that have recently been in the limelight for its vehicular applications.[46,47] Rechargeable lithium batteries have transformed portable electronic devices. Lithium ion batteries comprising a graphite-negative electrode (anode), a nonaqueous liquid electrolyte, and a positive electrode (cathode) formed from layered $LiCoO_2$ (Fig. 17.2). On charging process, lithium ions are deintercalated from the $LiCoO_2$ intercalation host, and pass through the electrolyte and are intercalated between the graphite layers in the anode. Discharge reverses this process. The electrons, of course, pass around the external circuit.[48] The major drawbacks of the graphite anode was its lower electrochemical potential for lithium intercalation leading to dendrite growth and low theoretical capacity (372 mA h g^{-1})[49], and it is not suitable for hybrid electric vehicles. Anode materials with relatively higher potential range (between 1.0 and 1.5 V vs. Li/Li$^+$) and very large capacity are highly favored as an alternative to the commercial graphite anode. Scientific community put forth much research efforts in replacing the existing graphite anode with alternate materials having enhanced energy and power densities. In this context, other anode materials were established, such as silicon (4200 mA h g^{-1}),[50] tin (994 mA h g^{-1}), and germanium (1600 mA h g^{-1}).[51] However, these materials exhibit low capacity, poor rate capability, and rapid capacity fading because of their volume expansion and low conductivity. To alleviate this drawback, carbonaceous materials like amorphous carbon, CNTs and graphene and its derivatives have been used in conjunction with metal/metal oxide nanoparticles for enhancing the performance of Li-ion batteries.[52]

In Li-ion batteries, graphene can serve as a binder material, eliminating the use of binding polymer materials such as poly(vinylidene fluoride). Also, the high conductivity associated with graphene sheets lends itself to rapid transport of electrons to and from the active material intercalation sites, particularly given the close physical association of the nanoparticles and the RGO sheets. Also the mechanical strength of the graphene has the potential to absorb some of the expansion and contraction of the anchored nanoparticles during the intercalation and deintercalation processes, which typically lead to mechanical failure of the electrode and performance reduction through the loss of intimate contact of the active material and the conductive carbon black mixed into the electrode material for enhanced conductivity. The electrode can ultimately be pulverized if the expansion is large enough; hence, the use of active materials exhibits small changes upon Li^+ intercalation. The storage capacity and cycling ability of graphene was shown to outperform graphite in a number of studies. Some have attributed this improvement to the higher surface area achieved in graphene relative to graphite.

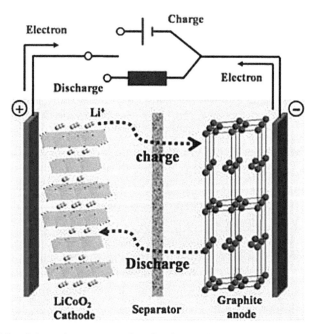

FIGURE 17.2 Schematic representation for the working mechanism of a lithium ion battery. Copyright reserved to the Royal Society of Chemistry.[53]

In one approach, Co_3O_4 nanoparticles wrapped with GO have significantly improved the specific capacity (1100 mA h g^{-1} at a low current density of 74 mA g^{-1}) in comparison with the highest value of the Co_3O_4-based electrodes (600–850 mA h g^{-1}).[54] Moreover, the Co_3O_4/GO hybrid-based electrode exhibits a very high and stable reversible capacity of about 1000 mA h g^{-1} after 130 cycles (about 94% retention). These properties confirm the supremacy of graphene/nanoparticle hybrids in Li-ion batteries. The GO layers connect together and maintain the electrical conductivity of the hybrid electrode and it facilitates the volume change of metal oxides during the charge–discharge cycles and it prevents the pulverization process of electrodes and improving the retention rate of the hybrid anode. Moreover, the graphene sheets have been wrapping around nanoparticles and it prevents the aggregation of nanoparticles. Wang et al. used anionic sulfate surfactants to stabilize graphene in aqueous solutions and make it possible the self-assembly of in situ grown nanocrystalline TiO$_2$ with graphene structure.[55] The surfactants provide the molecular template for controlled growth and nucleation of nanostructured materials. These approach schematically represented in Figure 17.3. The hybrid materials exhibited enhanced Li-ion insertion/extraction in TiO$_2$. The calculated specific capacity was more than doubled at high charge rates, as compared with the pure TiO$_2$ phase electrode. The improved capacity may be ascribed to the increased electrode conductivity due to the presence of a percolated graphene network presented in the metal oxide electrodes.

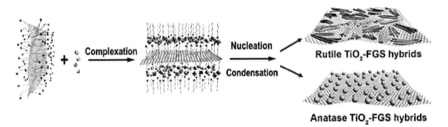

FIGURE 17.3 Anionic sulfate surfactant-mediated stabilization of graphene and growth of self-assembled TiO$_2$–FGS hybrid nanostructures. Copyright reserved to the American Chemical Society.[55]

Xie et al. were uniformly coated amorphous SnO$_2$ (a-SnO$_2$) thin films onto the surface of reduced graphene oxide (G) network by atomic layer deposition technique. The sheet structure of graphene provides an excellent support to tolerant volume change during charge/discharge cycle process. The a-SnO$_2$/G nanocomposites reached stable capacities of 800 mA h g^{-1} at 100 mA g^{-1} and 450 mA h g^{-1} at 1000 mA g^{-1}. The capacity from a-SnO$_2$ is higher than the bulk theoretical values (Fig. 17.4).[56] The extra high capacity is ascribed to additional interfacial charge storage resulting from the high surface area of the a-SnO$_2$/G nanocomposites and the effect of cystallinity on cycle stability.

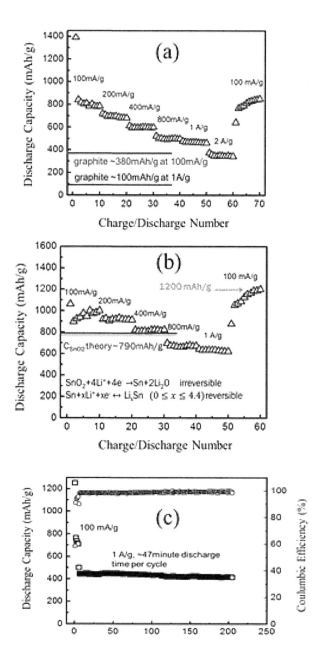

FIGURE 17.4 (a) Rate performance of SnO$_2$/G composites at various current densities, (b) rate performance of SnO$_2$ contribution only at various current densities, and (c) cycling performance and CE of SnO$_2$/G composites at 1000 mA g^{-1}. Copyright reserved to the American Chemical Society.[56]

Yao et al. established a facile hydrothermal approach combined with a preconditioning vacuum-assisted impregnation for in situ production of controlled anisotropic SnO_2 heterostructures inside graphene aerogel (GA). The 3D graphene aerogel provides an interconnected graphene framework which facilitates the electron transport across the electrode; numerous multi-dimensional channels in GA could effectively enhance the lithium diffusion within the bulk electrode, and an elastic matrix of GA could accommodate the volume expansion of metal oxides during long-term cycling. The SnO_2 and 3D GA synergistically led to enhanced lithium-storage properties (1176 mA h g^{-1} for the first cycle and 872 mA h g^{-1} for the 50th cycle at 100 mA g^{-1}) as compared with its two counterparts. Moreover, this hybrid material was able to deliver high specific capacity at rapid charge/discharge cycles (1044 mA h g^{-1} at 100 mA g^{-1}, 847 mA h g^{-1} at 200 mA g^{-1}, 698 mA h g^{-1} at 500 mA g^{-1}, and 584 mA h g^{-1} at 1000 mA g^{-1}).[57]

17.5.2 ELECTROCHEMICAL SUPERCAPACITORS

Among other energy storage systems, supercapacitors have received significant attention due to their fast charge–discharge process, very high power density, and superior cycle stability. Presently, supercapacitors are used together with batteries to give the additional power required in many applications.[58] Supercapacitors are electrochemical energy storage devices that combine the high energy-storage capability of conventional batteries with the high power-delivery capability of conventional capacitors.

Based on their storage mechanism, supercapacitors are mainly classified into two types, the pseudocapacitors and electrochemical double layer capacitors (EDLCs). The pseudocapacitors store their charges chemically by redox reaction at the surface of the material, while EDLCs store charges physically by reversible ion adsorption at the electrolyte–electrolyte interface. Normally, carbon-based materials (porous carbon, carbon nanotubes, grapheme, etc.) are used as electrode materials for EDLCs, while that for pseudocapacitors are conducting polymers (e.g., polyaniline, poly(3,4-ethylenedioxythiophene), polypyrrole, etc.) and transition metal oxides (e.g., MnO_2, RuO_2, NiO, Fe_2O_3, etc.).[58] Graphene electrodes offer great potential for double layer capacitors. Due to their high electrical conductivity, high-charge transport mobility, high mesoporosity, and high-electrolyte accessibility, graphene is attractive electrode material for developing high-performance supercapacitors. Besides, they were also found to be a versatile backbone to combine them with various metal oxides nanostructures such as

RuO$_2$, MnO$_2$, etc. to improve the performance, safety, and lifetime for supercapacitors. Graphene–metal oxide hybrids exhibit better total capacitance by an added pseudocapacitance from metal oxide nanostructures along with the double layer capacitance from graphene. The hybrid material possesses better chemical stability, capacity, and rate performance.[59,60] A lot of metal oxide nanostructures such as ZnO,[61] MnO$_2$,[62] RuO$_2$,[63] Co$_3$O$_4$,[64] and SnO$_2$[65] have been made composite with graphene for supercapacitor electrodes.

Geng et al. report a new and scalable method for the in situ synthesis of anatase TiO$_2$ nanospindles in a three-dimensional network structure of reduced graphene oxide (3D TiO$_2$@RGO) using cellulose as both an intermediate and structure-directing agent.[66] Due to the large aspect ratio of the TiO$_2$ nanospindles, and the exposed high-energy {0 1 0} facets of the TiO$_2$ crystals in the 3D network structure of the composite, the 3D TiO$_2$@RGO exhibited superior capacitive performance as an electrode material for supercapacitors, with a high specific capacitance (ca. 397 F g^{-1}), a high energy density (55.7 W h kg^{-1}), and a high power density (1327 W kg^{-1}) on the basis of the masses of RGO and TiO$_2$ (Fig. 17.5).

17.5.3 FUEL CELLS

Instead of burning fuel to create heat, fuel cells convert chemical energy directly into electricity. Although many different types can be constructed depending on the nature of the electrolyte materials used, they all work in the same principle.[67] They all in principle are an electrochemical cell consisting of the anode, the electrolyte, and the cathode. By pumping, for example, hydrogen gas onto one electrode (the anode), hydrogen is split into its constituent electrons and protons.[68] While the protons diffuse through the cell toward a second electrode (the cathode), the electrons flow out of the anode to provide electrical power. Electrons and protons both end up at the cathode to combine with oxygen to form water. The oxygen reduction reaction (ORR) at cathode can proceed either through a four-electron process to directly combine oxygen with electrons and protons into water as the end product or a less efficient two-step, two-electron pathway involving the formation of hydrogen peroxide ions as an intermediate.[69] The reduction would naturally happen very slowly without catalyst on the cathode to speed up the ORR, leading to insignificant production of electricity. Platinum nanoparticles have long been regarded as the best catalyst for the ORR, though the Pt-based electrode suffers from its susceptibility to time-dependent drift and CO deactivation.[69] Moreover, the high cost of the

Graphene-Based Hybrids for Energy Storage and Energy 267

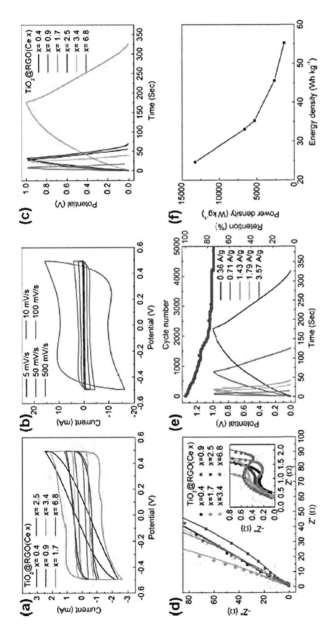

FIGURE 17.5 **(See color insert.)** (a) CV curves for 3D TiO$_2$@RGO composites measures at a scan rate of 50 mV s^{-1}. (b) CV curves collected for a 3D TiO$_2$@RGO(Ce 1.7) composite at different scan rates. (c) GV charge/discharge profiles for 3D TiO$_2$@RGO composites obtained at a current density of 0.36 A g^{-1}. (d) Nyquist plots for 3D TiO$_2$@RGO composites measured over a frequency range of 100 kHz to 0.01 Hz with an AC perturbation of 5 mV. (e) GV charge/discharge profiles collected for the 3D TiO$_2$@RGO composite at different current densities (left Y axis), and the long-term stability measured over nearly 5000 cycles at a current density of 1 A g^{-1} (right Y axis). (f) Relationship between power density and energy density for the 3D TiO$_2$@RGO composite. Copyright reserved to the American Chemical Society.[66]

platinum catalysts has been shown to be the major "showstopper" to mass market fuel cells for commercial applications. Pt-based catalysts are the most commonly used materials for low-temperature fuel cells.[70] Since Pt electrodes are very costly, Pt loading must be reduced; however, the fuel cell application cannot be compromised. Therefore, catalyst supports, such as graphene, carbon black, and CNTs, are being used to give better dispersity and thus provide very high specific surface area of the supported catalyst materials. Graphene–Pt hybrids have been also tried in the fuel cell applications, such as the oxygen reduction and methanol oxidation reaction.[71,72] Yoo et al. demonstrated that GNSs are used as a support for the formation of Pt clusters with small size and because these clusters are finely dispersed on the GNS surface and prevent from aggregation. The resulting hybrid exhibits very high activity for methanol oxidation than the Pt–carbon black catalyst. Doped graphene sheets show better conductivity than the undoped graphene which has also been used as the catalyst for fuel cells. The presence of nitrogen results in good dispersion and coverage for the in situ synthesis of Pt NPs. The N doping enhances the conductivity of the hybrid and exhibits improved electrocatalytic acivity.[73]

17.6 CONCLUSION

The collection of various valuable properties of graphene and its derivatives, together with the cost-effective nature and accessibility for large production, has made them potential building blocks in various multifunctional composites, incorporated with metals, metal oxides, etc. These stimulating research efforts include graphene synthesis, graphene functionalization, and the electrochemical storage and electrochemical conversion devices using graphene and graphene-based nanomaterials.

KEYWORDS

- graphene
- electrical conductivity
- thermal properties
- mechanical strength
- lithium-ion batteries

REFERENCES

1. Schatz, G. C. Nanotechnology for Next Generation Solar Cells. *J. Phys. Chem. C* **2009,** *113,* 15473–15475.
2. Sahaym, U.; Norton, M. G. Advances in the Application of Nanotechnology in Enabling a 'Hydrogen Economy'. *J. Mater. Sci.* **2008,** *43,* 5395–5429.
3. Kamat, P. V. Graphene-Based Nanoassemblies for Energy Conversion. *J. Phys. Chem. Lett.* **2011,** *2,* 242–251.
4. Guldi, D. M.; Martin, N. *Carbon Nanotubes and Related Structures: Synthesis, Characterization, Functionalization, and Applications. Carbon Nanotubes and Related Structures: Synthesis, Characterization, Functionalization, and Applications*; Wiley: Hoboken, NJ, 2010. DOI:10.1002/9783527629930.
5. Titirici, M.-M.; Antonietti, M. Chemistry and Materials Options of Sustainable Carbon Materials Made by Hydrothermal Carbonization. *Chem. Soc. Rev.* **2010,** *39,* 103–116.
6. Huang, X.; et al. Graphene-Based Materials: Synthesis, Characterization, Properties, and Applications. *Small* **2011,** *7,* 1876–1902.
7. Aricò, A. S.; Bruce, P.; Scrosati, B.; Tarascon, J.-M.; van Schalkwijk, W. Nanostructured Materials for Advanced Energy Conversion and Storage Devices. *Nat. Mater.* **2005,** *4,* 366–77.
8. Guo, Y.-G.; Hu, J.-S.; Wan, L.-J. Nanostructured Materials for Electrochemical Energy Conversion and Storage Devices. *Adv. Mater.* **2008,** *20,* 2878–2887.
9. Rolison, D. R.; et al. Multifunctional 3D Nanoarchitectures for Energy Storage and Conversion. *Chem. Soc. Rev.* **2009,** *38,* 226–252.
10. Liu, J.; et al. Oriented Nanostructures for Energy Conversion and Storage. *ChemSusChem* **2008,** *1,* 676–697.
11. Lee, C.; Wei, X.; Kysar, J. W.; Hone, J. Measurement of the Elastic Properties and Intrinsic Strength of Monolayer Graphene. *Science* **2008,** *321,* 385–388.
12. Geim, A. K. Graphene: Status and Prospects. *Science (80–)* **2009,** *324,* 1530–1534.
13. Balandin, A. A. Thermal Properties of Graphene and Nanostructured Carbon Materials. *Nat. Mater.* **2011,** *10,* 569.
14. Zhang, Y.; Tan, Y.-W.; Stormer, H. L.; Kim, P. Experimental Observation of the Quantum Hall Effect and Berry's Phase in Graphene. *Nature* **2005,** *438,* 201–204.
15. Bolotin, K. I.; et al. Ultrahigh Electron Mobility in Suspended Graphene. *Solid State Commun.* **2008,** *146,* 351–355.
16. Yang, N.; Zhai, J.; Wang, D.; Chen, Y.; Jiang, L. Two-Dimensional Graphene Bridges Enhanced Photoinduced Charge Transport in Dye-Sensitized Solar Cells. *ACS Nano* **2010,** *4,* 887–894.
17. Ji, L.; Meduri, P.; Agubra, V.; Xiao, X.; Alcoutlabi, M. Graphene-Based Nanocomposites for Energy Storage. *Adv. Energy Mater.* **2016,** *6,* 1502159.
18. Huang, X.; et al. Graphene-Based Materials: Synthesis, Characterization, Properties, and Applications. *Small* **2011,** *7,* 1876–1902.
19. Zhang, B.; Wang, Y.; Zhai, G. Biomedical Applications of the Graphene-Based Materials. *Mater. Sci. Eng. C* **2016,** *61,* 953–964.
20. Novoselov, K. S.; et al. Electric Field Effect in Atomically Thin Carbon Films. *Science* **2004,** *306,* 666–669.
21. Novoselov, K. S.; et al. Room-Temperature Quantum Hall. *Science (80–)* **2007,** *315,* 2007.

22. Nair, R. R.; et al. Fine Structure Constant Defines Visual Transperency of Graphene. *Science (80–)* **2008**, *320*, 2008.
23. Gusynin, V. P.; Sharapov, S. G.; Carbotte, J. P. Unusual Microwave Response of Dirac Quasiparticles in Graphene. *Phys. Rev. Lett.* **2006**, *96*, 256802.
24. Wang, F.; et al. Gate-Variable Optical Transitions in Graphene. *Science (80–)* **2008**, *320*, 206–209.
25. Balandin, A. A.; et al. Superior Thermal Conductivity of Single-Layer Graphene. *Nano Lett.* **2008**, *8*, 902–907.
26. Seol, J. H.; et al. Two-Dimensional Phonon Transport in Supported Graphene. *Science* **2010**, *328*, 213–216.
27. Tang, X.-Z.; et al. Synthesis of Graphene Decorated with Silver Nanoparticles by Simultaneous Reduction of Graphene Oxide and Silver Ions with Glucose. *Carbon N. Y.* **2013**, *59*, 93–99.
28. Choi, Y.; Gu, M.; Park, J.; Song, H. K.; Kim, B. S. Graphene Multilayer Supported Gold Nanoparticles for Efficient Electrocatalysts toward Methanol Oxidation. *Adv. Energy Mater.* **2012**, *2*, 1510–1518.
29. He, H.; Gao, C. Graphene Nanosheets Decorated with Pd, Pt, Au, and Ag Nanoparticles: Synthesis, Characterization, and Catalysis Applications. *Sci. China Chem.* **2011**, *54*, 397–404.
30. Kou, R.; et al. Enhanced Activity and Stability of Pt Catalysts on Functionalized Graphene Sheets for Electrocatalytic Oxygen Reduction. *Electrochem. Commun.* **2009**, *11*, 954–957.
31. Zhang, Y.; et al. Comparison of Graphene Growth on Single-Crystalline and Polycrystalline Ni by Chemical Vapor Deposition. *J. Phys. Chem. Lett.* **2010**, *1*, 3101–3107.
32. Luo, J.; Jiang, S.; Zhang, H.; Jiang, J.; Liu, X. A Novel Non-Enzymatic Glucose Sensor Based on Cu Nanoparticle Modified Graphene Sheets Electrode. *Anal. Chim. Acta* **2012**, *709*, 47–53.
33. Kim, H.; Moon, G.; Monllor-Satoca, D.; Park, Y.; Choi, W. Solar Photoconversion Using Graphene/TiO_2 Composites: Nanographene Shell on TiO_2 Core versus TiO_2 Nanoparticles on Graphene Sheet. *J. Phys. Chem. C* **2012**, *116*, 1535–1543.
34. Kim, H.; et al. SnO_2/Graphene Composite with High Lithium Storage Capability for Lithium Rechargeable Batteries. *Nano Res.* **2010**, *3*, 813–821.
35. Chen, S.; Zhu, J.; Wu, X.; Han, Q.; Wang, X. Graphene Oxide–MnO_2 Nanocomposites for Supercapacitors. *ACS Nano* **2010**, *4*, 2822–2830.
36. Akhavan, O. Photocatalytic Reduction of Graphene Oxides Hybridized by ZnO Nanoparticles in Ethanol. *Carbon N. Y.* **2011**, *49*, 11–18.
37. Dorgan, V. E.; Bae, M. H.; Pop, E. Mobility and Saturation Velocity in Graphene on SiO_2. *Appl. Phys. Lett.* **2010**, *97*.
38. Ji, Z.; Wu, J.; Shen, X.; Zhou, H.; Xi, H. Preparation and Characterization of Graphene/NiO Nanocomposites. *J. Mater. Sci.* **2011**, *46*, 1190–1195.
39. Teymourian, H.; Salimi, A.; Khezrian, S. Fe_3O_4 Magnetic Nanoparticles/Reduced Graphene Oxide Nanosheets as a Novel Electrochemical and Bioelectrochemical Sensing Platform. *Biosens. Bioelectron.* **2013**, *49*, 1–8.
40. Feng, M.; Sun, R.; Zhan, H.; Chen, Y. Lossless Synthesis of Graphene Nanosheets Decorated with Tiny Cadmium Sulfide Quantum Dots with Excellent Nonlinear Optical Properties. *Nanotechnology* **2010**, *21*, 75601.

41. Huang, J.; et al. Nanocomposites of Size-Controlled Gold Nanoparticles and Graphene Oxide: Formation and Applications in SERS and Catalysis. *Nanoscale* **2010**, *2*, 2733–2738.
42. Liu, J.; Fu, S.; Yuan, B.; Li, Y.; Deng, Z. Toward a Universal 'Adhesive Nanosheet' for the Assembly of Multiple Nanoparticles Based on a Protein-Induced Reduction. *Decor. Graphene* **2010**, 7279–7281.
43. Muszynski, R.; Seger, B.; Kamat, P. V. Decorating Graphene Sheets with Gold Nanoparticles Decorating Graphene Sheets with Gold Nanoparticles. *J. Phys. Chem. C* **2008**, *112*, 5263–5266.
44. Li, F.; et al. The Synthesis of Perylene-Coated Graphene Sheets Decorated with Au Nanoparticles and Its Electrocatalysis toward Oxygen Reduction. *J. Mater. Chem.* **2009**, *19*, 4022.
45. Yao, J.; et al. CoMoO$_4$ Nanoparticles Anchored on Reduced Graphene Oxide Nanocomposites as Anodes for Long-Life Lithium-Ion Batteries. *ACS Appl. Mater. Interfaces* **2014**, *6*, 20414–20422.
46. Tarascon, J. M.; Armand, M. Issues and Challenges Facing Rechargeable Lithium Batteries. *Nature* **2001**, *414*, 359–367.
47. Bruce, P. G.; Scrosati, B.; Tarascon, J.-M. Nanomaterials for Rechargeable Lithium Batteries. *Angew. Chem.* **2008**, *47*, 2930–2946.
48. Schalkwijk, W. Van & Scrosati, B. Advances in Lithium-ion Batteries. In *Seventeenth Annual Battery Conference on Applications and Advances Proceedings of Conference Cat No02TH8576*, **2002**. DOI:10.1007/b113788.
49. Goodenough, J. B.; Kim, Y. Challenges for Rechargeable Li Batteries. *Chem. Mater.* **2010**, *22*, 587–603.
50. Ge, M.; Rong, J.; Fang, X.; Zhou, C. Porous Doped Silicon Nanowires for Lithium Ion Battery Anode with Long Cycle Life. *Nano Lett.* **2012**, *12*, 2318–2323.
51. Liu, X. H.; et al. Reversible Nanopore Formation in Ge Nanowires During Lithiation-Delithiation Cycling: An In Situ Transmission Electron Microscopy Study. *Nano Lett.* **2011**, *11*, 3991–3997.
52. Nguyen, K. T.; Zhao, Y. Integrated Graphene/Nanoparticle Hybrids for Biological and Electronic Applications. *Nanoscale* **2014**, *6*, 6245–6266.
53. Xu, G.-L.; et al. Tuning the Structure and Property of Nanostructured Cathode Materials of Lithium Ion and Lithium Sulfur Batteries. *J. Mater. Chem. A* **2014**, *2*, 19941–19962.
54. Yang, S.; Feng, X.; Ivanovici, S.; Müllen, K. Fabrication of Graphene-Encapsulated Oxide Nanoparticles: Towards High-Performance Anode Materials for Lithium Storage. *Angew. Chemie—Int. Ed.* **2010**, *49*, 8408–8411.
55. Wang, D.; et al. Self-Assembled TiO$_2$–Graphene Hybrid Nanostructures for Enhanced Li-Ion Insertion. *ACS Nano* **2009**, *3*, 907–914.
56. Xie, M.; et al. Amorphous Ultrathin SnO$_2$ Films by Atomic Layer Deposition on Graphene Network as Highly Stable Anodes for Lithium-Ion Batteries. *ACS Appl. Mater. Interfaces* **2015**, *7*, 27735–27742.
57. Yao, X.; et al. In Situ Integration of Anisotropic SnO$_2$ Heterostructures Inside Three-Dimensional Graphene Aerogel for Enhanced Lithium Storage. *ACS Appl. Mater. Interfaces* **2015**, *7*, 26085–26093.
58. Yu, Z.; Tetard, L.; Zhai, L.; Thomas, J. Supercapacitor Electrode Materials: Nanostructures from 0 to 3 Dimensions. *Energy Environ. Sci.* **2015**, *8*, 702–730.
59. Huang, Y.; Liang, J.; Chen, Y. An Overview of the Applications of Graphene-Based Materials in Supercapacitors. *Small* **2012**, *8*, 1805–1834.

60. Wang, H.; et al. Advanced Asymmetrical Supercapacitors Based on Graphene Hybrid Materials. *Nano Res.* **2011**, *4*, 729–736.
61. Zhang, Y.; Li, H.; Pan, L.; Lu, T.; Sun, Z. Capacitive Behavior of Graphene–ZnO Composite Film for Supercapacitors. *J. Electroanal. Chem.* **2009**, *634*, 68–71.
62. Lee, H.; Kang, J.; Cho, M. S.; Choi, J.-B.; Lee, Y. MnO_2/Graphene Composite Electrodes for Supercapacitors: The Effect of Graphene Intercalation on Capacitance. *J. Mater. Chem.* **2011**, *21*, 18215.
63. Zhang, C.; et al. Synthesis of RuO_2 Decorated Quasi Graphene Nanosheets and Their Application in Supercapacitors. *RSC Adv.* **2014**, *4*, 11197–11205.
64. Wang, B.; Wang, Y.; Park, J.; Ahn, H.; Wang, G. In Situ Synthesis of Co_3O_4/Graphene Nanocomposite Material for Lithium-Ion Batteries and Supercapacitors with High Capacity and Supercapacitance. *J. Alloys Compd.* **2011**, *509*, 7778–7783.
65. Li, F.; et al. One-Step Synthesis of Graphene/SnO_2 Nanocomposites and Its Application in Electrochemical Supercapacitors. *Nanotechnology* **2009**, *20*, 455602.
66. Ding, Y.; et al. Cellulose Tailored Anatase TiO_2 Nanospindles in Three-Dimensional Graphene Composites for High-Performance Supercapacitors. *ACS Appl. Mater. Interfaces* **2016**, *8*, 12165–12175.
67. Carrette, L.; Friedrich, K. A.; Stimming, U. Fuel Cells: Principles, Types, Fuels, and Applications. *ChemPhysChem* **2000**, *1*, 162–193.
68. Barbir, F. PEM Fuel Cells. *Fuel Cell Technol.* **2006**, 27–51. DOI: 10.1007/1-84628-207-1_2.
69. Williams, M. C. *Fuel Cells: Technol. Fuel Process.* **2011**, 11–27. DOI: 10.1016/B978-0-444-53563-4.10002-1.
70. Song, D.; et al. A Method for Optimizing Distributions of Nafion and Pt in Cathode Catalyst Layers of PEM Fuel Cells. *Electrochim. Acta* **2005**, *50*, 3347–3358.
71. Kou, R.; et al. Enhanced Activity and Stability of Pt Catalysts on Functionalized Graphene Sheets for Electrocatalytic Oxygen Reduction. *Electrochem. Commun.* **2009**, *11*, 954–957.
72. Zhang, L.-S.; Liang, X.-Q.; Song, W.-G.; Wu, Z.-Y. Identification of the Nitrogen Species on N-Doped Graphene Layers and Pt/NG Composite Catalyst for Direct Methanol Fuel Cell. *Phys. Chem. Chem. Phys.* **2010**, *12*, 12055–12059.
73. Yoo, E.; et al. Enhanced Electrocatalytic Activity of Pt Subnanoclusters on Graphene Nanosheet Surface. *Nano Lett.* **2009**, *9*, 2255–2259.

CHAPTER 18

PIEZOELECTRIC POLYMER NANOCOMPOSITES FOR ENERGY-SCAVENGING APPLICATIONS

ANSHIDA MAYEEN[1] and NANDAKUMAR KALARIKKAL[1,2*]

[1]School of Pure and Applied Physics, Mahatma Gandhi University, Kottayam, Kerala, India

[2]International and Inter University Center for Nanoscience and Nanotechnology, Mahatma Gandhi University, Kottayam, Kerala, India

*Corresponding author. E-mail: nkkalarikkal@mgu.ac.in

ABSTRACT

Piezoelectric materials have gained considerable attention among the scientific community because of its profound potential applications. They are capable of generating electrical output when subjected to mechanical strain. Nowadays, piezoelectric materials are widely used to harvest energy. This chapter deals with the introduction of piezoelectric materials, history of piezoelectric materials, theory, classification, and applications of piezoelectric materials. Moreover, this chapter covers a detailed description of piezoelectric polymer nanocomposites, its advantages and applications over ceramic-based piezoelectric ceramics.

18.1 INTRODUCTION

Harvesting energy from the unused power has become a hot topic of discussion among the scientists in recent days. Nowadays, several energy scavenging techniques have been used; some of them are solar, thermoelectric, electromagnetic, piezoelectric, pyroelectric, etc. In electronics, researchers are trying to develop an efficient source that can charge the powering systems which will be cost-effective and efficient. Piezoelectric, pyroelectric, triboelectric, photovoltaic techniques energy harvesting are largely exploiting for energy scavenging due to its efficiency to produce the electrical output to response to various stimulus such as mechanical stress, temperature variation, friction, light, etc.[1]

Utilizing various kinds of unwanted energy from various sources and harvesting it offer tremendous potential for powering microelectronic devices.[1,2] In this chapter, we are discussing about the current trends and historical development in piezoelectric materials and their applications, mainly in the field of energy harvesting.

Polymer, polymer–ceramic composites, and polymer-based hybrid materials are used in micro electromechanical systems that have the following merits: high mechanical flexibility, low fabrication cost, biocompatible, etc.[2,3]

18.2 PIEZOELECTRICITY

The phenomenon piezoelectricity is the property shown in non-centrosymmetric crystals, which possess an electric polarization when it is subjected to an applied stress. This is termed as direct piezoelectric effect. Similarly, when this non-centrosymmetric material is subjected to an electric field, it creates an induced strain, which is termed as reverse piezoelectric effect. Direct piezoelectric effect is used to sense the variation of dynamic pressure, variation in acceleration due to vibration and changes in force, etc.; reverse piezoelectric effect is explored in actuators (Fig. 18.1).[4,8]

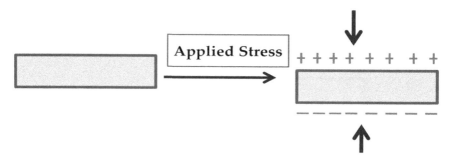

FIGURE 18.1 (See color insert.) The diagrammatic representation of a piezoelectric material.

Some of the piezoelectric materials are naturally occurring and some are not; naturally occurring piezoelectric crystals are quartz and Rochelle salt. Some polymers also possess piezoelectric property; they are polyvinyldene fluoride and its copolymers are trifluoroethylene, hexafluoropropylene, odd nylons, and polyvinyl chloride. Commercial applications like sensing piezoelectric ceramics such as lead zirconium titanate (PZT), barium titanate, etc. are used because of its advantages over single crystal materials. They possess good strength, ease of fabricating, generally into complex structure, etc.

Piezoelectric property of a material is measured in terms of two parameters: piezoelectric strain constant d and piezoelectric voltage coefficient g, where d and g are third rank tensor. The mathematical relation is as follows:

$$d_{ij} = \frac{dP_i}{d\sigma_j} \quad (18.1)$$

and

$$g_{ij} = \frac{dE_i}{d\sigma_j} \quad (18.2)$$

where $i = 1,2,3$ and $j = 1,2,3,4,5,6$.

Here, P_i and E_i are the polarization and the electric field vector, respectively, and σ_j represents the stress tensor, i and j represents the relative direction. When the piezoelectric property is expressed by d_{33} coefficient, then it means that induced polarization is in such a direction that polarization will be parallel to the direction where the material is polarized and it is expressed in pC/N. The coefficient g_{33} represents the induced electric field parallel to the direction in which the material is polarized per unit stress, and it is expressed in mV m/N.[4,6,9]

18.3 WHAT IS A FERROELECTRIC MATERIAL?

Materials that possess spontaneous polarization that can be reversed by an external electric filed are called ferroelectric materials. The necessary condition for a material to be ferroelectric is that it should possess a spontaneous polarization and the sufficient condition is that the polarization of the material can be reversed by the presence of an applied external electric field. Ferroelectrics are the electrical analogs to the ferromagnetic materials in magnetism. It is the property of certain nonconducting crystals (Fig. 18.2).[5,7]

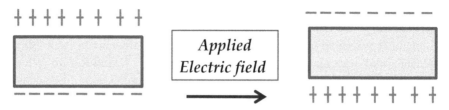

FIGURE 18.2 (See color insert.) The diagrammatic representation of a ferroelectric material.

18.4 WHAT IS A PYROELECTRIC MATERIAL?

Materials that have the capability to produce electrical output when it is subjected to a change in temperature are called pyroelectric materials. Pyro means temperature or heat, so in these materials, there exists a coupling between thermal and electrical parameters which gives rise to the phenomenon called pyroelectric effect. When the temperature is changed, it is sensed by the material and it may undergo change in polarization of the material; it will be reflected as the variation in electrical output, variation in voltage. Hence, corresponding to the variation in temperature or thermal gradient δT, these material shows variation in electrical output or electric voltage. Temperature-dependent electric polarization can be seen in pyroelectric materials. Crystals which possess 1, 2m, 2mm, 3, 3m, 4, 4mm, 6, and 6mm polar classes show pyroelectric property. These materials possess a unique polar axis which can be switched by an external electric field. Each unit cell within the pyroelectric material possess a dipole moment; the dipole moment of the material is called the spontaneous polarization.[8] When a pyroelectric material is subjected to a varying temperature, depending upon the temperature changes the dipole moment within the cell, and hence, spontaneous polarization will change.

Pyroelectric coefficient is represented as differential change in spontaneous polarization P due to the change in temperature (Fig. 18.3).

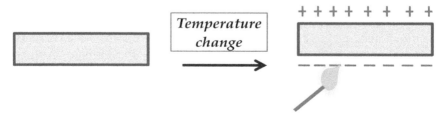

FIGURE 18.3 (See color insert.) The diagrammatic representation of pyroelectric materials.

There exists some difference between ferroelectric, piezoelectric, and pyroelectric materials. All ferroelectrics are piezoelectrics, but not all piezoelectrics are ferroelectrics. Due to the lack of center of symmetry, all ferroelectric shows piezoelectric property. The necessary condition for a material to be piezoelectric is that it should lack center of symmetry, that is, it should be non-centrosymmetric.

The necessary condition that a material to be ferroelectric is as follows:

1. It should possess polar axis and this polar axis can be reversed by the application of reverse electric field.
2. They possess a transition temperature called Curie's transition temperature beyond which the polarization disappears.
3. It should be non-centrosymmetric.

The necessary condition for the materials to be pyroelectric is as follows:

1. It should be non-centrosymmetric.
2. It should possess a unique polar axis in which polarization will be in that particular axis; this polar axis may or may not be reversed by reversible electric field (Fig. 18.4).

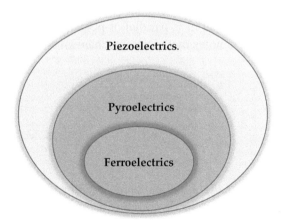

FIGURE 18.4 (See color insert.) Relationship between piezoelectric, pyroelectric, and ferroelectric materials.

So, a material is ferroelectric that will be piezoelectric and pyroelectric in nature, because a ferroelectric should satisfy all the criteria to be piezoelectric and to be pyroelectric, but all piezoelectric or pyroelectric not satisfy all the criteria to be ferroelectric; so every ferroelectrics is piezoelectric and pyroelectric, but every piezoelectric and pyroelectric are not ferroelectrics.

The characteristics of the pyroelectric materials are that they should possess a unique polar axis. These materials are non-centrosymmetric also.

18.5 THEORETICAL FORMULATION

Theoretical formulation of piezoelectricity in polymer-based materials was established by Wada and Hayakawa.

Mathematical relation between the electric displacement E and strain S of the piezoelectric material is given by the following equation:

$$S_\lambda = S^E_{\lambda\mu}T_\mu + d_{i\lambda}E_i \tag{18.3}$$

$$D_i = d_{i\lambda}T_\lambda + \varepsilon^T_{ik}E_i \tag{18.4}$$

where S, T, E, D corresponds to the strain, stress, electric field, electric displacement, respectively, and ε^T represents the permittivity at constant stress and d is the charge constant. There are two operating modes for the piezoelectric materials; 33 modes and 31 modes depend on whether the stress exerted on the material is parallel or perpendicular the electric field. In both the modes, k denotes that electromechanical coupling is given by

$$k^2_{ij} = \frac{\text{Energy transformed}}{\text{Incoming energy}} \tag{18.5}$$

and

$$K_{33} = \frac{d_{33}}{\sqrt{\varepsilon^T_{33}S^E_{33}}} \tag{18.6}$$

where K_{33} represents the longitudinal mode coupling coefficient.

Similarly, the transverse mode coupling coefficient is given as

$$K_{31} = \frac{d_{31}}{\sqrt{\varepsilon^T_{33}S^E_{11}}} \tag{18.7}$$

Extent of mechanical energy scavenged depends upon the coupling coefficient K. if K is higher, then more mechanical energy is scavenged. Higher deformation occurs more in d_{31} mode than from d_{33} mode.[10]

18.6 CLASSIFICATION OF PIEZOELECTRIC MATERIALS

Piezoelectricity can be seen only in nonconductive materials. Piezoelectric materials can be divided in two ways; they are

1. soft piezomaterials and hard piezomaterials and
2. crystals and ceramics.

18.6.1 SOFT PIEZOELECTRIC MATERIALS

These materials can be easily polarized at relatively low field strengths due to high domain mobility. They possess large piezoelectric coefficient, medium permittivity, and high coupling factors. These materials can be used in actuators for micropositioning and nanopositioning sensors such as conventional vibration detectors, ultrasonic transmitters, and receivers.

18.6.2 HARD PIEZOELECTRIC MATERIALS

These materials are ferroelectrically hard; its properties will change according to high mechanical and electrical stresses. These materials, under high mechanical loads and operating field strengths, possess moderate permittivity, low dielectric loss, large piezoelectric coupling constant, good mechanical properties, and good stability. These materials can be used in high power acoustic applications such as ultrasonic cleaning, machining of materials (ultrasonic bonding, drilling, and welding), ultrasonic processors (to disperse liquid media), surgical instruments, etc.

In actuators, ferroelectrically soft piezoceramic materials having low polarity reversal field strengths are used and for high-power acoustic applications, ferroelectrically hard lead zirconate were preferred.

18.6.3 PIEZOELECTRIC CRYSTALS

Quartz and Rochelle salt are naturally occurring piezoelectric single crystal and cannot grow synthetically. To attain better piezoelectric property in these crystals, it must be aligned and cut in a particular crystallographic direction. Quartz is used in accelerometers and Rochelle salt is used in sonic blast sensors. Other piezoelectric crystals are tourmaline, lithium sulfate used in commercial hydrophones, lithium tantalite and lithium niobate are used in high temperature acoustic sensors due to their sensitivity to withstand to higher temperature up to 400°C. High temperature piezoelectric crystals have high thermal stability which are the next class piezocrystals; they are layer-structured ferroelectric perovskite materials such as $Sr_2Nb_2O_7$ and $La_2Ti_2O_7$.

18.6.4 PIEZOELECTRIC CERAMICS

The majority of the piezoelectric materials found commercially today are inorganic ceramics, such as lead titanate, PZT, lithium tantalite, and barium titanate. These ceramics (perovskites) were discovered in the late 1940s and 1950s. First ferroelectric polycrystalline ceramic was barium titanate having perovskite structure which was discovered in 1943, independently by American and Japanese and Russian scientists.[11]

Out of the piezoelectric ceramics, the most significant one having high piezoelectric coefficient is PZT, which possesses good piezoelectric property due to its better polarization. Polarization of such a ferroelectric material can be enhanced by heating in a particular temperature which is above the Curie's temperature of the material and then allow it to cool slowly in the presence of a strong electric field up to 20 kV/cm. These ferroelectric ceramics can be poled by contact poling technique using two electrodes. Poling process enable the dipoles within the material to align in the direction of polarization.[11,12]

High d_{33} values in ceramics enable them to produce large electrical output, even to charge a phone, but its limitations are high stiffness, brittleness, etc. But piezoelectric polymers have good mechanical flexibility, so they emerged as a new class of elastically compliant electroactive materials.

Some examples of piezoelectric ceramics are listed below.

18.6.5 PZT

PZT is one most ferroelectric ceramic substance with a perovskite structure. Here, the lead atoms are located at the corners of the unit cell and oxygen at the face centers, both together forms a face-centered cubic array. Octahedrally, coordinate titanium ions or zirconium ions will be at the center of the unit cell. When cooling PZT from high temperatures, crystal structure undergoes phase transformations.

At the Curie's transition temperature, titanium-rich composition of PZT will undergo a phase transition from cubic (mm^3) to tetragonal phase (4 mm). PZT lies in the morphotrophic boundary region which separates the tetragonal and rhombohedral phase. At the morphotrophic boundary, PZT possess good dielectric and piezoelectric properties because of the coupling of two equivalent energy phases, tetragonal and rhombohedral, which allows moderate domain reorientation of the during the poling time. Doping PZT with donor and acceptor impurities will enhance the piezoelectric property.[11-13]

Since PZT is a polycrystalline ferroelectric ceramic, they contain randomly oriented polarized regions in each grain called domain, and these domains were cooled up to the Curie's transition temperature to reduce the total elastic energy in PZT. Due to this randomness in the polarized regions, in the normal condition, PZT do not show piezoelectric property. It can be made piezoelectric by applying a static electric filed at higher temperature, below the ferroelectric Curie's point, in which the field is larger than the saturation electric field of the material but lower than the breakdown field. Then, the domains within the material become easily aligned and this method in which domains were aligned with the application of the external high electric field is known as poling.

Poling aligns the domains either by reversal or by changing the angles which may be on the basis of the crystal structure. Hence, the spontaneous polarization will possess a component in the direction of poling. Some of the highly strained materials will revert back to the original position after the removal of the electric field but in most cases after the removal of the electric field, they will remain aligned or they will become permanently aligned.

PZT-based materials are used in high performance actuators and transducers because of its high piezo, dielectric, and electromechanical properties. But lead is toxic and not bio-compactable material; researchers are interested to make a substitute for lead-based material. In this aspect, the lead-free materials having piezoelectric properties have a great significance.

18.6.6 LEAD-FREE PIEZOELECTRIC CERAMICS

18.6.6.1 BARIUM TITANATE

Barium titanate belongs to the perovskite family of titanates having different crystal structure based on the variation of temperature. It is a ferroelectric, dielectric ceramic. It is one of the strong candidates for piezoelectric application and is the first developed ceramic for piezoelectric application. Depending on the variations in temperature, lattice parameters of BaTiO$_3$ change. At higher temperature, it will be in face-centered cubic structure and as the temperature decreases, it will distort to TiO$_6$ octahedra.[14,15]

BaTiO$_3$ materials prepared by chemical method show better dielectric and piezoelectric properties compared to the other lead-free materials. The piezoelectric coefficient in this material has very good stability and the d_{33} value is approximately given as 260 pC/N at 25°C, and at room temperature, it is found to be 190 pC/N. Barium titanate in tetragonal phase shows piezoelectric property, so sintering temperature has some role in controlling piezoelectric property of the material. In the case of BaTiO$_3$, samples sintered at 1190°C lies between two phases as in previous case of PZT, tetragonal and orthorhombic; at this region, better piezoresponse is obtained.

18.6.6.2 ZINC OXIDE

ZnO is a material which is both semiconducting and piezoelectric at the same time. It possess Wurzite structure and is non-centrosymmetric. Due to the lack of symmetry, it possess strong electromechanical coupling and thus strong piezoelectric and pyroelectric properties. So, it can be applicable in transducers and sensors. Zinc oxide has a wide band gap of 3.37 eV and is suitable for short wavelength optoelectronic application. Zinc oxide-based piezoelectric energy harvesting has particular advantages such as biocompatibility and can be used in biomedical applications. Zinc oxide can form diverse nanostructures compared to the other ceramics, such as nanowires, nanobelts, nanosprings, nanorings, nanohelics, nanoflowers, nanobows, etc. and a lot of work have been going on to find out which of the structures is more efficient for energy harvesting.[16–18]

18.7 PIEZOELECTRICITY IN POLYMERS

In 1880, Jaques and Pierre Curie described the direct piezoelectric effect; later, Lippmann theoretically predicted the existence of converse piezoelectric based on some thermodynamic arguments. The converse effect was confirmed by Curie brothers. Their work resulted in the creation of the first scientific instrument using the piezoelectric effect to measure force. In 1899 with this invention, the phonograph, Thomas Alva Edison was the first who used the piezoelectric effect in a commercial application. During the earlier period, piezoelectric effect is just a matter of curiosity, and at the time of First World War, scientific community showed more interest in piezoelectric properties, but during the Second World War, the study on piezoelectricity is increased significantly.[19]

Although some of the polymers exhibit piezoelectricity, during the initial periods of research, study of piezoelectric materials is restricted within crystals and ceramics. In polymers, the piezoelectric effect was first observed in polyvinyldene fluoride (PVDF) in 1969, Kawai discovered piezoelectric phenomenon in PVDF. He found that when PVDF is poled under an electric field, it shows strong piezoelectric property.

18.7.1 POLYVINYLDENE FLUORIDE

PVDF is a strong piezoelectric polymer, but there are some conditions that should be satisfied to show comparatively large piezoelectricity in this polymer; they are

1. Polymer chain should possess a large resultant dipole moment in a direction perpendicular to the chain axis.
2. Polymers have to crystallize into a polar crystal and its polar axis remains normal to the chain axis of the polymer.
3. In the case of film or fiber-based samples, molecular chain must be oriented in the direction perpendicular to the film or fiber dimension.
4. Crystallinity in the polymer should be high.
5. Polar axis will orient in a direction perpendicular to a direction perpendicular to the direction of thickness of the sample.

PVDF possess four crystalline phases; they are α, β, γ, and δ. Out of the four phases, β is the strong piezoelectric phase, α phase is not piezoelectric because it is a nonpolar phase. Remaining two phases γ-phase and δ-phase possess half the value of piezoelectric coefficient of the beta phase.

The α-phase of the polymer PVDF can be obtained by casting or melting it, chain containing *trans*-gauche–*trans*-gauche conformation in an antiparallel alignment. Hence, the resultant dipole moment is zero and this phase is not preferred for piezoelectric applications. The β-phase of PVDF is obtained by mechanical drawing and poling of the α-phase PVDF. When the crystal attains β-phase, it will possess a strong net dipole moment; hence, this preferred for piezoelectric application. In β-phase, crystal forms an orthorhombic structure in which chains are aligned in parallel direction and every chain will have planar all-*trans* (TTTT) conformation. The γ and δ phases are represented by polar derivative of the α-phase, that is, TTTGTTTG' (Fig. 18.5).[5,20]

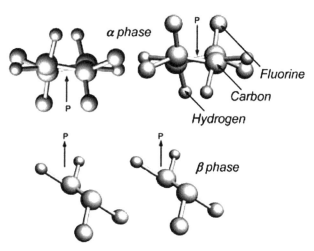

FIGURE 18.5 Molecular structure of the α and β phase of the polymer PVDF.

In PVDF, the degree of crystallinity is restricted up to 50–60% and the polarization value is continuously varying with temperature because of the disordered structure. To increase the crystallinity of the material copolymerizing, it with TrFE and hence resulting in the production of PVDF-TrFE copolymer has been studied by many authors.

18.7.2 POLYVINYLDENE FLUORIDE-TRIFLUOROETHYLENE

PVDF and PVDF-TrFE have been extensively studied for a wide range of applications, in nonvolatile memory in organic electronic devices. They possess ferroelectric, piezoelectric, and pyroelectric property. The copolymer TrFE exhibits more crystallinity than PVDF. This copolymer can be easily crystallized into the beta phase by heat treatment by heating it at a temperature between Curie's transition temperatures (T_m). Yagi et al. reported that since the structure of the hydrogen atom and fluorine atom, because of this, VDF and TrFE can randomly orient to form crystal structure and form a single crystalline phase analogous to the beta phase PVDF. In this copolymer, molar content of vinyldene fluoride will determine the values of spontaneous polarization value while the crystallization structure depends on the content of TrFE.

According to Holman and Kavarnos when this copolymer crystallizes from the melt, it attains all Trans' conformation, hence there is no need for cold drawing to make it in β-phase. In the case of polymers PVDF, PVDF-TrFE, etc., the piezoelectric charge constants are comparatively small to the charge constants of ceramics, polymers can be applicable in both sensors and actuators and not applicable in nanoenergy generators.[1,19–22]

18.7.3 POLYVINYLDENE CYANIDE-VINYL ACETATE

These polymers possess cyano groups, possess a large dipole moment, and possess strong piezoelectricity in copolymer of vinylidene cyanide (VDCN) and vinyl acetate having molecular formula $(CH_2-C(CN)_2-H_2-CHOCOCH_3)_n$. These kinds of polymers are belonging to amorphous group and piezoelectricity in such amorphous polymer is different from that of the crystalline polymers. In crystalline polymers, piezoelectricity is determined by the individual arrangement of the molecules. If the molecular rotational motion is along the direction of the polymeric chain, then it will possess large dipole moment and hence good piezoelectric property. Molecular arrangement can be varied by poling the material.[19–23]

18.7.4 POLYAMIDE 11 (NYLON 11)

Excellent thermal stability is the key characteristic of this polymer; beyond this property, it possess good ferroelectric, piezoelectric, and mechanical properties and they possess relatively low water absorption. Nylon 11 is more thermally stable than other piezoelectric polymers. In 1991, scientists Lee, Scheinbeim, and Newman found that when nylon 11 films were melt quenched, cold drawn and uniaxially stretched, and poled, they shows polarization vs. electric filed curve, that is, ferroelectric hysteresis loop, which confirms the ferroelectric behavior of Nylon 11. Ferroelectric behavior in the given polymer is due to the polar amide group. Here, all Trans conformation of the polymer and every amide groups will be oriented parallel.

Consider the case of a polyamide 11 film which is prepared by melt quenching; it forms doubly oriented hydrogen-bonded sheet structure in which hydrogen bonds are located at the plane of the film and molecular chain orients in the direction of drawing. When an electric field is applied to the film, amide group will reorient to the direction of the field, that is, a subsequent dipole switching by 180° takes place. Hydrogen-bonded networks will only randomize if the temperature exceeds melting point. So, after poling just anneal, the film below the melting will help to reduce the spacing between the hydrogen-bonded sheets and hence prevent the reorientation of the amide group with respect to the thermal motion. This is the way how polyamide 11 attains good thermal stability with the piezoelectric activity.[20,22,23]

18.8 POLING OF MATERIALS AND THE ENHANCEMENT OF THE PIEZOELECTRIC PROPERTY

Poling is a process in which reorientation of the crystallites or molecular dipoles within the bulk medium takes place by the application of an external high electric field at elevated temperature. If we want to retain orientation of the molecular dipoles, the temperature of the poled material is cooled in the presence of the electric field applied. There are two methods for poling contact poling and corona poling. In the case of ceramics, electrode poling or contact poling is used and in the case of polymeric films corona, poling is used.

Electrode poling or contact poling is easier than the corona poling; here, the conductive electrodes will be placed or deposited on both sides of the sample and apply high electric field; during the poling process the quality of the dipole alignment enhances.[23,24]

In corona poling, one side of the polymeric material should be covered with the electrode and the other layer is in contact with the air. A conductive needle is subjected to a very high voltage which is placed on the top a grid. After corona, poling the crystalline alignment within the polymer enhances.

18.9 WHY PIEZOELECTRIC POLYMER NANOCOMPOSITES

Ceramics having good piezoelectric properties have several limitations too; they are heavy and rigid, so researchers found that ferroelectric polymers have certain advantages over ceramics, they are light and flexible. They have a major role in the electronic industry. Continuous and ceaseless research in the field of flexible piezoelectric materials leads to the study of polymer–ceramics systems.

The addition of nanosized piezoelectric ceramic particles may influence the piezoresponse of the properties of polymer nanocomposites and comparatively very larger than that of the microcomposites, its mechanical and piezoresponse also increases. When nanomaterials were added to the polymeric matrix, it creates a new possibility to have extraordinary properties.

Piezoelectric materials are typically used to harvest mechanical energy as they have the ability to convert mechanical energy/vibrations into an electrical energy because of the presence of coupling between their mechanical and electrical domains. Piezoelectric nanocomposites in film and fiber forms are used for the device applications. The periodic use of the electronic devices and replacement of batteries can be avoided to a greater extent; certain ceramics having better piezoelectric coefficient than polymers but piezoelectricity produced will depend on the extent to which the sample can be stretched. In the case of ceramics, it is brittle so that we can't stretch it appropriately. So, the effective possible method is concentrating on the polymer–ceramic composites so that the resulting sample will show better piezoelectric coefficient and good mechanical response; it can be used for the fabrication of energy harvesting devices. This is the core reason why researchers are concentrating on ceramic–polymer piezoelectric nanocomposites, instead of choosing polymer, ceramics alone; here, we can exploit both the property of the polymer and ceramics also.

18.10 ENERGY HARVESTING USING PIEZOELECTRIC POLYMER NANOCOMPOSITES

Piezoelectric materials in various forms are available today for power harvesting applications. By modifying the materials, modification involves doping the host in the case of piezoceramics, or adding nano-sized ceramics into ferroelectric polymer, changing the electrode pattern, varying the poling condition and stress direction, or giving prestress to enhance the coupling and applied strain of the material, or by tuning the resonant frequency of the device, etc.; now the researchers are focusing to improve the efficiency of power harvesting using these piezoelectric materials and hence to generate maximum energy as possible.

Piezoelectric ceramics are $BaTiO_3$, PZT, $LiNbo_3$, $CaTiO_3$, ZnO, etc., and piezoelectric polymers are PVDF, odd Nylons (Nylon 11, Nylon 9, etc.), and PVC. Copolymers of PVDF show better piezoelectric property than neat PVDF, Nylons, PVCs, etc., and out of the copolymer family of PVDF, PVDF-TrFE shows better piezoelectric response. Researchers are trying to develop piezoelectric films, fibers and solid samples according to various applications.[19,24]

Piezoelectric polymers have been found useful for many different applications including sensing, energy harvesting and actuation. These applications, some of which are already in development, could include smart clothing, wristwatch straps, key straps, necklaces, belts, and footwear. For successful integration however, piezoelectric polymers must overcome some challenges:

- Torsion/Twisting and rough handling
- Decrease in piezoelectric properties with increase in temperature
- Washing (clothing)

For energy harvesting application, the global efficiency

Total = $\eta_{abs}\, \eta_{conv}\, \eta_{pow}$

$$\frac{e_{mech}}{e_{abs}} \times \frac{e_{pro}}{e_{abs}} \times \frac{e_{use}}{e_{pro}} = \frac{e_{use}}{e_{mech}},$$

where e_{abs} is the energy absorbed by the polymer, e_{mech} is the ambient mechanical energy, e_{pro} is the produced electrical energy, e_{use} is the useful electrical energy, and η_{conv}, η_{abs} correspond to the electromechanical efficiency.

18.11 APPLICATIONS

There were a lot of biomedical applications for the piezoelectric polymer-based nanocomposites. Since it possesses piezoelectric and pyroelectric properties, it can be applicable in the field of developing medical sensors and transducers.

18.11.1 PIEZOELECTRIC NANOCOMPOSITES IN BIOMEDICAL APPLICATIONS

Presently, piezoelectric materials are used to enhance the neural cell function activities, since these materials can produce the desired electrical output when it is mechanically stretched; these electrical output can provide a stimulus for axonal outgrowth and neural tissue regeneration.[25,27]

18.11.2 PIEZOELECTRIC NANOENERGY GENERATORS FOR FLEXIBLE ELECTRONICS

Piezoelectric nanoenergy generators are devices which convert external kinetic energy into an electrical energy based on the energy conversion by the nanostructured piezoelectric materials.

18.11.3 PIEZOELECTRIC TACTILE SENSORS

A device that can measure physical phenomenon through a contact or touch is called a tactile sensor. These kinds of sensors are widely used in the field of robotics, medicine, etc. Piezoelectric polymers and polymer-based nanocomposites have extensively used for the fabrication of tactile sensors, to measure force or pressure. Moreover, pyroelectric properties in these materials can also be exploited to measure the tactile temperature. Nowadays, a lot of works have been carried out on the piezoelectric and pyroelectric films for the fabrication of multifunctional sensors.

PVDF is a candidate used for these applications, which possess both piezoelectric and pyroelectric properties and have some merits like mechanical strength, light weight, low power consumption, biocompatible, etc. Moreover, PVDF has high dielectric permittivity so that the polarization and dipole moment will be high, thus piezo and pyro-properties. If we cooperate nanoparticles such as zinc oxide nanoparticles or $BaTiO_3$ nanoparticles, it may increase the piezoelectric property, mechanical strength, thermal stability, permittivity, etc.[9,10] Therefore, PVDF-based piezoelectric polymer nanocomposites have more significance in the fabrication of piezoelectric/pyroelectric sensor.

In these kinds of tactile sensors, two different and independent stimuli such as pressure and temperature can be measured simultaneously. We can find pressure from the change in electrical resistance through the piezoresistance of the material and other stimuli temperature can be measured based on the recovery time of the signal.

18.11.4 PIEZOELECTRIC VIBRATIONAL ENERGY HARVESTERS

Energy harvesting from mechanical vibrations have become an active field of research nowadays. A lot of piezoelectric materials have been used for this purpose. Among the three well-known vibrational to electrical energy transduction strategies such as electromagnetic, electrostatic, and piezoelectric methods, piezoelectric method was considered as the most efficient method, because piezoelectrics can cover largest power densities when compared to the lithium ion batteries and thermoelectric generators. Moreover, in other two modes of energy conversion can produce very weak voltage and it requires an additional upconversion to produce usable voltage; on the other hand, piezoelectric vibrational system can produce comparatively high voltage that can be generated directly.

Piezoelectric ceramics possess excellent piezoresponse when compared to piezoelectric polymers, but piezoceramics are fragile and possess lower stiffness, so that this alone cannot be chosen for flexible piezoelectric ceramics. On the other hand, polymers possess excellent mechanical stability. Due to high flexibility of the polymers, it prevents the energy harvesting device from fatigue and hence increases the life time of the device. Hence, a good solution is to incorporate these piezoceramics in piezoelectric polymers so that the advantages of both ceramics and polymers can be combined in a single system. The above strategy has been highly preferred/exploited by the researchers to develop piezoelectric energy harvesting systems.

To harvest more energy, now the focus is to generate high power over a broad band of frequencies than a single resonant frequency.

18.12 CONCLUSION

As the growing need of energy increases, researchers are searching to develop an appropriate way to harvest energy from day to day motions. Starting from the fundamental concept, scientific community have been continuously working with different methods to harvest energy from the environment, such as mechanical energy, thermal energy based on piezoelectric, and pyroelectric property. To meet the energy needs of the world, we need a sustainable energy resources; the mechanical energy can be easily converted to outdoor renewable energy at any time. Flexible energy harvesters called nanogenerators can produce electrical output from tiny physical motion such as inhalation, heartbeat, eye blinking which have been exploited for scavenging energy. These self-powered systems can be applicable in various fields, including sensing, medical science, defense, flexible electronics, etc.

KEYWORDS

- **piezoelectrics**
- **ferroelectrics**
- **polymer nanocomposites**
- **energy harvesting**

REFERENCES

1. Priya, S. Advances in Energy Harvesting Using Low Profile Piezoelectric Transducers. *J. Electroceram.* **2007**, *19* (1), 167–184.
2. Lee, M.; Chen, C. Y.; Wang, S.; Cha, S. N.; Park, Y. J.; Kim, J. M.; Chou, L. J.; Wang, Z. L. A Hybrid Piezoelectric Structure for Wearable Nanogenerators. *Adv. Mater.* **2012**, *24* (13), 1759–1764.
3. Anton, S. R.; Sodano, H. A. A Review of Power Harvesting Using Piezoelectric Materials (2003–2006). *Smart Mater. Struct.* **2007**, *16* (3), R1.
4. Mandal, D.; Yoon, S.; Kim, K. J. Origin of Piezoelectricity in an Electrospun Poly(Vinylidene Fluoride-Trifluoroethylene) Nanofiber Web-Based Nanogenerator and Nano-Pressure Sensor. *Macromol. Rapid Commun.* **2011**, *32* (11), 831–837.
5. Uchino, K. *Ferroelectric Devices*, 2nd ed. CRC Press: Boca Raton, FL, 2009.
6. Tressler, J. F.; Alkoy, S.; Newnham, R. E. Piezoelectric Sensors and Sensor Materials. *J. Electroceram.* **1998**, *2* (4), 257–272.
7. Bowen, C. R.; Kim, H. A.; Weaver, P. M.; Dunn, S. Piezoelectric and Ferroelectric Materials and Structures for Energy Harvesting Applications. *Energy Environ. Sci.* **2014**, *7* (1), 25–44.
8. Lingam, D.; Parikh, A. R.; Huang, J.; Jain, A.; Minary-Jolandan, M. Nano/Microscale Pyroelectric Energy Harvesting: Challenges and Opportunities. *Int. J. Smart and Nanomater.* **2013**, *4* (4), 229–245.
9. Damjanovic, D. Ferroelectric, Dielectric and Piezoelectric Properties of Ferroelectric Thin Films and Ceramics. *Rep. Progr. Phys.* **1998**, *61* (9), 1267.
10. Sirohi, J.; Chopra, I. Fundamental Understanding of Piezoelectric Strain Sensors. *J. Intell. Mater. Syst. Struct.* **2000**, *11* (4), 246–257.
11. Jaffe, B. *Piezoelectric Ceramics*; Elsevier: Amsterdam, 2012; vol. 3.
12. Hooker, M. W. *Properties of PZT-Based Piezoelectric Ceramics Between −150 and 250 C*. NASA/CR-1998-208708, NAS 1.26:208708, 1998.
13. Jung, W. S.; Lee, M. J.; Yoon, S. J.; Lee, W. H.; Ju, B. K.; Kang, C. Y. Flexible Nanocomposite Generator Using PZT Nanorods and Ag Nanowires. *Int. J. Appl. Ceram. Technol.* **2015**, *13* (3), 480–486.
14. Roberts, S. Dielectric and Piezoelectric Properties of Barium Titanate. *Phys. Rev.* **1947**, *71* (12), 890.
15. Chanmal, C. V.; Jog, J. P. Dielectric Relaxations in PVDF/BaTiO$_3$ Nanocomposites. *Express Polym. Lett.* **2008**, *2* (4), 294–301.

16. Lee, J. S.; Shin, K. Y.; Cheong, O. J.; Kim, J. H.; Jang, J. Highly Sensitive and Multifunctional Tactile Sensor Using Free-Standing ZnO/PVDF Thin Film with Graphene Electrodes for Pressure and Temperature Monitoring. *Sci. Rep.* **2015**, 5, 1–8..
17. Wang, Z. L.; Song, J. Piezoelectric Nanogenerators Based on Zinc Oxide Nanowire Arrays. *Science* **2006**, *312* (5771), 242–246.
18. Fu, Y. Q.; Luo, J. K.; Du, X. Y.; Flewitt, A. J.; Li, Y.; Markx, G. H.; Walton, A. J.; Milne, W. I. Recent Developments on ZnO Films for Acoustic Wave Based Bio-sensing and Microfluidic Applications: A Review. *Sens. Actuat.; B: Chem.* **2010**, *143* (2), 606–619.
19. Bohlén, M.; Bolton, K. Conformational Studies of Poly(Vinylidene Fluoride), Poly(Trifluoroethylene) and Poly(Vinylidene Fluoride-*co*-Trifluoroethylene) Using Density Functional Theory. *Phys. Chem. Chem. Phys.* **2014**, *16* (25), 12929–12939.
20. Baur, C.; Apo, D. J.; Maurya, D.; Priya, S.; Voit, W. Advances in Piezoelectric Polymer Composites for Vibrational Energy Harvesting. *Polymer Composites for Energy Harvesting, Conversion, and Storage*; American Chemical Society, 2014; pp 1–27.
21. Jo, Y. S.; Maruyama, Y.; Inoue, Y.; Chûjô, R.; Tasaka, S.; Miyata, S. Molecular Motions of Amorphous Piezoelectric Polymers Determined by 13C CPMAS NMR Spectroscopy. *Polym. J.* **1987,** *19* (6), 769–772.
22. Sharma, T.; Je, S. S.; Gill, B.; Zhang, J. X. Patterning Piezoelectric Thin Film PVDF-TrFE Based Pressure Sensor for Catheter Application. *Sens. Actuat.; A: Phys.* **2012**, *177*, 87–92.
23. Lang, S. B.; Muensit, S. Review of Some Lesser-Known Applications of Piezoelectric and Pyroelectric Polymers. *Appl. Phys. A* **2006**, *85* (2), 125–134.
24. Panda, P. K. Review: Environmental Friendly Lead-Free Piezoelectric Materials. *J. Mater. Sci.* **2009**, *44* (19), 5049–5062.
25. Fuh, Y. K.; Huang, Z. M.; Wang, B. S.; Li, S. C. Self-Powered Active Sensor with Concentric Topography of Piezoelectric Fibers. *Nanoscale Res. Lett.* **2017**, *12* (1), 44.
26. Nambiar, S.; Yeow, J. T. Conductive Polymer-Based Sensors for Biomedical Applications. *Biosens. Bioelectron.* **2011,** *26* (5), 1825–1832.
27. Shung, K. K.; Cannata, J. M.; Zhou, Q. F. Piezoelectric Materials for High Frequency Medical Imaging Applications: A Review. *J. Electroceram.* **2007**, *19* (1), 141–147.

CHAPTER 19

OPTIMIZATION OF BETANIN DYE FOR SOLAR CELL APPLICATIONS

APARNA THANKAPPAN[1*], V. P. N. NAMPOORI[2], and SABU THOMAS[1]

[1]*International and Inter University Centre for Nanoscience and Nanotechnology, Mahatma Gandhi University, Kottayam, India*

[2]*International School of Photonics, Cochin University of Science and Technology, Kochi, India*

*Corresponding author. E-mail: aparna.subhash@gmail.com

ABSTRACT

In this chapter, the authors report structure modification of ZnO via a simple facile two-step solution growth method using preexisting textured ZnO, ZnS, CdS, and TiO_2 seeds and the former was made to perform as photoanode in betanin-sensitized dye-sensitized solar cells (DSSC). The DSSC fabricated with TiO_2–ZnO composite structure photoanode achieved high solar to electricity conversion efficiency of 1.2%, whereas the CdS–ZnO composite structure exhibits higher IPCE (λ) than the other composites. The enhanced performance of the composite structures was credited to the parameters such as lasing emission, sulfur doping, morphology, and energy barrier.

19.1 INTRODUCTION

After Grätzel et al. developed a new type of solar cells, namely, dye-sensitized solar cell (DSSC), in 1991,[1] they have attracted considerable attention as environmental friendly and low-cost alternatives to conventional inorganic photovoltaic devices. Semiconductor hierarchical nanostructures with large surface areas have been favored as photoanodes in DSSC. A TiO_2 nanoparticle (NP) film has been widely investigated for DSSCs due to its high power efficiency. ZnO has similar band gap and electron affinity values to those of TiO_2 and it is also an environmentally friendly oxide semiconductors. DSSCs assembled from ZnO NPs play the second highest efficiencies after TiO_2.[2] Solar energy harvesting is a comparative fledged technology for outdoor applications. For indoor applications, it is necessary to note that the efficiency of the cell is very low as of its low light luminous intensity which is less than 10 W/m² as compared to 100–1000 W/m² under outdoor conditions depending on the type of light source and distance. It is known that the photosensitizer shows a significant role in determining the stability, light harvesting capability, and also the total cost of the DSSCs. Ruthenium-containing metalorganic complexes have been preferred in DSSCs due to their strong absorption of visible light, favorable spatial separation of highest occupied molecular orbital, and lowest unoccupied molecular orbital (LUMO), and their ability to get repetitively oxidized and reduced without degradation,[3] The best reported devices of this class which convert solar energy to electrical energy have an efficiency of 10–11%.[4-6] Recently, the efficiency of DSSCs has improved to 12.3% by the combination of cosensitization of porphyrin dyes and a cobalt-complex redox mediator.[7] However, synthetic dyes have offered problems as well, such as complicated synthetic routes, whereas natural dyes are more organic, better for the environment, or safer to use, than synthetic dyes. Due to their cost efficiency, nontoxicity, and environmental benefits, DSSCs based on natural dyes have been at the center of intense research.

Natural pigments from plants such as chlorophyll[8] and anthocyanin[9] have been extensively examined as sensitizers for the DSSC. Betalains are a class of vegetal pigments with favorable light absorption, existing in nature in association with various copigments such as the betalamic acid, indicaxanthin, and betanidin, which modify their light absorption properties and are capable of forming complex metal ions, possessing the requisite functional group (–COOH) to bind also to ZnO.[10] Although the betalain dyes fulfill the criteria mentioned above, they have not been properly studied as dye sensitizers.[11-13] The betalain pigments comprise the red–purple betacyanin, betanin(I) and betanidin(II), with maximum absorptivity at λ_{max} about 535 nm, and the yellow betaxanthins with λ_{max} near 480 nm.[14] Within these two categories, many subcategories are also present.[15]

Moreover, much effort has recently been invested to create new class of nanomaterials through surface coating such as core/shell type TiO_2/CdS nanowires,[16] Cu_2S/Au nanowires,[17] Zn/ZnO nanobelts[18] through different mixing morphologies.[19–21] These can modify the dynamic and interfacial properties of DSSC[22] which could enhance the solar cell conversion efficiency.

In this chapter, we propose a simple facile two-step solution growth method to build tailored nanomaterials on transparent conducting oxide (TCO) using preexisting textured ZnO, ZnS, CdS, and TiO_2 seeds. We have constructed DSSCs with the prepared composites as photoanodes comprising betanin natural dye and we compared their overall conversion efficiency, electron transport, and recombination properties. The results indicate that TiO_2–ZnO composite structure can achieve an excellent balance between dye adsorption and electron transport, thus resulting higher efficiency than the others.

19.2 EXPERIMENTAL

19.2.1 SYNTHESIS OF ZnO NANOSTRUCTURES WITH DIFFERENT MORPHOLOGIES

ZnO nanocomposite films were grown by a simple facile two-step solution method. For this, we have synthesized ZnO, TiO_2, ZnS, and CdS NPs using different methods and the details of the fabrication process are discussed below. First, ZnO NPs were synthesized according to the method described by Pacholski.[23] In brief, 1 M zinc acetate in methanol was added dropwise to a well-stirred solution of 3 M NaOH in methanol maintained at 60°C. After 2 h growth, the solution was centrifuged to separate the NPs of size 72 nm. The fresh methanol was infused to suspend the particles. ZnO nanostructures were grown on the glass substrate that was seeded with the use of preexisting textured ZnO, ZnS, CdS, and TiO_2 seeds as reported in the work by Greene et al.[24] Second, ZnS nanocrystals are synthesized by using wet chemical precipitation at room temperature. 1 M $ZnCl_2$ and 1 M Na_2S in water are used as sources of Zn and S, respectively, and 2 mL of triethanolamine (TEA) is used as capping agent and the solution was centrifuged to separate the NPs. The X-ray diffraction (XRD) study reveals that the average size of the nanocrystals is about 2.6 nm.

Third, CdS NP (28 nm) was synthesized using the chemical bath deposition technique as described in literature.[25] Briefly, 1 M $CdAc_2$ and 1 M thiourea are dissolved in distilled water, using TEA as the capping agent. The NPs are separated from the medium by centrifugation, washed, and then dried at room temperature. Lastly, the TiO_2 sol was prepared via sol–gel method and synthesized as follows: Titanium tetra-isopropoxide (TTIP) was used as the precursor to prepare TiO_2 sol. A mixture of HCl and isopropyl alcohol was added to a mixture of TTIP and isopropyl alcohol under continuous magnetic agitation at room temperature. The TiO_2 sol composition was of molar ratio TTIP/isopropanol/water = 1:26.5:1.5. The [H⁺]/[TTIP] molar ratio ranged from 0.02 to 1.1. The resultant TiO_2 sol was clear, yellow, and stable. The as-obtained NPs were treated thermally at 100°C to remove the excess moisture content. Thus, amorphous TiO_2 NPs of 35 nm were obtained.

The as-prepared NPs were deposited as seeded layer onto indium tin oxide (ITO) by a dip-coated method. Following dip coating, ITO substrates were annealed at 200°C in air for 10 min. The nanostructures were grown on the seeded substrates by placing them face down in a closed vessel containing equimolar concentration (0.025 M) of zinc nitrate hexahydrate and hexamine in deionized water at 80°C for 3 h. Subsequently, the substrates were washed with water and annealed at 300°C for 30 min for giving good porosity and to remove the surface defects.

19.2.2 DYE SENSITIZATION AND CELL ASSEMBLY

The composite photoelectrodes were impregnated with filtered fresh juice from red beets without further purification. Dye adsorption was carried out for 24 h at room temperature, after which the photoelectrodes were removed from the dye solution and rinsed several times with water to remove weakly adsorbed dye molecules and finally dried in a hot plate. The platinized ITO glass was used as the counter electrode. For this, ITO conducting glass was dipped into a 10 mM chloroplatinic acid hexahydrate solution in ethanol and then annealed at 450°C for 15 min. The counter electrode was then placed on top of the photoelectrode and filled with electrolyte by using capillary force through the sides. Binder clips were used to hold the electrodes together. The electrolyte was made with the ferrocene, in which 0.05 M ferrocenium hexafluorophosphate mixed with 0.1 M ferrocene and 0.5 M *tert*-butylpyridine (TBP) in acetonitrile. The TBP was used in the electrolyte solution to improve the photovoltage. This Fe^{2+}/Fe^{3+} redox couple does not induce corrosion because it does not interact with oxygen during the fabrication process.[26]

19.2.3 CHARACTERIZATION

The morphologies and sizes of the as-prepared samples were examined by scanning electron microscopy (JEOL/EO and JSM6390) and XRD method.

The XRD data were collected on an AXS Bruker D% diffractometer using Cu Kα-radiation ($\lambda = 0.1541$ nm, the operating conditions were 35 m A and 40 kV at a step of 0.020° and step time of 29.5 s in the 2θ range from 30° to 70°). The electrical characteristics (J–V) of the fabricated solar cells were examined by using a standard solar irradiation of 1000 W xenon arc lamp as light source. The cell surface was exposed to light through a water column. The input intensity was measured using a light meter (Metravi 1332). Incident monochromatic photo-to-current conversion efficiency (IPCE) measurements were carried out using narrow band-pass filters to create monochromatic light.

The energy conversion efficiencies (η) were calculated by

$$\eta = \frac{P_{max}}{P_{in}}, \quad \eta = \frac{J_{max} \cdot V_{max}}{P_{in}}, \quad \eta = \frac{J_{sc} \cdot V_{oc} \cdot FF}{P_{in}}$$

where P_{in} is the incident power, P_{out} is the output power, FF is the fill factor, η is the efficiency, J_{sc} is the short-circuit current density, and V_{oc} is the open circuit voltage.

Incident photon-to-electron conversion efficiencies, IPCE (λ), were recorded as a function of incident wavelength for the betanin-sensitized solar cell and are calculated using the formula[27]:

$$\text{IPCE}(\lambda) = \frac{1240 \cdot J_{sc}}{\lambda . P_{in}}$$

where λ is the wavelength of the incident light in nm, P_{in} is the incident power in watts, and J_{sc} is the maximum current in amps.

19.3 RESULTS AND DISCUSSION

The morphological features of ZnO nanostructures deposited on different preexisting textured ZnO, ZnS, CdS, and TiO_2 seeds are identified by SEM images (Fig. 19.1). Figure 19.1a exhibits the highly densed nonaligned branched ZnO nanowires properly grown on ZnO seeded substrates. The growth of nanowires appears to be from a common nucleus, and these structures might be due to the initial aggregation of several ZnO particles and are growing along the preferential direction of the crystalline network. We have repeated the same synthesis method with various seeds, and the XRD pattern of seeded layers used for the synthesis is shown in Figure 19.2. With the TiO_2 and ZnS seed, a new morphology of nanomaterial appears, namely, vertical microrods (top view) and vertical nanorods that are shown in Figure 19.1b and c. Nevertheless a small proportion of ZnS NPs do not allow for the exclusive growth of nanorods. In addition to these structures, randomly distributed nanoplate and nanofibre structures are also observed in Figure 19.1c (i.e., ZnS/ZnO composite). But with CdS NPs, we have obtained honey bee structure with several nanoballs. The schematic representation of the synthesis of tailored growth of ZnO is shown in Figure 19.3.

FIGURE 19.1 (a) SEM images of ZnO nanowire, (b) TiO_2–ZnO composites, (c) ZnS–ZnO composites, and (d) CdS–ZnO composites.

In order to investigate the effect of ZnO morphologies on the photovoltaic properties of DSSCs, their photoelectrochemical behaviors were studied. Figure 19.4 compares the current density–voltage (J–V) characteristics of DSSCs based on different morphologies measured by using 1000 W xenon arc lamp. The related physical values, such as J_{sc} (short-circuit current), V_{oc} (open circuit voltage), FF (fill factor), and η (light to electricity conversion efficiency), are summarized in Table 19.1. The performance of cells with ZnO composites can be influenced by factors such as the dye-adsorption capability, light harvesting, sulfur doping, and the existence of energy barrier.

Optimization of Betanin Dye for Solar Cell Applications

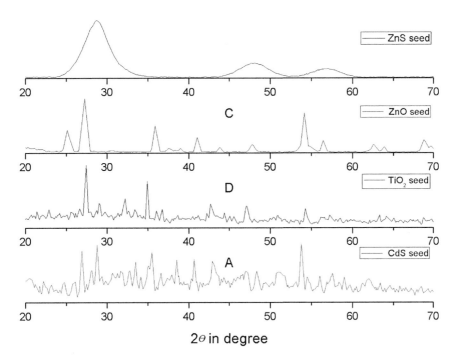

FIGURE 19.2 XRD patterns of different seeds: ZnS, ZnO, TiO$_2$, and CdS.

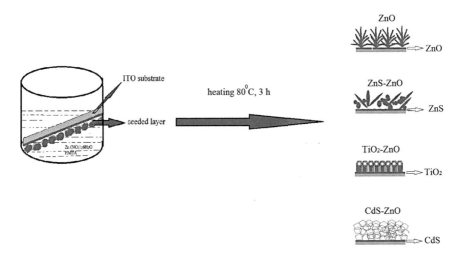

FIGURE 19.3 Schematic representation of tailored growth of ZnO nanostructures.

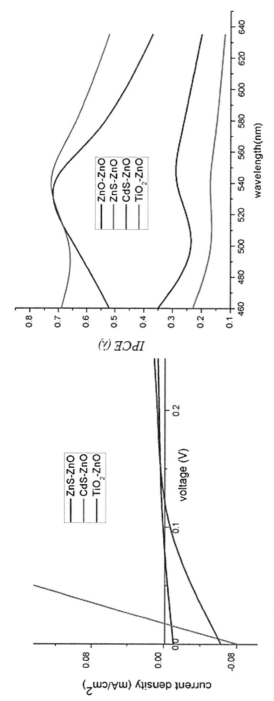

FIGURE 19.4 (See color insert.) (a) Current density against voltage characteristics of different composites and (b) their IPCE (λ).

TABLE 19.1 Comparison of DSSC Based on Semiconductor Composites with Ferrocene Electrolyte.

	P_{in} (klx)	J_{sc} (mA/cm²)	V_{oc} (V)	FF	η (%)
ZnO–ZnO	2.62	0.06	0.07	0.46	0.48
ZnS–ZnO	0.8	0.01	0.08	0.23	0.21
CdS–ZnO	0.8	0.09	0.02	0.21	0.28
TiO$_2$–ZnO	0.8	0.06	0.09	0.26	1.2

Self-seeded ZnO-based DSSCs showed low efficiency compared to TiO$_2$–ZnO due to the formation of Zn^{2+}/dye complex by dissolving ZnO onto the –COOH group of dye, which blocks the electron injection from the dye to ZnO.

The points of zero charge of ZnO and TiO$_2$ are reported to be for pH = 8–9[28] and pH = 5.5–6.5,[29,30] respectively, whereas the pH value for the dye sensitization process is approximately 5, which inactivates some dye molecules and decreases the efficiency. But TiO$_2$ has similar pH value to that of dye solution and the formation of ion/dye complexes does not take place. The open circuit voltage (V_{oc}) of TiO$_2$–ZnO is higher than that of self-seeded ZnO branched nanowire even with a small input flux due to less charge recombination that reflects in higher IPCE which is shown in Figure 19.4b. This reduced recombination occurring in the TiO$_2$–ZnO composite is due to the existence of energy barrier formed in the TiO$_2$–ZnO interface (work function of ZnO = 5.6 eV and that of TiO$_2$ = 4.8 eV) and also due to higher electron mobility of ZnO.[31] The presence of TBP suppresses the recombination of conduction band electrons with the electrolyte and raises the conduction band energy of TiO$_2$. The betanin adsorbs strongly on TiO$_2$ film through the two ortho–OH groups on the aromatic ring so that the driving force for the electron injection is greater in TiO$_2$ leading to the faster electron injection and the faster rate of reverse electron transfer leads to the increase in the photocurrent by suppressing recombination. Above all, the efficiency enhancement in TiO$_2$–ZnO composite is also ascribed to the enrichment of light harvesting and the reduction of electron back reaction due to direct conduction pathway.

Further, a multifold enhancement of efficiency in TiO$_2$–ZnO composites photoanode is achieved in comparison with ZnS–ZnO composites. Due to hetero-hierarchical structure, it is expected that ZnS–ZnO composite will show higher photovoltaic performance, but on the contrary, it shows the less efficiency, which may be due to the formation of metal–ion/dye complex, that is, between constituents and the carboxylic group of the dye, that hinders its dye-loading ability. It is assumed that the semiconductor starts to leak electrons before reaching the quasi-Fermi level and most of the photons do not fall into the acceptance cone of the integrating sphere and get scattered out;[32] this is mainly due to poor dye adsorption. The S atoms in the ZnS layer are able to fill with the oxygen vacancies on ZnO layer which can reduce the recombination and the tunneling of electrons from the LUMO energy levels of the betanin dyes to the conduction band the ZnO and to the ZnS. This process is prevented by the higher band gap of ZnS, but the quasi-Fermi level of ZnO increases with the increase of the density of electrons injected from ZnS.[33]

We believe that increased photoabsorption of CdS–ZnO is the main reason for the enhancement of conversion efficiency compared with ZnS–ZnO. With the absorption of sulfur atoms the formation of new defects or surface state such as Zn–CdS or Cd–Zn–O in the interface region of the CdS–ZnO heterojunction improves the conductance leading to the higher photocurrent[34] and higher IPCE (λ). The major problem for using CdS as a seed layer is the trapping/detrapping events at the grain boundaries of the pores of honey bee structure reduce driving force of electrons and slows down the electron transport which can results in an increasing probability of charge recombination with the oxidized dye molecules and/or redox species, and this is obvious from the low value of V_{oc}. We suggest that great improvement in IPCE (λ) results from the sufficient light harvesting in the visible region which arise from the increased overall surface area due to the random multiple scattering within the structure leading to photon localization that increases the probability of interaction between the dye molecules and photons. Even though we did not quantitatively measure, we could verify the higher dye adsorption as the color of CdS–ZnO was much deeper than the other composites as seen by naked eye. The absorption and reflection of incident light by the conductive glass which could reduce the incident light power by as much as 20%[35] may also be responsible for the higher value of IPCE (λ). The higher value of IPCE (λ) suggests that betanin can inject more than one electron per photon, as betanin is capable of expending up to two electrons without being irreversibly oxidized.[36] The absorbed two electrons per photon would explain how recombination could limit the maximum attainable photovoltages despite the high photocurrents obtained. The sulfur atoms can diffuse into the inner part of the ZnO as results in a sulfur-doping effect in ZnO, and this sulfur doping could be considered for the conductance enhancement. Another factor is the electron injection from the ZnO to CdS through the staggered gap (for ZnO, E_g =3.37 eV and for CdS, E_g = 2.42 eV at 300 K). In addition, multiple scattering is responsible for the higher IPCE in CdS–ZnO and it may be due to the lasing emission on the composite. The formation of the oxidation product, that is, melanin-like polymer of cyclo-DOPA along with the possible formation of indicaxanthin and the heat-induced cleavage of betanin into betalamic acid may play a crucial role in the cell performance. We assume that the betalain-derived melanin adsorbed on the TiO_2–ZnO film is very stable.

19.4 CONCLUSION

In summary, four types of ZnO nanostructures were synthesized via a simple two-step growth method and they were used as photoanodes to construct DSSCs. Dye adsorption capability, sulfur doping, morphology, energy barrier, and the effective surface area show remarkable influence on the performance of DSSCs. The DSSC with the TiO_2–ZnO composite structure exhibits the highest conversion efficiency among the four ZnO structures. This study provides a facile method to tailor the morphologies of nanomaterials and the work indicates that the present ZnO nanostructures are favorable to further improving the conversion efficiency.

ACKNOWLEDGMENT

The author Aparna Thankappan acknowledges the financial support from UGC-DSKPDF.

KEYWORDS

- betanin
- DSSC
- nanoparticles
- ZnO
- solar cell

REFERENCES

1. O'Regan, B.; Grätzel, M. A Low-cost, High-efficiency Solar Cell Based on Dye-sensitized Colloidal TiO$_2$ Films. *Nature* **1991**, *353*, 737–740.
2. Keis, K.; Magnusson, E.; Lindstrom, H.; Lindquist, S. E.; Hagfeldt, A. *Sol. Energy Mater. Sol. Cells* **2002**, 735.
3. Nazeeruddin, M. K.; De Angelis, F.; Fantacci, S.; Selloni, A.; Viscardi, G.; Liska, P.; Ito, S.; Takeru, B.; Grätzel, M. Combined Experimental and DFT–TDDFT Computational Study of Photoelectrochemical Cell Ruthenium Sensitizers. *J. Am. Chem. Soc.* **2005**, *127*, 16835–16847.
4. Nazeeruddin, M. K.; Kay, A.; Rodicio, I.; Humphry-Baker, R.; Muller, E.; Liska, P.; Vlachopoulos, N.; Gratzel, M. *J Am. Chem. Soc.* **1993**, *115*, 6382.
5. Argazzi, R.; Iha, N. Y. M.; Zabri, H.; Odobel, F.; Bignozzi, C. A. *Coord. Chem. Rev.* **2004**, *248*, 1299.
6. Polo, A. S.; Itokazu, M. K.; Iha, N. Y. M. *Coord. Chem. Rev.* **2004**, *248*, 1343.
7. Yella, A.; et al. Porphyrin-sensitized Solar Cells with Cobalt(II/III)-Based Redox Electrolyte Exceed 12 Percent Efficiency. *Science* **2011**, *334*, 629–634.
8. Wanga, X.-F.; Koyama, Y.; Kitao, O.; Wada, Y.; Sasaki, S.; Tamiaki, H.; Zhou, H. Significant Enhancement in the Power-conversion Efficiency of Chlorophyll Co-sensitized Solar Cells by Mimicking the Principles of Natural Photosynthetic Light-harvesting Complexes. *Biosens. Bioelectron.* **2010**, *25*, 1970–1976.
9. Zhu, H.; Zeng, H.; Subramanian, V.; Masarapu, C.; Hung, K. H.; Wei, B. Anthocyanin-sensitized Solar Cells Using Carbon Nanotube Films as Counter Electrodes. *Nanotechnology* **2008**, *19*, 465204.
10. Dumbravă, A.; Enache, I.; Oprea, C. I.; Georgescu, A.; Gîrţu, M. A. Toward a More Efficient Utilisation of Betalains as Pigments for Dye-sensitized Solar Cells. *Dig. J. Nanomater. Bios.* **2012**, *7*, 339–351.
11. Zhang, D.; Lanier, S. M.; Downing, J. A.; Avent, J. L.; Lum, J.; McHale, J. L. *J. Photochem. Photobiol. A: Chem.* **2008**, *195*, 72.
12. Zhang, D.; Yamamoto, N.; Yoshida, T.; Minoura, H. *Trans. Mater. Res. Soc. Jpn.* **2002**, *27*, 811.
13. Calogero, G.; Di Marco, G.; Cazzanti, S.; Caramori, S.; Argazzi, R.; Di Carlo, A.; Bignozzi, C. A. *Int. J. Mol. Sci.* **2010**, *11*, 254.
14. Zhang, D.; Laniera, S. M.; Downing, J. A.; Avent, J. L.; Lumc, J.; McHalea J. L. Betalain Pigments for Dye-sensitized Solar Cells. *J. Photochem. Photobiol. A: Chem.* **2008**, *195*, 72.
15. Thankappan, A.; Thomas, S.; Nampoori, V. P. N. Solvent Effect on the Third Order Optical Nonlinearity and Optical Limiting Ability of Betanin Natural Dye Extracted from Red Beet Root. *Opt. Mater.* **2013**, *35*, 2332–2337.
16. Cao, J.; Sun, J. Z.; Li, H. Y.; Hong, J.; Wang, M. *J. Mater. Chem.* **2004**, *14*, 1203.
17. Wen, X.; Yang, S. *Nano Lett.* **2002**, *2*, 451.
18. Ding, Y.; Kong, X. Y.; Wang, Z. L. *J. Appl. Phys.* **2004**, *95*, 306.
19. Jiang, C. Y.; Sun, X. W.; Lo, G. Q.; Kwong, D. L.; Wang, J. X. *Appl. Phys. Lett.* **2007**, *90*, 263501.
20. Singh, D. P. *Sci. Adv. Mater.* **2010**, *2*, 245–272.
21. Akhtar, M. S.; Cheralathan, K. K.; Chun, J. M.; Yang, O. B. *Electrochim. Acta* **2008**, *53*, 6623.

22. Ito, S.; Kitamura, T.; Wada, Y.; Yanagida, S. *Sol. Energy Mater. Sol. Cells* **2003**, *76*, 3.
23. Pacholski, C.; Kornowski, A.; Weller, H. *Angew. Chem. Int. Edn.* **2002**, *7*, 411188.
24. Greene, L. E.; Law, M.; Goldberger, J.; Kim, F.; Johnson, J. C.; Zhang, Y. F.; Saykally, R. J.; Yang, P. D. *Angew. Chem. Int. Edn.* **2003**, *42* (*26*), 3031.
25. Prabhu, R. R.; Abdul Khadar, M. *Pramana J. Phys.* **2005**, *65*, 801–807.
26. Sonmezoglu, S.; Akyurek C.; Akin, S. High-efficiency Dye-sensitized Solar Cells Using Ferrocene-based Electrolytes and Natural Photosensitizers. *J. Phys. D: Appl. Phys.* **2012**, *45*, 425101.
27. Zhang, D.; Lanier, S. M.; Downing, J. A.; Avent, J. L.; Lum, J.; McHale, J. L. *J. Photochem. Photobiol. A: Chem.* **2008**, *195*, 72–80.
28. Blok, L.; DeBruyn, P. L. *J. Colloid. Interf. Sci.* **1970**, *32*, 518.
29. Kosmulski, M. *Adv. Colloid. Interf. Sci.* **2002**, *99*, 255.
30. Kallay, N.; Babic, D.; Matijevic, E. *Colloids. Surf.* **1986**, *19*, 375.
31. Yu, X.-L.; Song, J.-G.; Fu, Y.-S.; Xie, Y.; Song, X.; Sun, J.; Du, X.-W. ZnS/ZnO Heteronanostructure as Photoanode to Enhance the Conversion Efficiency of Dye-sensitized Solar Cells. *J. Phys. Chem. C* **2010**, *114*, 2380–2384.
32. Baxter, J. B.; Waker, A. M.; van Ommering, K.; Aydil, E. S. Synthesis and Characterisation of ZnO Nanowires and Their Integration into Dye Sensitized Solar Cells. *Nanotechnology* **2006**, *17*, 304–312.
33. Chung, J.; Myoung, J.; Oh, J.; Lim, S. Synthesis of a ZnS Shell on the ZnO Nanowire and Its Effect on the Nanowire-based Dye-sensitized Solar Cells. *J. Phys Chem. C* **2010**, *114*, 21360–21365.
34. Gao, T.; Li, Q.; Wang, T. Sonochemical Synthesis, Optical Properties, and Electrical Properties of Core/Shell-type ZnO Nanorod/CdS Nanoparticle Composites. *Chem. Mater.* **2005**, *17*, 887–892.
35. Sandquist, C.; McHale, J. L. *J. Photochem. Photobiol. A* **2011**, *221*, 90.
36. Butera, D.; Tesoriere, L.; Gaudio, R. D.; Bongiorno, A.; Allegra, M.; Pintaudi, A. M.
37. Kohen, R.; Livrea, M. A. *J. Agric. Food Chem.* **2002**, *50*, 6895–6901.

CHAPTER 20

POLYMER ELECTROLYTE MEMBRANE-BASED ELECTROCHEMICAL CONVERSION OF CARBON DIOXIDE FROM AQUEOUS SOLUTIONS

P. SURESH, K. RAMYA[*], and K. S. DHATHATHREYAN

Centre for Fuel Cell Technology, International Advanced Research Centre for Powder Metallurgy and New Materials (ARCI), 2nd Floor, IIT-M Research Park, No. 6, Kanagam Road, Taramani, Chennai 600113, India

[*]Corresponding author. E-mail: ramya.k.krishnan@gmail.com

ABSTRACT

This chapter reports the study of a polymer electrolyte membrane fuel cell-type reactor for the electrochemical reduction of CO_2. Hydrogen gas and carbon dioxide dissolved in water/acid have been used as reactants with copper oxide and platinum as catalysts. Formate formation occurred with very high efficiency in a period of 1 h in acidified solutions when operated at potentials <1.1 V. The cumulative faradaic efficiency decreased when operated for a long period of time in recirculation mode. An operating temperature of 27°C was found to present a very high faradaic efficiency compared to operation at subambient and high temperatures. Further, the efficiency was found to be higher when hydrogen gas was used instead of water.

20.1 INTRODUCTION

The emission of greenhouse gases and the resultant global warming is an emerging problem faced due to extended use of fossil fuels to satisfy the various energy needs. Various methods are being tried for storing CO_2, one of the major greenhouse gases. Carbon capture and storage of the major greenhouse gas CO_2 and sequestration of CO_2 using molecular hydrogen to form formic acid or methanol is an attractive method to solve the problems. The usage of CO_2 to synthesize HCOOH is preferred[1] as the formation of HCOOH requires only one equivalent of H_2 compared to three equivalents of H_2 for every methanol molecule (as one equivalent is used for the formation of water). Formic acid (being a liquid) stores 4.3% of hydrogen and can be easily transported at ambient conditions. Several routes such as electrochemical, biochemical, thermochemical, photochemical, chemical, and radiochemical are being tried out for storing CO_2 as formic acid or methanol.[2,3]

A hybrid method of solar cell and electrochemical reduction may be advantageous industrially to convert CO_2 into useful products. Table 20.1 lists the different types of electrochemical reductions such as electrocarboxylation, direct electroreduction, etc. being investigated for conversion of CO_2 into useful products.[4–9] It has been reported that high faradaic efficiency or selectivity for CO_2 electrochemical conversion is achieved on metals such as Cd, In, Pb, Hg, Au, or Cu which have high overpotential for hydrogen generation. These metals provide the best environment for electron to be transferred to assist in the electrochemical reduction of CO_2.[10] Parameters such as kinetics of electron transfer, adsorption/desorption at the electrode surface, and diffusion of CO_2 to the electrode govern the reaction rate.[11]

TABLE 20.1 Electrochemical CO_2 Conversion Techniques.

Method	Catalysts/solvents	Products	References
Electrocarboxylation	Aprotic solvents	Carboxylate, polycarboxylates, olefins, etc.	4–7
Electroreduction with heterogeneous catalysts	Bulk metals, e.g., In, Hg, Sn, Au, Ag, Pt, Cu, Pb, etc.	CO, CH_4, C_2H_4, alcohols, formate, H_2	8
Electroreduction with homogeneous catalysts	Transition metal complexes of Ni, Pd, Cu, Co, Rh, Ru, etc.	CO, HCOO–	9

Copper (Cu) has been widely used in the study of electrochemical reduction of CO_2, forming hydrocarbons and alcohols in aqueous electrolytes at ambient temperature and pressure. Cu has intermediate adsorption strength for CO_2 and an intermediate hydrogen evolution capability. Table 20.2 lists some of the products that have been obtained using Cu under various

TABLE 20.2 Electroreduction Using Cu as Catalysts.

Catalyst	Potential (V)	Current density (mA/cm^2)	Hydrocarbons	Alcohol	CO	Formate/lactate	H$_2$	Remarks	References
Cu	−1.44 vs. NHE	5.0	58.8	8.7	1.3	9.4	20.5	KHCO$_3$ electrolyte 18.5 ± 0.5°C	8
Cu	−1.18 vs. NHE	11	50	11	1	1	17	0.5 M KHCO$_3$ saturated CO$_2$.	12
Cu net electrode	−0.6 V vs. Ag/AgCl		12.9	1.9	0.2	48.8	30.3	0.5 M KCl solution at pH 3.0	13
Cu net electrode	−2.4 vs. Ag/AgCl		46.5	7.1	25.0	2.4	25.6	0.5 M KCL solution at pH 3.0	13
Anodized Cu foil	−1.9 V vs. SCE		—	240	Trace		Trace	0.5 M KHCO$_3$, pH 7.6	14
Cu		375	Hydrocarbon					Aq. KHCO$_3$ electrolyte Pressure > 30 atm	15
Cu dinuclear complex	−0.03 V vs. NHE					12 equivalents of oxalate			16
Cu	−2.3 V Ag quasi-ref. electrode	436	6.0		46.6	34.6	4	Nonaqueous methanol. 40 atm	17
Cu oxide	1.0 V vs. SCE	11 mA						0.5 M KHCO$_3$	18
Cu/Zn/Al								0.1 M KHCO$_3$, MEA with H$_2$ gas reactant	19
Cu/CuO	−3.0 V vs. Ag/AgCl		∼100%					Pulsed mode reduction 0.1 M Na$_2$SO$_4$	20

conditions. The selectivity of this system and the product formed could be tuned by modification of parameters such as cathode potential, current density, temperature, pressure, and the supporting electrolyte (aqueous and nonaqueous), catalyst form, crystal structure, etc.[12–20]

Further, it has been reported that high current densities can be achieved with gas diffusion-based electrodes. Ikeda et al.[21] employed Cu-loaded gas-diffusion electrodes (GDEs) for mass reduction of CO_2. The products (CH_4, C_2H_4, C_2H_5OH, CO, and HCOOH) were formed with high current densities with GDEs than Cu plate electrodes. The solid polymer electrolyte method may be employed for electrochemical reduction of CO_2 in gas phase without solvent because no supporting electrolyte is required. Electroless plating methods have also been employed to prepare Cu/Nafion® electrodes for electroreduction of CO_2.[22,23] The efficiency of the product depended on the type of SPE and was found to be 17% in Nafion (C_2H_4) and 27% in selemion (HCOOH), respectively.[24] Narayanan et al.[25] have studied the conversion of CO_2 to formate ion using a polymer electrolyte membrane sandwiched between two catalyzed electrodes as this type of cell allows higher conversion efficiencies. The product formate was monitored by ultravoilet–visible (UV–vis) spectroscopy. The efficiency was found to be dependent on the concentration at the surface of the electrodes. A maximum cumulative faradaic efficiency of 80% was achieved at 300 s with 1 M bicarbonate solutions.

Thus, for CO_2 electroreduction, hydrogen evolution and CO_2 reduction occur simultaneously on the same electrode with the anodic process being the generation of oxygen by water oxidation. If the process is aimed such that the anodic reaction is not water oxidation, hydrogen oxidation then potentially can be reduced and the product formed may be different. Kobayashi and Takahashi[19] have studied the reduction of CO_2 to methanol with CO_2 and H_2 as the cathode and anode reactant. Electron generated on the anode transfers to the cathode through the external circuit. CO_2 is reduced by the transferred proton and electron on the cathode. This chapter utilizes the similar concept and a copper-coated GDE has been used as the cathode and Pt-coated GDE used as the anode. Pure hydrogen and aqueous CO_2 solution are used as the reactants. CO_2 reduction has been carried out using PEM configuration using acidic and neutral solutions saturated with CO_2. The effect of electrolyte and temperature on the efficiency of conversion of CO_2 to product using copper-based GDEs has been studied using hydrogen as a proton donor and electron source.

20.2 EXPERIMENTAL

20.2.1 CELL CONFIGURATION

The heart of the electrochemical cell is the Nafion membrane sandwiched between the catalyst-coated electrodes. The anode is formed by coating 40% Pt/C (0.5 mg/cm^2) on the GDE formed with microporous layer coating on carbon cloth and the cathode is formed by coating CuO powder (10 mg/cm^2) coated on carbon cloth. The area of both electrodes is 6.75 cm^2. The Nafion112 membrane was placed between the anode and cathode and then hot pressed at 170 kg/cm^2 and 130°C.

20.2.2 REACTANT SOLUTIONS

H_2 gas at a flow rate of 0.14 LPM was passed along the anode side. In the cathode side, two different solutions were circulated at the flow rate of 2 mL/min using a peristaltic pump. They are deionized water saturated with carbon dioxide prepared by continuous bubbling of the pure gas at a pressure of 1 atm or CO_2 dissolved in 0.1 M H_2SO_4 solution. In total, 200 mL of the solution was taken for every experiment.

20.2.3 ELECTROCHEMICAL REDUCTION PROCESS

The electrochemical reduction of carbon dioxide was conducted under galvanostatic conditions. The current–voltage performance of the cell was investigated and a maximum current density of ~160 mA/cm^2 was obtained at voltages less than 1.1 V. The constant current experiments to measure Faradaic efficiency were conducted by passing the current continuously for at least 1 h. During the studies, to measure the Faradaic efficiency, the product was allowed to accumulate in the cathode reservoir. Product was identified using UV–vis spectrophotometry using the broad absorption band in the region of 200–250 nm. The absorption at 230 nm was the wavelength chosen for the formate concentration determination as described by Narayanan et al.[25] Calibration curves for formate at 230 nm obeyed Beer's law with an extinction coefficient of 15.665 mol^{-1} cm^{-1}.

20.3 RESULTS AND DISCUSSION

20.3.1 ELECTROCHEMICAL CELL DESIGN

The electrochemical cell has been configured with perfluorosulphonic acid-based polymer flanked on either side with CuO-coated carbon cloth and a GDE coated with Pt catalyst. Although copper has been widely studied as an electrode catalyst material for CO_2 electroreduction, wide variety of products have been obtained based on the operating conditions such as temperature, pressure, potential, crystal structure, etc. Thus, Cu can be tuned to give different products based on the operating conditions. The reduction of CO_2 to formate takes place at a potential of −0.43 V with respect standard hydrogen electrode.

$$CO_2 + H^+ + e^-\ HCOO^- \qquad (20.1)$$

Literature studies cite that operation at low potentials (above −1.1 V) has resulted in the production of CH_3OH.[18] In the present investigation, to operate the cell below the water oxidation potential, the electrochemical cell has been configured with H_2 on the anode instead of water. The schematic representation of the polymer electrolyte membrane-based electrochemical cell for CO_2 reduction has been shown in Figure 20.1. As indicated in Figure 20.1, the cell is supplied with saturated solution of CO_2, hydrogen, and

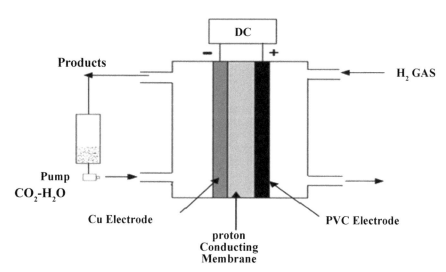

FIGURE 20.1 Schematic representation of the experimental arrangement for the production of organic products from carbon dioxide using a membrane cell.

electricity. The reduction products formed at the cathode are collected and analyzed. At the anode, hydrogen is oxidized with Pt catalyst and the protons migrate into the cathode side through the membrane. The protons, CO_2, and the electrons combine to give reduced organic product. The advantage of using a membrane electrode assembly similar to that of PEMFC (polymer electrolyte membrane fuel cell) include separation of the products of CO_2 electroreduction from the anode compartment and scalability due to the use of porous electrodes of low resistance. Thus, in the present chapter, we compare the efficiencies of formate production by operating the cell in acidic and neutral media at low voltages (0.4–1.2 V) and compare the same with the cell operated with water as the anode reactant. Gas chromatography studies carried out with the gaseous product of the cathode side indicated only the presence of H_2 confirming the absence of methane, CO, etc. in cathode effluent.

20.3.2 EFFECT OF ANODE REACTANT

In the present chapter, CO_2 was reduced by application of potentials using hydrogen as the anode reactant rather than the use of water, which is the anode reactant in most of the cells reported. Activated hydrogen species useful for CO_2 reduction are supplied to the cathode by the use of hydrogen as the anode reactant. When water is used as the anode reactant, oxygen evolution occurs at the anode and hydrogen evolution and CO_2 reduction occurs simultaneously at the cathode. Table 20.3 compares the results of the experiments carried out with water and hydrogen as the reactants. The experiments were carried out at a constant current of 1 A (corresponding to a current density of ~160 mA/cm²). In order to maintain a current density of 160 mA/cm², a potential of 2.5–3.5 V had to be applied to the cell with water as the reactant compared to a potential of 0.5–1.1 V for hydrogen as the reactant. It can be seen that the use of hydrogen as the anode reactant results in reduction of CO_2 at lower potentials and higher efficiency of conversion to formic acid.

TABLE 20.3 Reactions with Different Anode Reactants.

Anode catalyst	Cathode catalyst	Anolyte	Catholyte	Current (A)	Voltage (V)	Time (s)	Faradaic efficiency (%)
Pt	CuO	H_2O	CO_2/H_2O	1	2.5–3.5	10,800	14.9
Pt	CuO	H_2 gas	CO_2/H_2O	1	0.5–1.1	10,800	26.5

20.3.3 EFFECT OF CATHODE REACTANT

Two different cathode solutions were used in this study. They were water solution obtained by dissolution of CO_2 in water and CO_2 dissolved solutions with 0.1M H_2SO_4. The anode reactant in all the experiments was pure hydrogen gas. The current while using both the cells was kept at 1 A (corresponding to a current density of ~160 mA/cm²). Table 20.4 gives the performance characteristics of the cell when operated with different catholyte solutions. It can be seen that the performance improves in the presence of acid in the catholyte solution.

TABLE 20.4 Performance Comparison in Different Cathode Reactants.

Anode catalyst	Cathode catalyst	Anolyte	Catholyte	Current (A)	Voltage (V)	Time (s)	Faradaic efficiency (%)
Pt	CuO	H_2 gas	CO_2/H_2O	1	0.5–1.1	3600	50
Pt	CuO	H_2 gas	0.1 M H_2SO_4/CO_2/H_2O	1	0.5–1.1	3600	73

The differences in the performance between the two studies arise due to the differences in the ionic contact between the electrode and the solution. The membrane electrode assembly is formed by bonding the electrodes containing Nafion binder with Nafion membrane. At the membrane catalyst interface, contact is improved further in the presence of sulfuric acid, thereby reducing the resistance further. The concentration of CO_2, carbonate, and bicarbonate in the solution is dependent on the pH of the solution.

$$HCO_3^- + H_3O^+ = CO_2 + H_2O \quad K_1 = 2.5 \times 10^{-4} \tag{20.2}$$

$$HCO_3^- + H_2O = CO_3^{2-} + H_3O^+ \quad K_2 = 5.6 \times 10^{-11} \tag{20.3}$$

The performance of the cell is dependent on the amount CO_2 adsorbed on the electrode surface. From eq 20.2, it can be seen that due to the increased availability of CO_2 near the electrode surface, faradaic efficiency is higher when acidified solutions are used. Further, the presence of acid solution at the cathode helps in presenting the active form of hydrogen for reducing the adsorbed CO_2 at the electrode surface. Similar results have been obtained by Whipple et al.[26] who observed that operating at acidic pH resulted in a significant increase in faradaic and energy efficiencies to 89% and 45%, respectively, with a current density of 100 mA/cm².

20.3.4 EFFECT OF TIME

All the experiments conducted in the present study were in recirculation mode by circulating CO_2-rich solutions through the cathode. An experiment to study the cumulative formic acid production as a function of time was carried out with the conditions for a reaction tabulated in Table 20.4. Figure 20.2 represents the cumulative faradaic efficiency at various intervals of time. It can be found that there is an increase in formic acid production upto a certain time and this is followed by a decline in product formation for both neutral and acidified solution of CO_2. This suggests that as the reaction proceeds, the concentration of formic acid builds up and an increase in product concentration with time is seen. This cumulative concentration effect seems to decrease after a certain period time. As the reaction proceeds, the concentration of CO_2 near the electrode gets depleted. It is transported by diffusion through the electrode to the catalyst site. As the concentration gradient builds up, availability of the CO_2 near the electrode decreases. Faradaic efficiency and product formation, hence, decreases. Further, it has been found that the diffusion coefficient of CO_2 in water is about 1.1×10^{-5} cm^2/s. When diffusion of CO_2 takes place through a porous electrode coated with catalyst and ionomer, the diffusion coefficient decreases further leading to a concentration gradient with depleted CO_2 near the catalyst.[25] Hence, the cumulative faradaic efficiency decreases as a function of time.

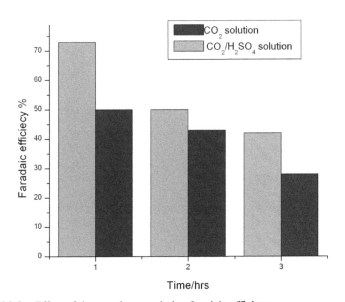

FIGURE 20.2 Effect of time on the cumulative faradaic efficiency.

20.3.5 EFFECT OF TEMPERATURE

Two important considerations for electrochemical reduction of CO_2 are temperature and electrocatalytic activity. Even though catalytic activity is known to improve with increased temperature, the amount of CO_2 dissolved in the solution decreases, and hence, faradaic efficiency decreases. To anlyze the effect of temperature, a study was carried out to evaluate the faradaic efficiency at different temperatures of electroreduction. Figure 20.3 gives the faradaic efficiency at different temperature. Hydrogen evolution on copper electrodes is further reduced at low temperatures and the efficiency for catalytic reduction is expected to be high. However from the results, it was found that formation of formic acid is improved at room temperatures as at low temperature, catalytic activity is low and at high temperatures, CO_2 concentration in the solution is decreased.

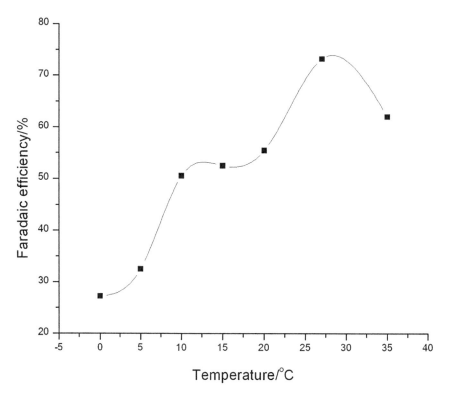

FIGURE 20.3 Effect of temperature on faradaic efficiency.

Formic acid formed may be obtained from the following intermediates formed during the electrocatalytic reduction of CO_2.

1. Electron transfer step to CO_2 adsorbed on the cathode

$$CO_{2\,ads} + e^- \rightarrow CO_{2\,ads}^-$$

2. Proton transfer forming adsorbed formic acid

$$CO_{2\,ads}^- + 2H^+ \rightarrow HCOOH_{ads}$$

Formation of hydrogen bonds with the solution and circulation of the reactants removes the HCOOH acid formed and prevents further oxidation of the products formed. We can thus conclude that product selectivity and activity of the electrocatalyst to form any product is influenced both by electrode materials and the operation conditions such as electrode potentials, duration, pH, temperature, etc.

20.4 CONCLUSIONS

A PEMFC-type cell was designed for electrochemical reduction of CO_2. The cell used gas diffusion-based electrodes, Nafion membrane, copper oxide, and platinum catalysts for the production of formate with high efficiency. Hydrogen was used as the anode reactant and CO_2 reduction was performed at voltages less than water oxidation potential (<1.1 V). Operation of the cell with acidified CO_2 solutions resulted in higher faradaic efficiency of 73% compared to operation with neutral solutions (53%). Faradaic efficiency was the highest at 27°C (73%) as at low temperatures, catalytic activity was low and at high temperatures, there is a decrease in CO_2 concentration in the solution. Faradaic efficiency values were found to be high (73%) for the conversion of CO_2 to formate at low intervals of time (1 h). However, the cumulative faradaic efficiency decreased over a period of 3 h due to the buildup of concentration gradient caused by the depletion of CO_2 as a result of product formation.

ACKNOWLEDGMENTS

The authors would like to thank Dr. G. Sundararajan, Director, ARCI, Hyderabad for supporting this work and the Department of Science and Technology (DST), Government of India for financial support.

KEYWORDS

- CO_2 utilization
- electrochemical method
- gas diffusion layer based
- Cu catalyst
- HCOOH formation

REFERENCES

1. Williams, R.; Crandall, R. S.; Bloom, W. Use of Carbon Dioxide in Energy Storage. *Appl. Phys. Lett.* **1978**, *33*, 381.
2. Sun, H.; Wang, S. Research Advances in the Synthesis of Nanocarbon-based Photocatalysts and Their Applications for Photocatalytic Conversion of Carbon Dioxide to Hydrocarbon Fuels. *Energy Fuels* **2014**, *28*, 22.
3. Nikulshima, V.; Hirsch, D.; Mazzotti, M.; Steinfeld, A. CO_2 Capture from Air and Co-production of H_2 via $Ca(OH)_2$–$CaCO_3$ Cycle Using Concentrated Solar Power–Thermodynamic Analysis. *Energy* **2006**, *31*, 715.
4. Wawzonek, S.; Duty, R. C.; Wagenknecht, J. H. Polarographic Studies in Acetonitrile and Dimethylformamide, VIII. Behavior of Benzyl Halides and Related Compounds. *J. Electrochem. Soc.* **1964**, *111*, 74.
5. Simonet, J. Conditions for a Large Cathodic Sequestration of Carbon Dioxide into Glassy Carbon. Generation of a Carbon Poly-carboxylate Versatile Material. *Electrochem. Comm.* **2012**, *21*, 22.
6. Dano, C.; Simonet, J. Cathodic Reactivity of Graphite with Carbon Dioxide: An Efficient Formation of Carboxylated Carbon Materials. *J. Electroanal. Chem.* **2004**, *564*, 115.
7. Derien, S.; Clinet, J. C.; Dunach, E.; Perichon, J. Electrochemical Incorporation of Carbon Dioxide into Alkenes by Nickel Complexes. *Tetrahedron* **1992**, *48*, 5235.
8. Hori, Y.; Wakebe, H.; Tsukamoto, T.; Koga, O. Electrocatalytic Process of CO Selectivity in Electrochemical Reduction of CO_2 at Metal Electrodes in Aqueous Media. *Electrochim. Acta* **1994**, *39*, 1833.
9. Christensen, P. A.; Higgins, S. J. The Electrochemical Reduction of CO_2 to Oxalate at a Pt Electrode Immersed in Acetonitrile and Coated with Polyvinylalcohol [Ni(dppm)2Cl2]. *J. Electroanal. Chem.* **1992**, *387*, 127.
10. Ogura, K.; Farrell, J. R.; Cugini, A. V.; Smotkin, E. S.; Villapando, M. D. S. CO_2 Attraction by Specifically Adsorbed Anions and Subsequent Accelerated Electrochemical Reduction. *Electrochim. Acta* **2010**, *56*, 381.
11. Watanabe, M.; Shibata, M.; Kato, A. Design of Alloy Electrocatalysts for CO_2 Reduction: III. The Selective and Reversible Reduction of CO_2 on Cu Alloy Electrodes. *J. Electrochem. Soc.* **1991**, *138*, 3382.

12. Jaramillo, T. F.; Norskov, J. K. In *Global Climate Energy Project*, Sixth Annual Research Symposium Creating a Sustainable Energy System for the 21st Century and Beyond, Stanford University, California, September 2010.
13. Yano, H.; Shirai, F.; Nakayama, M.; Ogura, K. Efficient Electrochemical Conversion of CO_2 to CO, C_2H_4, and CH_4 at a Three-phase Interface on a Cu Net Electrode in Acidic Solution. *J. Electroanal. Chem.* **2002**, *519*, 93.
14. Frese Jr., K. W. Electrochemical Reduction of CO_2 at Intentionally Oxidized Copper Electrodes. *J. Electrochem. Soc.* **1991**, *138*, 3338.
15. Hara, K.; Tsuneto, A.; Kudo, A.; Sakata, T. Electrochemical Reduction of CO_2 on a Cu Electrode Under High Pressure Factors That Determine the Product Selectivity. *J. Electrochem. Soc.* **1994**, *141*, 2097.
16. Angamuthu, R.; Byers, P.; Lutz, M.; Spek, A. L.; Boumann, E. Electrocatalytic CO_2 Conversion to Oxalate by a Copper Complex. *Science* **2010**, *327*, 313.
17. Saeki, T.; Hashimoto, K.; Fujishima, A.; Kimura, N.; Omata, K. Electrochemical Reduction of CO_2 with High Current Density in a CO_2–Methanol Medium. *J. Phys. Chem.* **1995**, *99*, 8440.
18. Le, M.; Ren, M.; Zhang, Z.; Sprunger, P. T.; Kurtz, R. I.; Flake, J. C. Electrochemical Reduction of CO_2 to CH_3OH at Copper Oxide Surfaces. *J. Electrochem. Soc.* **2011**, *158*, E45.
19. Kobayashi, T.; Takahashi, H. Novel CO_2 Electrochemical Reduction to Methanol for H_2 Storage. *Energy Fuels* **2004**, *18*, 285.
20. Yano, J.; Yamasaki, S. Pulse-mode Electrochemical Reduction of Carbon Dioxide Using Copper and Copper Oxide Electrodes for Selective Ethylene Formation. *J. Appl. Electrochem.* **2008**, *38*, 1721.
21. Ikeda, S.; Ito, T.; Azuma, K.; Ito, K.; Noda, H. Electrochemical Mass Reduction of Carbon Dioxide Using Cu-loaded Gas Diffusion Electrodes I. Preparation of Electrode and Reduction Products. *Denki Kagaku* **1995**, *63*, 303.
22. DeWulf, D. W.; Jim, T.; Bard, A. J. Electrochemical and Surface Studies of Carbon Dioxide Reduction to Methane and Ethylene at Copper Electrodes in Aqueous Solutions. *J. Electrochem. Soc.* **1989**, *136*, 1686.
23. Cook, R. L.; MacDuff, R. C.; Sammels, A. F. Reduction of Carbon Dioxide to Methane at High Current Densities. *J. Electrochem. Soc.* **1987**, *134*, 1873.
24. Sciboh, M. A.; Viswanathan, B. Electrochemical Reduction of Carbon Dioxide: A Status Report. *Proc. Indian Natl. Sci. Acad.* **2004**, *70*, 1.
25. Narayanan, S. R.; Haines, B.; Soler, J.; Valdez, T. I. Electrochemical Conversion of Carbon Dioxide to Formate in Alkaline Polymer Electrolyte Membrane Cells. *J. Electrochem. Soc.* **2011**, *158*, A167.
26. Whipple, D. T.; Finke, E. C.; Kenis, P. J. A. Microfluidic Reactor for the Electrochemical Reduction of Carbon Dioxide: The Effect of pH. *Electrochem. Solid State Lett.* **2010**, *13*, B109.

CHAPTER 21

GRAPHENE-BASED NANOSTRUCTURED MATERIALS FOR ADVANCED ELECTROCHEMICAL WATER/WASTEWATER TREATMENT

EMMANUEL MOUSSET[1*] and MINGHUA ZHOU[2,3]

[1]Laboratoire Réactions et Génie des Procédés, UMR CNRS 7274, Université de Lorraine, 1 rue Grandville BP 20451, 54001 Nancy cedex, France

[2]Key Laboratory of Pollution Process and Environmental Criteria, Ministry of Education, College of Environmental Science and Engineering, Nankai University, Tianjin 300071, China

[3]Tianjin Key Laboratory of Urban Ecology Environmental Remediation and Pollution Control, College of Environmental Science and Engineering, Nankai University, Tianjin 300071, China

*Corresponding author. E-mail: emmanuel.mousset@univ-lorraine.fr

ABSTRACT

The emerging development of graphene-based electrodes is critically reviewed. These promising materials have shown to enhance the electrochemical advanced oxidation processes (EAOPs) efficiency during wastewater treatment. Thorough investigations, the main factors responsible of the success of such cathode in undivided electrolytic cell have been identified as follows: (1) carbonaceous raw material, (2) nanostructured-coated materials, (3) high-porosity materials, and (4) nitrogen-doped materials. These parameters participate to the enhancement of electroactive surface area and conductivity of electrode, two main material's properties that affect its efficiency. By combining all these characteristics, graphene-based electrode opens up new exciting EAOPs setups and should bring them closer to real applications.

21.1 INTRODUCTION

Nanomaterials have gained many interests in the two last decades and especially graphene that was discovered in 2004[1] and drawn tremendous research attention due to its extraordinary properties such as high electron mobility (up to 200,000 cm^2 V^{-1} s^{-1})[2], high thermal conductivity (up to 5000 W m^{-1} K^{-1}),[3] high mechanical strength (Young's modulus around 1.0 TPa),[4] high optical transparency (up to 97.7%),[5] large theoretical specific surface area (2630 m^2 g^{-1}),[6] and many other promising characteristics. This ideal one-atom-thick fabric of sp^2-hybridized carbon atoms arranged in a honeycomb structure is remarkably attractive for numerous applications that include energy generation and storage, sensors, composite materials, paints and coating, and electronics and photonics. In recent past, the benefit from its large surface area was implemented in the treatment of water and wastewater such as heterogeneous photocatalysis,[7] desalination,[8] and adsorption.[9] Combined with an extremely high conductivity, graphene was more recently suggested to be used as an electrode material either in a galvanic mode (e.g., microbial fuel cells)[10] or in an electrolytic cell (e.g., electrochemical advanced oxidation processes, EAOPs).[11–17] The later technologies constitute a hot topic of research in water treatment area due to the clean reagent that is involved to produce extremely high oxidizing agents such as hydroxyl radicals ($^{\bullet}$OH).[18–22] Such processes are able to degrade and mineralize a wide range of organic load present in water even those containing persistent organic pollutants (POPs) that are bioresistant.

In this context, the present chapter aims at giving a critical review of the research papers dealing with graphene-based electrodes employed in EAOPs processes by presenting the following consecutive steps as summarized in Figure 21.1: (1) synthesis of graphene and its characterization, (2) manufacture of graphene-based electrode and assessment of its properties, and (3) its application in EAOPs for water/wastewater treatment.

21.2 SYNTHESIS OF GRAPHENE

There exist many production routes of graphene that can be grouped into either bottom-up or top-down approaches.[23] Both of them have been used

Graphene-Based Nanostructured Materials

to produce graphene-based electrode for EAOPs application and they are described in the Sections 21.2.1 and 21.2.2.

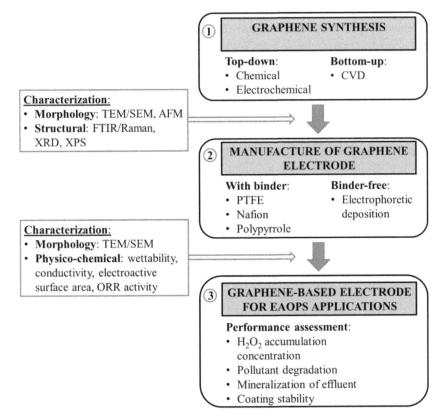

FIGURE 21.1 Three consecutive steps for graphene-based nanostructured materials applications in EAOPs treatment of water/wastewater.

21.2.1 BOTTOM-UP METHODS

Bottom-up methods generate graphene by assembling small-molecular building blocks into a single layer or a few layers of graphene.[24] Among them, chemical vapor deposition (CVD) has been widely developed for the very high purity graphene production.[25] It allows growing graphene either on copper (Cu) or nickel (Ni) substrates and then the graphene is transferred onto another substrate, depending on the application. The performance of CVD graphene in the form of graphene monolayer (Gmono), graphene multilayer (Gmulti), and three-dimensional (3D) graphene foam (Gfoam) have been previously studied in EAOPs applications.[13] To synthesize those graphene materials, hydrogen and a carbon source (methane) gases were introduced into a furnace that heat at high temperature (around 1000°C). The high thermal oxidation condition leads to methane decomposition in the presence of a transition metal (Cu or Ni) on which carbon atoms are deposited and rearranged into sp^2-hybridized carbon structures. In the case of Gmono and Gmulti, a thin copper film is used as a substrate[26] while a Cu or Ni foam is employed to grow Gfoam.[27] The Gmono is then transferred onto quartz material,[26] while the metal skeleton is etched away to get graphene sheets (Gmulti) and a porous Gfoam structure.[27] Large-scale CVD synthesis can be obtained using large substrates and large tube furnace according to the substrate-rolling synthesis strategy.[28] For example, Gfoam area of 170 × 220 mm^2 could be obtained with a furnace equipped with a quartz tube of 71 mm inner diameter.[27]

However, the transfer process and metal removal usually cause significant degradation of the graphene quality due to the formation of wrinkles or structural damage from tearing and ripping.[24] As a consequence, CVD graphene displays poorer electronic and structural properties than the ideal one,[29] which makes competitive the top-down approach as presented thereafter (Section 21.2.2).

21.2.2 TOP-DOWN METHODS

The top-down technique aims at weakening the van der Waals forces between the layers of graphite used as raw material to progressively obtain one single-layer graphene. Two methods have been tested to produce graphene for EAOPs application: (1) the well-known and conventional Hummers method followed by reduction of graphene oxide (GO)[11,12,15–17,30] and (2) the more recent electrochemical exfoliation technique.[14]

The first approach consists on applying strong chemical oxidation conditions (KMnO$_4$, NaNO$_3$, and H$_2$SO$_4$) to get GO from exfoliated graphite.[31] To avoid generation of toxic gas and to better control the temperature, a modified Hummers method has been proposed, namely, Tour's method, by increasing the amount of KMnO$_4$, excluding the use of NaNO$_3$, and performing the reaction in a 9:1 mixture of H$_2$SO$_4$/H$_3$PO$_4$.[13,32] The GO is then reduced to get reduced graphene oxide (rGO).[14,33]

Alternatively, the electrochemical exfoliation method has been developed. It consists of applying a bias voltage (around 10 V) between the cathode (graphite) and the anode (inert counter electrode such as platinum, Pt) both immersed in a sulphated electrolyte (around 0.1 mol L^{-1}).[14,34–37] SO$_4^{2-}$ ions can then be intercalated within graphite layer leading to its subsequent exfoliation. This method presents many advantages over the chemical method as, first, it is faster with production rate of graphene that averaged 30.0 mg h^{-1} against 1.2 mg h^{-1} chemically.[14,24] Second, it can be operated under ambient conditions while the chemical one needs to heat at least at 50°C, which further increases the energy requirements. Third, it is more environmental friendly and safer, as it requires only power and a diluted electrolyte (0.01–1 mol L^{-1}), whereas the rGO production method needs highly concentrated (0.3–20 mol L^{-1}) acids, oxidizing agents, and reducing agents. Additionally, it can be precisely tweaked by controlling the applied potential or current on the contrary to chemical approach.

Graphene needs to be then characterized to further compare the efficiency of methods to produce graphene as discussed in the following section.

21.3 CHARACTERIZATION OF GRAPHENE

To investigate the characteristics of the graphene produced via CVD, chemical, or electrochemical methods, the surface morphology as well as structural and physicochemical properties are commonly assessed.

21.3.1 SURFACE MORPHOLOGY

The morphology (e.g., roughness, sheet size, number of sheets, interlayer distance, and pore size) of graphene is commonly assessed by microscopy: (1) atomic force microscope (AFM), (2) (high resolution) transmission electron microscope ((HR)TEM), and/or (3) scanning electron microscope (SEM).

Examples of images depicting CVD Gmono, electrochemically exfoliated graphene (EEG), CVD Gmulti, and CVD Gfoam are shown in Figure 21.2(a–d), respectively.[13,14] AFM results (Fig. 21.2a) show that the average roughness of CVD Gmono was 0.37 nm, which is very close to 0.34 nm—the theoretical thickness of a single layer of graphene.[13] The slight difference of 0.03 nm could be attributed to the presence of functional groups on the graphene surface (Section 21.3.2). Comparatively, the roughness of GO produced by Hummers method was 1 nm,[11] demonstrating the presence of numerous oxygen functional groups on the graphene surface (Section 21.3.2). Regarding the CVD Gmulti, an average roughness of 227 nm highlighted an average number of layer around 667.

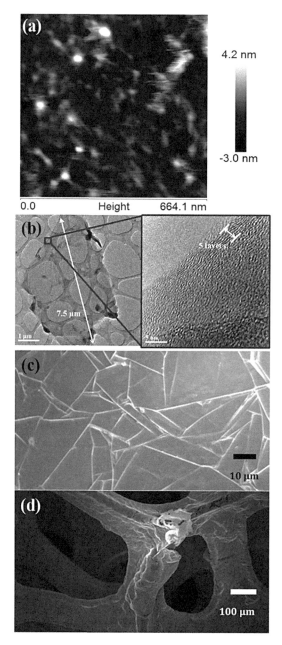

FIGURE 21.2 Morphology of different graphene materials use in EAOPs applications and given by (a) AFM images of CVD Gmono, (b) TEM (left) and HRTEM (right) images of EEG, SEM images of (c) CVD Gmulti, and (d) CVD Gfoam. Adapted with permission from Refs. [13,14]. Copyright 2016 Elsevier.

In Figure 21.2b (left), an EEG sheet was distinctly identified with a size around 7.5 µm, in the range of EEG sheet (5–10 µm) obtained by statistical analysis.[34,37] A micrograph on the edge of the sheet (Fig. 21.2b, right) highlights the number of layers (5) with an interlayer distance of 0.4 nm (2 nm thickness in total). Considering the theoretical interlayer distance of 0.34 nm, the 0.06-nm difference could be attributed to the few oxygen-containing groups on the graphene surface,[14,35] as discussed in Section 21.3.2. Still, the high degree of transparency in TEM micrographs of single or few graphene sheets made by CVD, electrochemical, or chemical method indicated the very low thickness of this nanomaterial.[13,15,30]

SEM picture in Figure 21.2c displayed the flat surface of Gmulti and numerous graphene sheets less ordered than in graphite material.[13] Contrastingly, the 3D Gfoam (Fig. 21.2d) clearly showed large open pores (100–600 µm) that are expected to increase the surface area and facilitate molecule diffusion inside the pores.[13]

Further analyses are usually performed to better assess the quality of produced graphene as mentioned in the following section.

21.3.2 PURITY AND STRUCTURAL PROPERTIES

Spectroscopy techniques such as X-ray diffraction (XRD), Fourier transform infrared (FTIR), Raman, as well as photoelectron spectroscopy (XPS) have been commonly used to assess the crystalline structure (by XRD), the functional groups (by FTIR, Raman, and XPS), and the surface elemental composition (by XPS) of graphene materials.[11,13–15,17,30]

XRD patterns of Gmono, Gmulti, and Gfoam produced by CVD method are displayed in Figure 21.3a. A string and sharp diffraction peak (002) at $2\theta = 26.5°$ and a weaker sharp peak (004) at $2\theta = 54.6°$ were observed for all CVD graphene materials. Those peaks could be ascribed to the crystalline graphitic structure.[13,38] It revealed the high crystalline degree of Gmono and Gmulti while background noise in Gfoam patterns could be attributed to amorphous characteristics and lesser ordered structure.[13] On the contrary, XRD of GO does not exhibited a peak at 26°, whereas a reflection at 12.5°[11] was noticed confirming the high degree of oxidation and certifying that all the raw graphite was converted into GO.[11] The reflection of a peak at 47.8° exhibited the hexagonal phase (012) of GO.[30]

Raman spectra of graphene materials usually display three kinds of peaks, namely, (1) the D peak at around 1350 cm^{-1} that represents the number of defects, (2) the G peak at around 1580 cm^{-1} that is relevant to the sp²-hybridized carbon–carbon bonds in graphene, and (3) the 2D peak at around 2690 cm⁻¹ that provides information on the stacking order.[39,40] Two main ratios have employed as indicator to assess and compare the quality of graphene, that is, the intensity of D peak (I(D)) over the intensity of G peak (I(G)) (I(D)/I(G)) that expresses the number of defects and I(D) over the intensity of 2D peak (I(2D)) (I(2D)/I(G)) that exemplifies the graphitization degree for C=C sp² bonds in graphitic carbons. Both ratios have been plotted in Figure 21.3b and allow comparing the GO,[15,30] EEG,[14] and CVD productions of Gmono, Gmulti, and Gfoam.[13] I(D)/I(G) ratio of EEG was lower (0.68) than for GO (0.90). Knowing that lower the I(D)/I(G) ratio, lower the defects, it shows that EEG had less number of defects than GO. It could be accredited to the oxygen-functional groups of GO. Regarding the I(2D)/I(G) ratio values, they were ranked as follows: CVD Gmono (3.66) > CVD Gfoam (1.58) > CVD Gmulti (0.97) > EEG (0.92) > Graphite (0.59). CVD provided higher quality of C=C bond recovery when the number of layer decreases.[13,36] In addition, the ratio value of Gfoam (3.66)[13] was close to the theoretical one (4).[40] Graphite had the lower quality while ratio value of EEG was close to CVD Gmulti.[13,14]

To further determine the surface elements and functional groups that could be present on graphene materials, XPS analyses are traditionally performed. C1s deconvolution graphs allow indentifying and qualifying C=C/C–C/C–H bonds present in aromatic rings (peak at 284.5 eV), epoxy and alkoxy bonds (C–O) (peak at 286 eV), C=O bond (peak at 287.5 eV), and O–C=O (peak at 289 eV) functional group.[13,33] The sp² carbons were present in majority in all graphene materials whatever the production method.[11,13,15] Oxygen-containing group such as C–O was still distinguished in CVD graphene materials, GO, and rGO.[11,13,15] C=O and O–C=O functional groups were found in a lesser extent in GO and rGO materials.[11,15] The presence of oxygen in the graphene material is further assessed by the carbon atomic percentage over oxygen atomic percentage (C/O) ratio. The higher the C/O ratio, the higher the quality of graphene. C/O ratio of GO,[11] CVD graphene,[13] and graphite[13] have been compared in Figure 21.3c and ranked as follows: CVD Gmono (32) > CVD Gmulti (25) > CVD Gfoam (22) > graphite (19) > GO (2.75). First, it can be considered that graphite was completely converted into GO since the C/O ratio was higher than 2.0.[41] Second, it again emphasized the higher quality of graphene obtained with CVD technique. Few oxygen atoms remained in CVD Gmono which corroborated the difference of thickness (0.37 nm) with the theoretical value (0.34 nm) (Section 21.3.1).

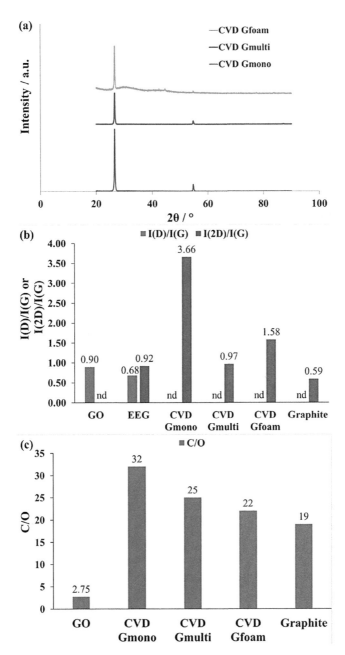

FIGURE 21.3 Purity and structural properties of different graphene materials use in EAOPs applications: (a) XRD patterns of CVD Gmono, Gmulti, and Gfoam, (b) I(D)/I(G) ratio obtained by Raman results (nd: not determined), and (c) C/O ratio obtained by XPS results. Source: Adapted with permission from Ref. [13]. Copyright 2016 Elsevier.

21.3.3 SHEET RESISTANCE

The quality of graphene can be further assessed by sheet resistance measurements usually expressed in ohm per square ($\Omega\ \square^{-1}$). Two- or preferentially four-point probe devices connected to a multimeter are commonly used to quantify the sheet resistance of graphene sheets.[42] The EEG was compared to a modified Hummer's method (Tours technique) to produce graphene. A sixfold higher sheet resistance was noticed with the chemical method (30 $\Omega\ \square^{-1}$) against the electrochemical (5.1 $\Omega\ \square^{-1}$). The typical range of rGO (14–52 $\Omega\ \square^{-1}$)[43] made it less conductive than the EEG. Knowing that CVD method could be reached a sheet resistance as low as 4.4 $\Omega\ \square^{-1}$ for a single-layer graphene, the electrochemical exfoliation method opens up an environmental friendly alternative to produce graphene.[14]

21.4 MANUFACTURE OF GRAPHENE-BASED ELECTRODES

Once the graphene is synthesized and characterized, it is either tested as an electrode itself (CVD Gmono, Gmulti, and Gfoam)[13] or it is transferred onto a raw conductive material,[11,12,14–17,30] which is the subject of this section. Several transfer techniques have been proposed and they have been categorized into two ways of electrode manufacturing: (1) methods requiring the use of binder and (2) methods without binder requirement.

21.4.1 MANUFACTURING METHODS USING BINDER

Most of the studies developed methods requiring the use of additional binder to coat the graphene on raw conductive material to be used in EAOPs applications (Table 21.1). Three binding agents have been employed: (1) polytetrafluoroethylene (PTFE),[15,16] (2) Nafion,[14,30] and (3) polypyrrole.[17] Those polymers have good electrochemical stability and they are not electroactive in the range of potentials applied in EAOPs.[14,44–46] In addition, it has been laid out that PTFE delivers two functions in the electrode material namely to bind the high surface graphene onto a cohesive layer and to impart some hydrophobic characteristics.[14,15,47] Moreover, Nafion is an ionomer consisting of hydrophobic fluoro-backbone and a hydrophilic sulfonic acid group Nafion can strongly be adsorbed onto graphene surface by hydrophobic interactions followed by steric and increased electrostatic repulsions between graphene sheets by inducing a negative charge on the

Graphene-Based Nanostructured Materials

TABLE 21.1 Different Methods to Manufacture Graphene-based Electrodes for EAOPs Applications.

Nature of graphene	Binder	Raw materials	Operating conditions	References
GO*	PTFE	Stainless steel mesh	Mixture of graphite/graphene/PTFE (optimal ratio: 8:1:2) ultrasonicated for 15 min at room temperature; heated at 80°C; pressed at 15 MPa for 2 min; heated at 300°C for 2 h	[16]
GO*	PTFE	Stainless steel mesh	Graphene-based solution mixed with 10 wt% of PTFE in ethanol; ultrasonicated for 20 min; heated at 70°C; paste cold-pressed at 10 Mpa for 8 min; heated at 360°C for 1 h. N-rGO@CNT preparation: mixed rGO and CNT with ammonia (25 wt%) and heated at 180°C for 12 h (rGO@CNT same without ammonia addition)	[15]
rGO*	Polypyrrole	Polyester fabric membrane/stainless steel mesh	Polyester membrane immersed and sonicated in graphene solution for 10 min; ammonium persulphate sprayed on it; vapor deposition of pyrrole at 90°C for 15 min followed by in situ polymerization	[17]
GO*	Nafion	Glassy carbon	Microwave-assisted hydrothermal treatment of graphene mixture at 160°C for 15 min; ethanol (1 mL) and Nafion (0.5%) were added to graphene mixture (5 mg) and dropped on glassy carbon; dried by lyophilization for 24 h	[30]
EEG	Nafion	Carbon cloth	Nafion/graphene (optimal ratio: 1:4 (w/w)) solution mixed in water/ethanol (1:1 v/v) and ultrasonicated for 1 h; cloth soaked in graphene ink and heated at 55°C for 3 h and at 250°C for 1 h	[14]

*GO obtained by (un)modified Hummers method; subsequent reduction of GO (rGO) is considered to occur by heat treatment during electrode manufacture, except in Zhao et al.[17] study in which GO is reduced before electrode manufacture.

graphene surface. Both repulsive reactions allowed overcoming the van der Waals force between the graphene sheets. Therefore, the amphiphilic property of Nafion favors graphene dispersion and stabilization in solution. Nafion also acts as interfacial binder and favor the hydrophobic interactions between the graphene dispersions and the raw material. Moreover, the sole presence of Nafion caused a slight improvement in electrocatalytic activity,[14] which could be ascribed to its electrical conductivity that could possibly increase the electrochemical rate of reactions occurring at the surface of the electrode.[48]

The concentration of interfacial binder as well as the binder/graphene ratio play a major role in the electrocatalytic efficiency of electrode. An increase of hydrophobic binder (e.g., PTFE) has shown to minimize the cathode flooding and facilitating gas distribution and subsequent electrochemical reaction such as O_2 reduction into H_2O_2.[47] However, an excess of binder (e.g., Nafion) has demonstrated to reduce the porosity and the available surface area.[14,45] Similarly, an increase of graphene concentration exposed a raise of catalytic activity by increasing the specific surface area and the electrical conductivity until a maximal graphene concentration.[14] At such high amount, graphene starts aggregating due to insufficient quantity of dispersant, which decreases surface area and porosity of the electrode material.[14] After varying the concentrations of Nafion and graphene, a Nafion to graphene ratio of 1:4 (w/w) was found optimal when coated on carbon cloth,[14] whereas an optimal graphite/graphene/PTFE ratio was determined to be 8:1:2 when stuck on stainless steel mesh.[16]

The graphene loading achieved by the mass transfer of graphene to the raw material is another important parameter to determine. It allows getting accurate binder to graphene mass ratio and normalizing the amount of graphene to the surface of the raw material employed. Though this is fundamental to allow comparing graphene cathodes, most papers do not report the graphene loading except Mousset et al.[14] who calculated the weight of graphene transferred to the cloth. The average graphene loading equaled 0.27 mg cm^{-2} at optimal Nafion and graphene concentrations.[14]

The nature of raw materials employed as current collector is further capital. Two kinds of primary materials have been applied, for example, stainless steel mesh[15–17] and carbon materials (glassy carbon[30] and carbon cloth[14]). Carbonaceous materials are usually preferred for EAOPs applications as it catalyzes the H_2O_2 electrogeneration by having (1) high H_2 evolution overpotential, (2) low catalytic activity for H_2O_2 decomposition, (3) good conductivity, (4) high chemical resistance, and (5) by being nontoxic for water/wastewater applications.[18]

Following the wake of graphene, recent applications have been devoted to the assembly of 2D materials.[49] A 3D carbon network has been developed with alternate connections of aligned carbon nanotubes (CNTs) and rGO nanosheets (rGO@CNT) doped with nitrogen (N-rGO@CNT) to get advanced multifunctional structures.[15] Such carbonaceous material has been employed as cathode in EAOPs application.[15]

Though binder properties have shown to be positive in the graphene-based electrode manufacture, some authors suggested alternative method to produce electrode without the need of binders. Such techniques are presented hereafter.

21.4.2 BINDER-FREE MANUFACTURING TECHNIQUES

Some authors[12,50,51] assume that the presence of binder hinder the properties of graphene, which result in lower conductivity of the composite and lower graphene loading. It also requires the addition of chemicals, which makes the method less simple.[51] Le et al.[12,50] suggested an electrophoretic deposition of GO (1.5 mg mL^{-1})—produced by modified Hummers method—followed by its electrochemical reduction to form rGO on a raw material (e.g., carbon felt, CF).[50] The essential parameters affecting the modification such as current density value, time of GO deposition on the substrate, and conditions of electrochemical reduction were investigated. Finally, a current density of −1.5 mA cm^{-2} for 10 min of electrolysis was determined to be optimal.[50] Furthermore, GO could serve as weak supporting electrolyte as it is conductive, meaning that sodium sulfate or potassium chloride were not necessary to add in the solution.[50] The constant potential reduction of GO on CF raw material (CPrGO-CF) method was compared to the chemical reduction technique (CrGO-CF) by adding a mixture of hydrazine (35 wt%) and ammonia (28 wt%) and the thermal reduction (TrGO-CF) approach by pyrolizing at 1000°C for 1 h.[12]

Upon all, these techniques implemented to manufacture graphene-based electrodes, their properties need to be established to compare the different approaches as discussed in the following section.

21.5 PROPERTIES OF GRAPHENE-BASED ELECTRODES

The manufactured graphene-based electrode can be characterized according to their (1) surface morphology, (2) wettability, (3) electrical conductivity, (4) electroactive surface area, and (5) oxygen reduction reaction (ORR) activity.

21.5.1 SURFACE MORPHOLOGY

To investigate the coating efficiency of the graphene on primary materials, SEM analyzes are first conducted as a qualitative approach.[11,12,14,16,17] Samples of images[14] are displayed in Figure 21.4. From the micrographs obtained at ×1000 magnification (Fig. 21.4a), it is shown that carbon cloth consisted of an entwined network of carbon microfilaments with an average diameter of 10 μm.[14] It is further illustrated that the fibers of the uncoated cloth (left) were clean, whereas the fibers of graphene-coated cloth (right) displayed microscale clusters of graphene sheets intertwined in the carbon network, indicating successful transfer onto the primary material.[14] At ×5000 (Fig. 21.4b) magnification, the image on the left revealed the uniform and smooth surface, whereas details of the graphene clusters (right) ranging from 1 to 20 μm in length appended on the cloth. A feature confirmed by Le et al.[12] and Zhao et al.[17] studies. The reduction treatment of GO impacted on the surface morphology as well. Though raw CF had a soft surface, the chemical reduction makes increase the number of pores and clusters of graphene deposited on CF.[12] The electrochemical and thermal reduction of GO could even higher the amount of asperity.[12]

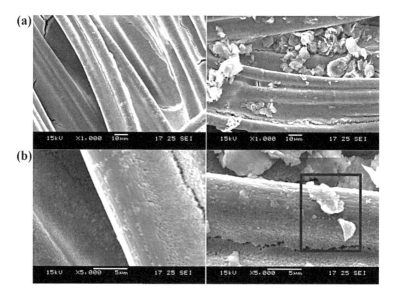

FIGURE 21.4 Surface examination of the uncoated cloth (left) and graphene-coated cloth (right) by SEM: (a) ×1000 and (b) ×5000 magnification. Optimal ink-coating conditions: Nafion = 0.025% (w/v), and [graphene]$_{cloth}$ = 1.0 mg mL^{-1}. Adapted with permission from Ref. [14]. Copyright 2016 Elsevier.

The polished surface of the uncoated material is expected to lower surface area, whereas the rough surface forecasts an increase of the area as further noticed.[12,14,17] This particularity would impact on the electrocatalytic activity of the electrode material as further discussed.

21.5.2 WETTABILITY

The hydrophilicity/hydrophobicity of electrode material is an important criterion that has shown to have double-contradictory effects in literature: (1) a high hydrophilicity could improve dissolved O_2 diffusion and adsorption at the cathode surface and can then be easily reduced into H_2O_2 (eq 21.1),[12] whereas (2) a high hydrophobicity could favor the O_2 gas absorption on the cathode material before being reduced into H_2O_2 (eq 21.1).[47] Both mechanisms lead to the generation of ·OH through Fenton reaction (eq 21.2):

$$O_2 + 2H^+ + 2e^- \rightarrow H_2O_2 \qquad (21.1)$$

$$H_2O_2 + Fe^{2+} \rightarrow \cdot OH + Fe^{3+} + HO^- \qquad (21.2)$$

To assess the hydrophilicity/hydrophobicity on the surface of graphene-based electrode materials contact angles measurements of water droplets have been performed.[12,13] Figure 21.5 depicts contact angles measurement done with CVD Gmono, CVD Gmulti, and CVD Gfoam materials.[13] The carbon materials were ranked as follows, from the least hydrophobic: Gfoam (128.3°) > Gmulti (87.1°) > Gmono (82.6°). It is clearly discerned that the more layers of graphene, the more hydrophobicity of graphene, which corroborates the increasing apolar aromatic rings.[13,52] Contact angles measurements have been also performed with uncoated CF and graphene-coated CF (CrGO-CF, CPrGO-CF, and TrGO-CF)[12] as displayed in Figure 21.6a. It is obvious that the reduction method affected the hydrophilicity of the material in the following (from the least hydrophobic): CF (89.9°) > CrGO-CF (68.3°) > CPrGO-CF (41.8°) > TrGO-CF (~ 0). The water droplets adsorbed immediately after contact with TrGO-CF material. The thermal reduction affected more the raw material by bringing a lot of open pores—as inspected on SEM pictures—that helped the quick water adsorption.[12] The same phenomenon could also occur with the two other reduction methods (e.g., chemical and electrochemical) but in a lesser extent. It is also interesting to note that CF-based materials were more hydrophilic than the CVD graphene electrodes. It could be attributed to the higher oxygen content in

CrGO-CF (C/O = 4.66),[12] CPrGO-CF (C/O = 7.72),[12] and TrGO-CF (C/O = 15.39)[12] than in CVD graphenes (C/O ranging from 22 to 32).[13] It means that the numerous oxygen functional groups in rGO-coated CF participated to increase the polarity of the material as compared to the nonpolar sp^2 carbon of pristine graphene.

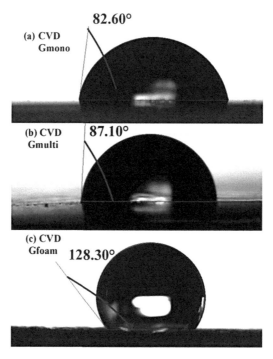

FIGURE 21.5 Contact angle measurements of (a) CVD Gmono, (b) CVD Gmulti, and (c) CVD Gfoam. Adapted with permission from Ref. [13]. Copyright 2016 Elsevier.

21.5.3 ELECTRICAL CONDUCTIVITY

The electrical conductivity of electrode material is an important parameter as it plays roles (1) in the kinetics of electrochemical reaction occurring at electrode surface and (2) on the energy consumption by reducing the cell potential. Electrochemical impedance spectroscopy (EIS) is an efficient tool for studying the interface properties and electron transfer mobility between the electroactive substance and the electrode material. EIS has been used to assess the electrical conductivity of the graphene-based electrode by using a three-electrode system connected to a potentiostat, in which an electroactive specie ($K_3[Fe(CN)_6]$)—sensitive to the surface chemistry of carbon-based electrodes—and an inert support electrolyte (KCl) are added in aqueous solution.[12–14] The impedance spectra can be then recorded at open circuit voltage (OCV) with frequency ranging from 50 kHz to 100 mHz and a voltage amplitude of 10 mV.[12–14] The Nyquist plot allows then determining the interfacial charge transfer resistance (R_{ct}) of the material that corresponds to the intercept of the semicircle with the real axis (X axis).[12–14]

R_{ct} values of pristine CVD graphene materials have been represented in Figure 21.6b and ranked as follows: Gmono (~3600 Ω) > Gmulti (123 Ω) > Gfoam (1.6 Ω). A significant increase of electrode conductivity was noticed with the increase of graphene thickness as reported by further authors.[13,53] The graphene layers in Gfoam are randomly assembled, unlike the ordered stacking observed with graphite.[53] The arbitrary stacking decouples the adjacent graphene sheets so that the electronic property of each layer is similar to that of single graphene.[54] Consequently, the overall resistivity of multilayer graphene decreases with the number of graphene sheets.[53,54] This is further consistent with the lower conductivity of graphite (92 Ω) as compared to Gfoam (1.6 Ω). Interestingly, the conductivity of CVD Gfoam was slightly higher than CF, whose R_{ct} value equaled 2.4 Ω[12] as shown in Figure 21.6b. CF had larger number of defects caused by the presence of oxygen functional group that alters the electrical conductivity.[13,43] Since the high conductivity of pristine graphene rely on its sp^2-hybridized honeycomb structure, the presence of oxygen functional groups disrupt its structure and localize π-electrons which therefore decreased their mobility, that is, decreased the conductivity of rGO compared to pristine graphene.[43]

Moreover, the reduction method further improved the conductivity of raw CF with R_{ct} values decreasing from 0.5 Ω (CrGO-CF) to around 0.21 Ω (CPrGO-CF and TrGO-CF). This was in agreement with the increasing C/O values from raw CF to TrGO-CF as mentioned in Section 21.5.2. Furthermore, the presence of EEG on carbon cloth could reduce the R_{ct} from 81.1 Ω (raw cloth) to 2.45 Ω, a 33-fold improvement (Fig. 21.6b).[14] Contrastingly, a three-times lower enhancement of CPrGO-CF against raw CF was noticed. This could be accounted for the use of rGO which was reported to contain defects related to oxygen functionalities that could reduce its electrical conductivity.[13,43]

21.5.4 ELECTROACTIVE SURFACE AREA

The electroactive surface area is a key point in advanced electrocatalysis as it increases the number of active site for molecule reduction. This is the case of O_2 adsorption on the carbon cathode surface before its subsequent reduction into H_2O_2, the adsorption stage being the limiting step according to the following reaction sequence (eqs 21.3a–3e) that is equivalent to the global eq 21.1[55,56]:

$$O_2 \rightarrow O_{2(ads)} \tag{21.3a}$$

$$O_{2(ads)} + e^- \rightarrow O_{2\ (ads)}^- \tag{21.3b}$$

$$O_{2\ (ads)}^- + H^+ \rightarrow HO_{2\ (ads)}^{\cdot} \tag{21.3c}$$

$$HO_{2\ (ads)}^{\cdot} + e^- \rightarrow HO_2^- \tag{21.3d}$$

$$HO_2^- + H^+ \rightarrow H_2O_2 \tag{21.3e}$$

Electroactive surface area of carbon materials can be obtained from cyclic voltammetry (CV) of $[Fe(CN)_6]^{3-}/[Fe(CN)_6]^{4-}$ reversible redox couple in which the anodic peak current value (I_p in A) of the cycle is used to calculate the surface (A in cm²) according to Randles–Sevcik equation (eq 21.4)[12–14,57,58]:

$$I_p = 2.69 \times 10^5 \times AD^{1/2}n^{1/2}\gamma^{1/2}C \tag{21.4}$$

where n is the number of electrons participating in the redox reaction ($n = 1$), D is the diffusion coefficient of the molecule in solution (7.60×10^{-6} cm²s⁻¹ for $[Fe(CN)_6]^{3+}/[Fe(CN)_6]^{4-}$ redox couple), C is the concentration of the probe molecule in the bulk solution (mol cm⁻³), and γ is the scan rate of the potential perturbation (V s⁻¹).

Generally, the applied potential is ranging from −0.4 to 0.8 V versus Ag/AgCl at a scan rate of 10 mV s⁻¹.[12–14] From Randles–Sevcik equation, it can be deduced that higher the I_p, the higher electroactive surface area. Electroactive surface area of different graphene-based electrode materials have been calculated and displayed in Figure 21.6c. CVD graphene materials could be ranked as follows: Gmono (0.015 cm²) < Gmulti (4.99 cm²) < Gfoam (55.02 cm²). Gfoam acquired higher surface area, which could be due to its 3D porous structure, as observed by SEM (Section 21.3.1).[13] CF-based electrodes were ranked as follows: raw CF (13.80 cm²) < CrGO-CF (27.59 cm²) < TrGO-CF (124.17 cm²) < CPrGO-CF (137.97 cm²).[12] Again, thermal and electrochemical reduction of GO make increase the surface area higher, with an improvement ranging from 9 to 10 times as compared to the uncoated CF, which is in accordance with SEM pictures (Section 21.5.1). A 11.5-fold

Graphene-Based Nanostructured Materials 339

enhancement could be further noticed between EEG-cloth (6.31 cm²) and the raw cloth (0.57 cm²). This is slightly higher gain of surface as compared to CPrGO-CF with CF.

To better compare the materials, it has been previously suggested to normalized the electroactive surface area either by the geometric surface of 2D electrodes or the volume of electrode for 3D materials.[13] The results are shown in Figure 21.6c. The following rank was noticed for 2D materials: CVD Gmono (0.0024 cm² cm⁻²) < cloth (0.024 cm² cm⁻²) < CVD Gmulti (0.125 cm² cm⁻²) < EEG-cloth (0.26 cm² cm⁻²),[13,14] whereas for 3D materials, it was ordered as follows: CVD Gfoam ≈ CF (27.5 cm² cm⁻³) < CrGO-CF (55.2 cm² cm⁻³) < TrGO-CF (248 cm² cm⁻³) < CPrGO-CF (276 cm² cm⁻³).[12,13] EEG-cloth and CPrGO-CF accounted for the best existing graphene-based 2D and 3D materials, respectively, in terms of electroactive surface area. It is still important to note that porous 3D materials remain preferable in their use as cathode materials in EAOPs processes.[59–62]

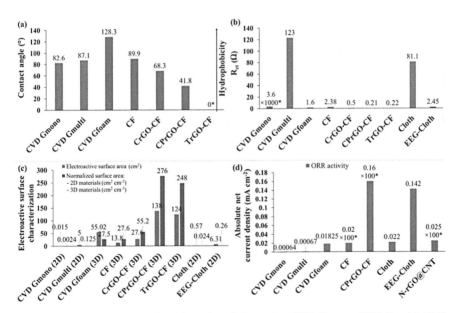

FIGURE 21.6 Assessment of graphene-based electrodes (CVD Gmono, CVD Gmulti, CVD Gfoam, CF, CrGO-CF, CPrGO-CF, TrGO-CF, cloth, and EEG-cloth) properties: (a) contact angle (°) measurements (*water droplets adsorbed immediately after contact with TrGO-CF material), (b) interfacial charge transfer resistance (R_{ct}) by EIS technique (value to be multiplied by 1000 for better reading convenience), (c) electroactive surface characterization by CV method, and (d) absolute net current density (ORR activity) from LSV curves (values to be multiplied by 100 for better reading convenience).

21.5.5 ORR ACTIVITY

ORR activity is responsible for H_2O_2 electrogeneration at cathode surface through the two-electron reaction pathway (eq 21.1). H_2O_2 is part of the Fenton's reagent that promotes the production of OH radicals (eq 21.2) and degrades organic effluent from water/wastewater when implemented in specific EAOPs.

Linear scan voltammetry (LSV) is an efficient device to investigate the electrochemical reaction kinetics on electrode materials and specially to carry out the ORR activity.[63–67] LSV can be performed with a potentiostat using a static working electrode[12–14] or a rotating ring-disk electrode (RRDE)[15] in a mixture of an electroactive specie ($K_3[Fe(CN)_6]$) and an inert supporting electrolyte (KCl).

This method allows getting the polarization curves that reflect the transient current response with respect to the applied cathodic potential. Generally, LSV is first performed in N_2-saturated solution, that is, free of dissolved O_2, meaning that ORR is inhibited so that only H_2 evolution reaction (HER) occurs (eq 21.5)[18]:

$$2H^+ + 2e^- \rightarrow H_{2(g)} \qquad (21.5)$$

Then, LSV measurements are done in O_2-saturated solution and both HER and ORR activity are monitored. The net current density can therefore be defined as the difference between the currents recorded in O_2- and N_2-saturated solutions, determining the ORR activity of the working electrode. Thus, a higher net current density would suggest a higher ORR activity, that is, a higher H_2O_2 electrogeneration efficiency (eq 21.1) and in the meantime a greater production of competitive waste four-electron ORR pathway that produce H_2O (eq 21.6)[12,14,68]:

$$O_2 + 4H^+ + 4e^- \rightarrow 2H_2O \qquad (21.6)$$

Due to the existence of competitive reactions, there exists an optimal applied cathodic potential that was reported to be in the range of −0.5 to −0.75 V versus saturated calomel electrode for carbonaceous materials (e.g., graphite, CVD Gmono, CVD Gmulti, CVD Gfoam, cloth, EEG-cloth, N-rGO@CNT, CF, and CPrGO–CF).[12–15] At respective optimal cathodic potential value, the net current density (in absolute value) of each material have been compared in Figure 21.6d. The following net current rank was noticed (expressed in mA cm^{-2}): CVD Gmono (0.00064) < CVD Gmulti (0.00067) < CVD Gfoam (0.018) < cloth (0.022) < EEG–cloth (0.14) < CF (2.00) < N-rGO@CNT

(2.50) < CPrGO–CF (16). It is interesting to note that CVD graphene materials portrayed lower ORR activity than other carbonaceous cathodes. The higher presence of oxygen-functional groups on rGO-based electrodes and raw cloth and CF materials could increase the ORR activity according to the following reactions (eqs 21.7a–d)[60]:

$$C=O + e^- \rightarrow C=O^- \qquad (21.7a)$$

$$C=O^- + O_2 \rightarrow C=O + O_2^- \qquad (21.7b)$$

$$2O_2^- + H_2O \rightarrow O_2 + HO_2^- + OH^- \qquad (21.7c)$$

$$O_2^- + H_2O + e^- \rightarrow HO_2^- + OH^- \qquad (21.7d)$$

Still, 3D materials such as Gfoam and CPrGO-CF electrodes could enhance the ORR activity as compared to 2D materials. It has been previously reported that electrochemical processes occur at the carbon spaces rather than at the outer planar surface, resulting in a three-dimensional electrochemical activity,[69] which was consistent with the better electrocatalytic activity gained with 3D materials.

In addition, EEG and CPrGO increased the ORR activity of the cloth by 6.4-fold[14] and of the CF by eightfold,[12] respectively. The slight difference of efficiency between EEG and CPrGO could be referred to the numerous oxygen functionalities present on the later one.

Liu et al.[15] further stated that N-rGO@CNT had higher ORR activity than rGO and CNT separately. The enhanced ORR activity could be ascribed to two main factors: (1) the easier electron transfer benefited from the bridge between CNT and graphene in the 3D assembly composite and (2) the nitrogen doping. The presence of nitrogen implies a dipole effect between N–C bonds, meaning that the electron density of the three adjacent carbon atoms is lowered—the electronegativity of N (3.04) being higher than C (2.55)—thus favoring the chemisorption of O_2 on positively charged carbon atoms and the subsequent ORR.[70] Therefore, the synergy of graphene/CNT composite and the presence of N-doped carbons are both parameters that appear to be an interesting feature to take into account for electrode characterization and further development of novel graphene-based cathode materials.

After characterizations and comparisons of the different enhanced properties benefited from graphene-based electrodes, it appears interesting to evaluate their efficiency in real applications such as EAOPs for water/wastewater treatment as discussed in the following section.

21.6 WATER/WASTEWATER TREATMENT APPLICATIONS OF GRAPHENE-BASED ELECTRODES

21.6.1 CONTEXT AND DEVELOPMENT OF EAOPs

Water scarcity is becoming a global concern that can be mainly imputed to the ever increasing water demand and the overall climate change. Additionally, the continuous release of POPs from municipal and industrial wastewater treatment plant that pollute the natural water bodies along with more and more stringent regulations makes water/wastewater treatment one of the greatest challenges of the century.[71] These POPs such as pesticides, pharmaceuticals, cosmetics, etc. also characterized as micropollutants and/or emerging contaminants are known to be not only toxic to the environment and human health but also refractory to the current biological treatments. Regarding this context there is a need to develop advanced physico-chemical treatment able to eliminate such POPs from water. Advanced oxidation processes (AOPs) have been developed for this purpose.[72] All AOPs rely on the production of ·OH that has a very high oxidation potential (2.8 V vs. standard hydrogen electrode [SHE]),[73] second after fluorine that have to be avoided in a context of water remediation. This radical is non-selective and its reaction kinetics are even quicker with aromatic rings and double C=C bonds (10^8–10^{10} L mol^{-1} s^{-1}), major components of the xenobiotic compounds.[18] Recently, EAOPs have emerged as they require only electrical power and optionally a supporting electrolyte in too low conductivity effluent.[18–22] In contrast to the traditional AOPs (e.g., ozonation, Fenton technologies, photocatalysis, etc.), the oxidants are produced in situ and continuously in mild conditions (room temperature and pressure) through the use of adequate catalytic electrode materials.[18] Other advantages include high energy efficiency, easy handling, safety, modularity, and versatility as EAOPs can treat a very wide range of effluents (i.e., from micropollution to highly loaded industrial effluent).[20]

In such processes, the electrode material is a key element of the system. Especially, carbonaceous cathode material can promote the Fenton reaction (eq 21.2) accountable for the ·OH formation in an electro-Fenton process. In this technology, H_2O_2 is electrogenerated through O_2 reduction (eq 21.1) while ferrous ion (Fe^{2+}) is regenerated from ferric ion (Fe^{3+}) reduction (eq 21.8)[74,75]:

$$Fe^{3+} + e^- \rightarrow Fe^{2+} \tag{21.8}$$

The superiority of electro-Fenton over traditional chemical Fenton process remains also in the addition of only a catalytic amount (0.05–0.2 mM) of iron source—if not already present in the effluent,[76,77]—which restrain sludge production. In addition, the degradation and mineralization rates and yields can be very high (>99%), even with refractory compounds.[78–82]

As explained in Section 21.4.1, carbon-based cathode (e.g., carbon felt, graphite felt, PTFE-gas diffusion electrode, GDE) manifested excellent performance in electro-Fenton process, but still the yield of H_2O_2 production and the rates of Fe^{2+} regeneration were not satisfying enough to scale up the technology.[13,14] Therefore, extensive efforts have been made to elevate the catalytic efficiency of carbonaceous cathodes via chemical modification,[60] electrochemical modification,[66] acidic treatment,[83] or surface modification with organic compounds and rare earth-derived compounds.[84–88] More recently, the benefits that could be gained from graphene-based materials due to their exceptional properties (Section 21.5) have been investigated for EAOPs applications.[11–17] The results are presented and compared hereinafter by taking into account five criterion: (1) the H_2O_2 accumulation efficiency, (2) the pollutant degradation efficiency, (3) the mineralization efficiency, (4) the stability of electrode, and (5) the energy requirements.

21.6.2 H_2O_2 ACCUMULATION CONCENTRATION

Being performed in undivided cell, the kinetics of H_2O_2 accumulation concentration that is produced in an electro-Fenton treatment display always the same trend whatever the applied current density, that is, a transient phase followed by a steady state. This behavior can be explained by the competition between H_2O_2 electrogeneration at the cathode (eq 21.1) and the H_2O_2 decomposition at the cathode (eq 21.9), at the anode (eqs 1.10a–10b), and in a lesser extent in the bulk solution (eq 21.11)[18]:

$$H_2O_2 + 2H^+ + 2e^- \rightarrow 2H_2O \quad (21.9)$$

$$H_2O_2 \rightarrow HO_2^\cdot + H^+ + e^- \quad (21.10a)$$

$$HO_2^\cdot \rightarrow O_2 + H^+ + e^- \quad (21.10b)$$

$$2H_2O_2 \rightarrow O_2 + 2H_2O \quad (21.11)$$

Varying the applied current density, and therefore the applied cathodic potential, further exhibited an optimal H_2O_2 accumulation concentration value.[13,14] This is ascribed to the parasitic reactions, that is, HER (eq 21.5) and H_2O_2 decomposition (eqs 21.9–11), that become dominating when the applied current density is too high (i.e., too low cathodic potential).

The maximal amounts of H_2O_2 accumulated in the cell using different graphene-based cathode material have been represented in Figure 21.7a.[13–16] The concentration values have been taken at optimal applied cathodic potential and normalized to the cathode area to be comparable. The following rank of maximal H_2O_2 accumulation concentration was given (expressed in mg H_2O_2 L^{-1} cm^{-2}): CVD Gmono (0.032) < CVD Gmulti (0.048) < CVD Gfoam (0.213) < rGO (0.5) < cloth (1.43) < rGO@CNT (2.25) < EEG-cloth (2.81) < N-rGO@CNT (5) < rGO-graphite GDE (15.7). Among the CVD graphenes, 3D Gfoam highlighted again much better performance, which could be due to its higher electroactive surface area, conductivity, and ORR activity. EEG-cloth could increase by two times the H_2O_2 production from the raw cloth, indicating the benefit got from graphene. In total, 4.6- and 6.7-fold improvements were noticed with N-rGO@CNT as compared to rGO@CNT and rGO, respectively. It brings to light the synergy obtained with 3D assembly of rGO@CNT and the N-doped material. Interestingly, rGO-graphite GDE gave the better H_2O_2 production efficiency. GDE are known to generate high concentrations of H_2O_2 thanks to O_2 bubbling through the carbon material which ensures enhanced contact between the cathode surface and O_2.[75] However, GDE cathodes typically suffer from reduced surface area as compared to 3D carbon materials, leading to lower Fe^{2+} regeneration and therefore, Fe^{3+} accumulation and lower production of ·OH.[75] Moreover, Xu et al.[16] did not compare the degradation and mineralization efficiencies of their graphene material with an uncoated cathode which does not allow for complete comparison in terms of the coating efficiency.

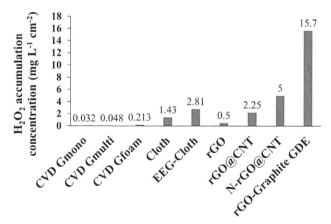

FIGURE 21.7 Maximal H_2O_2 accumulation concentration obtained with graphene-based electrodes: CVD Gmono, CVD Gmulti, CVD Gfoam, CF, cloth, EEG-cloth, N-rGO@CNT, and rGO-graphite GDE.

21.6.3 POLLUTION DEGRADATION EFFICIENCY

EAOPs are able to degrade many kinds of pollutants and mainly dyes (brilliant blue,[16] methylene blue,[17] and acid orange 7[11,12]), phenolic compounds,[13,14] and plastic additives (dimethyl phthalate)[15] have been treated by electro-Fenton using graphene-based cathode (Table 21.2). Electro-Fenton process successfully degraded all monitored contaminants thanks to the very high oxidation power of ˙OH that react mainly through two mechanisms: (1) hydrogen atom abstraction (e.g., aliphatics, R) (eqs 21.12a–12c) or (2) by electrophilic addition to an unsaturated bond (e.g., aromatics, Ar) (eqs 21.13a–13c) to initiate a radical oxidation chain[18,20]:

$$RH + \cdot OH \rightarrow H_2O + R\cdot \quad (21.12a)$$

$$R\cdot + O_2 \rightarrow ROO\cdot \quad (21.12b)$$

$$ROO\cdot + RH \rightarrow ROOH + R\cdot \quad (21.12c)$$

$$ArH + \cdot OH \rightarrow ArH(OH)\cdot \quad (21.13a)$$

$$ArH(OH)\cdot + O_2 \rightarrow [ArH(OH)OO]\cdot \quad (21.13b)$$

$$[ArH(OH)OO]\cdot \rightarrow ArH(OH) + HO_2\cdot \quad (21.13c)$$

Most of the time, at the beginning of the oxidation treatment, the pollutant decay follows a pseudo-first-order kinetic model, assuming a quasi-stationary state for ˙OH concentration as it does not accumulate into the solution ([˙OH]) (eq 21.14):

$$\frac{d[\text{pollutant}]}{dt} = -k_{abs} [\cdot OH][\text{pollutant}] = k_{app} [\text{pollutant}] \quad (21.14)$$

where k_{abs} is the absolute rate constant of pollutant decay and k_{app} is the pseudo-first-order rate constant of pollutant decay.

After integration of eq 21.14, the following relation is obtained (eq 21.15):

$$\ln\left(\frac{[\text{pollutant}]_0}{[\text{pollutant}]}\right) = k_{app} t \quad (21.15)$$

Then k_{app}, values can be deduced from linear regression of the semilogarithmic plots. k_{app} value is an efficient parameter that allows comparing the degradation efficiencies of different operating conditions and processes. This constant is often converted into the half-life time ($t_{1/2}$), which is the time to degrade 50% of the monitored contaminant. $t_{1/2}$ values have been reported in Table 21.2 for each electro-Fenton treatment employing a graphene-based cathode. CVD graphene setups allowed degrading 50% of phenol (1 mM) after 86 min (Gfoam), 315 min (Gmulti), and 866 min (Gmono).[13] The rate of phenol decay with Gfoam was 3.7 and 10 times faster than Gmulti and Gmono, respectively. This trend was in agreement with the higher conductivity (Section 21.5.3), higher electroactive surface area (Section 21.5.4), and higher ORR activity (Section 21.5.5) of Gfoam.[13] It allows electrogenerating the Fenton reagent (especially H_2O_2, Section 21.6.2) at higher rates and yield.[13] In similar operating conditions, EEG-cloth could decrease by 3.1-fold, the half-life time of phenol (44 min) as compared to the raw cloth (136 min). Again, the higher conductivity, electroactive surface area and ORR activity led to higher amount of ˙OH produced and quicker contaminant degradation rates. Comparatively, CPrGO-CF increased by 2.2 times the degradation rates efficiency of acid orange 7 as compared to the raw CF. Interestingly, Liu et al.[15] further exhibited the gain (4–16.2 times of increase) that could be obtained using a N-rGO@CNT-based cathode ($t_{1/2}$ = 21.4 min) as compared to a rGO@CNT ($t_{1/2}$ = 86.6 min) and rGO-based cathode ($t_{1/2}$ = 347 min) when degrading dimethyl phthalate.

21.6.4 MINERALIZATION EFFICIENCY

EAOPs and especially electro-Fenton offer the possibility to mineralize quasi-completely the effluent after an extended treatment, that is, to reach the quasi-total conversion of organic compounds into CO_2, H_2O, and inorganic ions when heteroatoms are present. The oxidation pathway leads to the ring-opening reactions (eq 21.16) of the hydroxylated aromatic compounds (ArOH) initially formed (eqs 1.13a–13c)[89,90]:

$$Ar(OH) + \cdot OH/O_2 \rightarrow CA + CO_2 \quad (21.16)$$

Carboxylic acids are then degraded to form shorter chain CAs (eq 21.17)[89–91]:

$$CA_1 + \cdot OH/O_2 \rightarrow CA_2 + CO_2 \quad (21.17)$$

These low molecular weight CAs, for example, oxalic acid, formic acid, and acetic acid, constitute generally the latest organic products because they are directly converted into CO_2 and H_2O (eq 21.18)[89–91]:

$$CA_2, CA_3, \ldots CA_n + \cdot OH/O_2 \rightarrow CO_2 + H_2O \quad (21.18)$$

Total organic carbon (TOC) measurements are usually performed to obtain the mineralization degree, considering that mineralization yields are equivalent to TOC removal.[13] TOC decay have been previously monitored during an electro-Fenton treatment with different graphene-based electrodes.[11–14] It is interesting to note that in all the cases, the TOC decay rates decreased toward the end of treatment. This trend is related to the increased formation of CAs and especially short-chain CAs that are more recalcitrant to mineralization, with rate constants of reaction with ˙OH ranging from 1.4×10^6 L mol^{-1} s^{-1} (oxalic acid) to 1.3×10^8 L mol^{-1} s^{-1} (acetic acid).[13,91] In the meantime, several waste reactions such as between ˙OH and Fe^{2+} (eq 1.19) and between ˙OH and H$_2$O$_2$ (eq 21.20) become predominant and compete with the Fenton reaction (eq 21.2):

$$Fe^{2+} + ˙OH \rightarrow Fe^{3+} + HO^- \qquad (21.19)$$

$$H_2O_2 + ˙OH \rightarrow H_2O + HO_2˙ \qquad (21.20)$$

Regarding the mineralization yield, CVD graphene could reach 1.9%, 5.3%, and 14.2% with Gmono, Gmulti, and Gfoam, respectively, after 2 h of electro-Fenton treatment (Table 21.2), highlighting again the superiority of the 3D Gfoam.[13] The presence of EEG on carbon cloth could reach higher mineralization yield (16.0%), a 1.5-times improvement as compared to the raw cloth (10.6%).[14] A lower enhancement was noticed (1.3 times) between CPrGO-CF (74%) and CF (56%).[12] From 2 to 10 times improvement of rGO@CNT (19% of mineralization yield) and rGO materials (4% of mineralization yield) was reached with N-rGO@CNT electrode (40% of mineralization yield).[15] These trends corroborated the results given in Sections 21.6.3 and 21.6.4.

21.6.5 STABILITY OF ELECTRODE

The stability of graphene-based electrodes is a crucial factor that needs to be assessed before the development of EAOPs for environmental applications. In this context, electro-Fenton treatments are repeated several times in a row, from 5 successive batches[14] to 7 cycles,[17] 10 cycles,[12,13] and 20 cycles,[15] according to the literature. The values represented in Table 21.2 reproduce the percentage of 2 h-electro-Fenton degradation[15,17] or mineralization[12–14] after the first and the fifth cycle. The stability of the graphene-based cathode can be ranked as follows from the least stable (expressed in % of stability

TABLE 21.2 Performance Comparison of Graphene-Based Cathodes Applied in Lab-Scale Electro-Fenton for Wastewater Treatment.

Nature of graphene	Raw material for cathode	Immersed cathode surface area (cm^2)	Operating conditions[a]	Degradation efficiency ($t_{1/2}$) (min)[b,*]	Mineralization yield after first cycle (%)[c,*]	Stability over 5 cycles (%)	Energy consumption (kW h g^{-1} pollutant)[d,*]	Energy consumption (kW h g^{-1} TOC)[e,*]	References
rGO	Stainless steel mesh	12	TOC_{init} = 18 mg-C L^{-1}; [Reactive brilliant blue] = 0.13 mmol L^{-1}; [Fe^{2+}] = 0.75 mmol L^{-1}; pH = 3; [Na$_2$SO$_4$] = 50 mmol L^{-1}; j = 2.0 mA cm^{-2}, V = 200 mL, anode: Pt, electrode distance: 3 cm	nd/80	nd/25	nd	nd	nd	[16]
rGO	Stainless steel mesh	4 (2 cm × 2 cm)	TOC_{init} = 31 mg-C L^{-1}; [Dimethyl phthalate] = 0.26 mmol L^{-1}; [Fe^{2+}] = 0.5 mmol L^{-1}; pH = 3; [Na$_2$SO$_4$] = 50 mmol L^{-1}, E_{cat} = −0.2 V, V = 100 mL, anode: Pt	347/21.4[g] (93.8%)	4/40[g] (900%)	97/97[f,g] (0%)	1.42 × 10^{-2}/3.28 × 10^{-3} (77%)	nd	[15]
rGO	Polyester fabric membrane/ stainless steel mesh	40 (2 cm × 10 cm)	TOC_{init} = 3000 mg-C L^{-1}; [Methylene blue] = 15.6 mmol L^{-1}; [Fe^{2+}] = 0.2 mmol L^{-1}; pH = 3; [Na$_2$SO$_4$] = 50 mmol L^{-1}, E_{cat} = −1 V, V = 50 mL, anode: stainless steel, electrode distance: 1 cm	nd/nd (8%)	nd	90/90[f] (0%)	nd	nd	[17]
CPrGO	Carbon felt	2	TOC_{init} = 19 mg-C L^{-1}; [Acid orange 7] = 0.10 mmol L^{-1}; [Fe^{2+}] = 0.2 mmol L^{-1}; pH = 3; [Na$_2$SO$_4$] = 50 mmol L^{-1}; j = 20 mA cm^{-2}, V = 30 mL, anode: Pt, electrodes distance: 3 cm	2.7/1.2 (55%)	56/74 (33%)	74/[f,c] (6%)	nd	0.74/0.53 (29%)	[11,12]

Graphene-Based Nanostructured Materials

TABLE 21.2 *(Continued)*

Nature of graphene	Raw material for cathode	Immersed cathode surface area (cm²)	Operating conditions[a]	Degradation efficiency ($t_{1/2}$) (min)[b,c]	Mineralization yield after first cycle (%)[c]	Stability over 5 cycles (%)	Energy consumption (kW h g⁻¹ pollutant)[d,e]	Energy consumption (kW h g⁻¹ TOC)[c,e]	References
EEG	Carbon cloth	24 (3 cm × 5 cm)	TOC$_{init}$ = 100 mg-C L⁻¹; [Phenol] = 1.4 mmol L⁻¹; [Fe²⁺] = 0.1 mmol L⁻¹; pH = 3; [K₂SO₄] = 50 mmol L⁻¹; j = 1.25 mA cm⁻², V = 80 mL, anode: Pt, electrodes distance: 3 cm	136/44 (68%)	10.6/16.0 (51%)	16.0/12.7[c] (21%)	nd	0.31/0.20 (36%)	[14]
CVD Gmono	—	(2.5 cm × 2.5 cm)	TOC$_{init}$ = 72 mg-C L⁻¹; [Phenol] = 1 mmol L⁻¹; [Fe²⁺] = 0.1 mmol L⁻¹; pH = 3; [K₂SO₄] = 50 mmol L⁻¹; E_{cat} = −0.6 V vs. Ag/AgCl (3 mol L⁻¹), V = 150 mL, anode: Pt, electrodes distance: 3 cm	−/866	−/1.9	1.9/0[c] (100%)	nd	−/0.072	[13]
CVD Gmulti	—	40 (5 cm × 4 cm)		−/315	−/5.3	5.3/4.9[c] (7.5%)	nd	−/0.053	[13]
CVD Gfoam	—	40 (5 cm × 4 cm)		−/86	−/14.2	14.2/12.3[c] (13.4%)	nd	−/0.039	[13]

*Before/After cathode modification (percentage change).

nd: Not determined.

[a]TOC$_{init}$: Initial TOC of solution; E_{cat}: cathodic potential; j: current density; V: volume of solution.

[b]$t_{1/2}$ corresponds to the half-life time value obtained after 50% of pollutant degradation, considering a pseudo-first order electro-Fenton degradation rate of the pollutant.

[c]Percentage of mineralization after 2 h of electro-Fenton treatment, except in Liu et al.[15] study in which 3 h of mineralization is considered.

[d]Energy requirement for 50% of pollutant degradation by electro-Fenton.

[e]Energy requirement for 40% electro-Fenton mineralization.

[f]Percentage of 2 h-electro-Fenton degradation after the first and the fifth cycle.

[g]N-rGO@CNT is compared to rGO as primary material.

decrease): CVD Gmono (100%) < EEG-cloth (21%) < CVD Gfoam (13.4%) < CVD Gmulti (7.5%) < CPrGO-CF (6%) < rGO (~0%) ≈ N-rGO@CNT (~0%). It was clear that CVD Gmono was not stable at all, which rule out its use as a single cathode. The stability of EEG-cloth was decreasing with the number of cycles but still showed 5.3% higher mineralization yield than the uncoated cloth (10.6%) after five cycles. CVD Gfoam and CVD Gmulti depicted better stability but the percentage of stability decrease is not completely representative to the stability of these materials. Five more successive cycles demonstrated the existence of a plateau, highlighting the stabilization behavior of these CVD materials. Contrastingly, the decrease percentage was 6% with CPrGO-CF; however, pursuing the successive treatments over 10 cycles makes decrease again the stability until a 14% decrease of stability. Interestingly, rGO-based cathode from Zhao et al.[17] and N-rGO@CNT material displayed very good stability. Nevertheless, the assessment of stability was performed by monitoring the degradation of pollutant, unlike the other works. It overestimates the real stability as it has been established previously. For instance, decrease percentages of stability equaled 2.5% with Gmulti and 0.8% with Gfoam after 10 cycles of 2 h of electro-Fenton degradation, which was much better than monitoring the mineralization.[13] It can be further perceived that the coating procedure of EEG-cloth and CPrGO-CF needs to be improved.

21.6.6 ENERGY REQUIREMENTS

In EAOPs, the electrical power is an important feature as it corresponds to the main operating cost. Therefore, reducing the energy consume ($E_{consumption}$) is a decisive step before implementing such processes in industrial applications. Energy requirements are often considered per amount of TOC removed and can be calculated according to the following relation (eq 21.21):

$$E_{consumption}\left(kW\,h\,g^{-1}TOC\right) = \frac{E_{cell}It}{(\Delta TOC)_t V_s} s \qquad (21.21)$$

where E_{cell} is the average cell voltage (V), I is the applied current (A), t is the electrolysis time (h), V_s is the solution volume (L), and $\Delta(TOC)_t$ is the TOC decay (mg C L^{-1}).

The energies consumed to reach 40% of mineralization were ranked as follows (Table 21.2): CVD Gfoam (0.039 kW h g^{-1} TOC) < CVD Gmulti (0.053 kW h g^{-1} TOC) < CVD Gmono (0.072 kW h g^{-1} TOC) < carbon cloth

(0.31 kW h g^{-1} TOC) < EEG-cloth (0.20 kW h g^{-1} TOC) < CF (0.74 kW h g^{-1} TOC) < CPrGO-CF (0.53 kW h g^{-1} TOC). Interestingly, the CVD graphene materials required much lower energy to be performed than the other carbon-based cathodes. Still, a 1.6-fold improvement of the EEG-cloth against the raw cloth could be obtained while a 1.4-times enhancement was procured by the CPrGO-CF over the uncoated CF. The positive role of graphene is again emphasized and could be even surpassed as it has been showed a 4.3 decrease of energy when using N-rGO@CNT instead of rGO.

21.7 CONCLUDING REMARKS

The use of graphene-based electrodes has been reviewed from the synthesis and characterization of graphene to the manufacture of electrode and finally to their applications in EAOPs for water/wastewater treatment. Different factors have shown to affect the performance (in terms of H_2O_2 accumulation concentration, pollutant degradatio, and mineralization efficiency) of the graphene-based cathode in EAOPs treatment. Those parameters have been listed in Table 21.3 and the main impacted materials properties have been defined.

The comparison of these different factors allows identifying the two major properties of the materials that influence its efficiency in EAOPs applications (Fig. 21.8): (1) the conductivity of materials that increase the kinetics rates of electrochemical reaction and (2) the specific surface area that raises the number of active sites for molecules reduction.

The purity of graphene materials has demonstrated to have adverse effects; lower defects make increase the conductivity while it makes decrease the ORR activity. Similarly, the hydrophilicity has contrasting impacts; higher hydrophilicity can favor the kinetics of gaseous O_2 reduction while it can decrease the dissolved O_2 reduction rate in the meanwhile. Therefore, the purity and hydrophilicity of materials could be considered having a secondary role in the electrocatalytic efficiency of the electrode.

Finally, the way to improve the electrocatalytic properties of nanostructured cathodes in EAOPs applications (Fig. 21.8) is to (1) employ raw carbonaceous materials (such as graphene), (2) which can be coated with nanomaterials (graphene and/or CNT), (3) by using preferentially porous 3D primary materials, and (4) that can be doped with nitrogen. Such electrodes need to be developed in the near future to proceed to real applications.

TABLE 21.3 Influencing Parameters and Impacted Properties of Graphene-Based Electrodes Employed in EAOPs Applications.

Influencing parameters	Results	Main impacted materials properties
Coated vs. uncoated materials	Graphene-coated carbon > uncoated carbon materials	Higher surface area
		Higher conductivity
		Higher hydrophilicity
		Lower purity
		Higher ORR activity
EEG vs. rGO	EEG > rGO	Higher surface area
		Higher conductivity
		Similar purity
		Lower ORR activity
Binder vs. binder-free coating methods	Binder ≈ binder-free coating methods	Similar stability
		Higher conductivity with binder
		Lower hydrophilicity with binder
3D vs. 2D materials	3D > 2D materials	Higher surface area
		Higher conductivity
		Lower hydrophilicity
		Lower purity
		Higher ORR activity
N-doped vs. undoped materials	N-doped > undoped materials	Higher number of active sites
		Higher conductivity
		Lower purity
		Higher ORR activity

```
        MANUFACTURING
ADVANCED NANOSTRUCTURED ELECTRODES
       FOR EAOPS APPLICATIONS
```

INFLUENCING PARAMETERS:
- Employing raw carbonaceous materials
- Coating electrode with nanomaterials
- Increasing porosity with 3D materials
- Doping materials with Nitrogen

MAIN MATERIALS PROPERTIES TO GET:
- High Conductivity
- High Active Surface Area

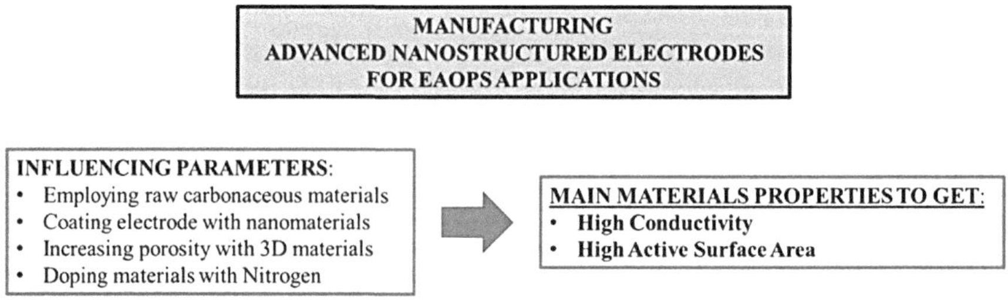

FIGURE 21.8 Routes proposed to manufacture advanced nanostructured cathodes for EAOPs applications.

ACKNOWLEDGMENT

This work was partially supported by Key Project of Natural Science Foundation of Tianjin (no. 16JCZDJC39300) and Natural Science Foundation of China (51178225 and 21273120).

KEYWORDS

- graphene
- oxidation
- reduction
- sp^2 hybridization
- electro-Fenton

REFERENCES

1. Geim, A.; Novoselov, K. The Rise of Graphene. *Nat. Mater.* **2007,** *9*, 183–191.
2. Morozov, S. V.; Novoselov, K. S.; Katsnelson, M. I.; Schedin, F.; Elias, D. C.; Jaszczak, J. A.; Geim, A. K. Giant Intrinsic Carrier Mobilities in Graphene and Its Bilayer. *Phys. Rev. Lett.* **2008,** *100*, 016602(1)–016602(4).
3. Balandin, A. A.; Ghosh, S.; Bao, W.; Calizo, I.; Teweldebrhan, D.; Miao, F.; Lau, C. N. Superior Thermal Conductivity of Single-layer Graphene. *Nano Lett.* **2008,** *8*, 902–907
4. Lee, C.; Wei, X.; Kysar, J. W.; Hone, J. Measurement of the Elastic Properties and Intrinsic Strength of Monolayer Graphene. *Science* **2008,** *321*, 385–388.
5. Nair, R. R.; Blake, P.; Grigorenko, A. N.; Novoselov, K. S.; Booth, T. J.; Stauber, T.; Peres, N. M. R.; Geim, A. K. Fine Structure Constant Defines Visual Transparency of Graphene. *Science.* **2008,** *320*, 1308.
6. Stoller, M. D.; Park, S.; Yanwu, Z.; An, J.; Ruoff, R. S. Graphene-based Ultracapacitors. *Nano Lett.* **2008,** *8* (10), 3498–3502.
7. Chowdhury, S.; Balasubramanian, R. Graphene/Semiconductor Nanocomposites (GSNs) for Heterogeneous Photocatalytic Decolorization of Wastewaters Contaminated with Synthetic Dyes: A Review. *Appl. Catal. B Environ.* **2014,** *160–161*, 307–324.
8. Goh, P. S.; Ismail, A. F. Graphene-based Nanomaterial: The State-of-the-art Material for Cutting Edge Desalination Technology. *Desalination* **2014,** *315* (2013), 115–128.
9. Chowdhury, S.; Balasubramanian, R. Recent Advances in the Use of Graphene-family Nanoadsorbents for Removal of Toxic Pollutants from Wastewater. *Adv. Colloid. Interface. Sci.* **2014,** *204*, 35–56.
10. Filip, J.; Tkac, J. Is Graphene Worth Using in Biofuel Cells? *Electrochim. Acta* **2014,** *136*, 340–354.

11. Le, T. X. H.; Bechelany, M.; Champavert, J.; Cretin, M. A Highly Active Based Graphene Cathode for the Electro-Fenton Reaction. *RSC Adv.* **2015**, *5* (53), 42536–42539.
12. Le, T. X. H.; Bechelany, M.; Lacour, S.; Oturan, N.; Oturan, M. A.; Cretin, M. High Removal Efficiency of Dye Pollutants by Electron-Fenton Process Using a Graphene-based Cathode. *Carbon* **2015**, *94*, 1003–1011.
13. Mousset, E.; Wang, Z.; Hammaker, J.; Lefebvre, O. Physico-Chemical Properties of Pristine Graphene and Its Performance as Electrode Material for Electro-Fenton Treatment of Wastewater. *Electrochim. Acta* **2016**, *214*, 217–230.
14. Mousset, E.; Ko, Z. T.; Syafiq, M.; Wang, Z.; Lefebvre, O. Electrocatalytic Activity Enhancement of a Graphene Ink-coated Carbon Cloth Cathode for Oxidative Treatment. *Electrochim. Acta* **2016**, *222*, 1628-1641.
15. Liu, T.; Wang, K.; Song, S.; Brouzgou, A.; Tsiakaras, P.; Wang, Y. New Electro-Fenton Gas Diffusion Cathode Based on Nitrogen-doped Graphene@carbon Nanotube Composite Materials. *Electrochim. Acta* **2016**, *194*, 228–238.
16. Xu, X.; Chen, J.; Zhang, G.; Song, Y.; Yang, F. Homogeneous Electro-Fenton Oxidative Degradation of Reactive Brilliant Blue Using a Graphene Doped Gas-diffusion Cathode. *Int. J. Electrochem. Sci.* **2014**, *9*, 569–579.
17. Zhao, F.; Liu, L.; Yang, F.; Ren, N. E-fenton Degradation of MB During Filtration with Gr/PPy Modified Membrane Cathode. *Chem. Eng. J.* **2013**, *230*, 491–498.
18. Brillas, E.; Sirés, I.; Oturan, M. A. Electro-Fenton Process and Related Electrochemical Technologies Based on Fenton's Reaction Chemistry. *Chem. Rev.* **2009**, *109* (12), 6570–6631.
19. Rodrigo, M. A.; Oturan, N.; Oturan, M. A. Electrochemically Assisted Remediation of Pesticides in Soils and Water: A Review. *Chem. Rev.* **2014**, *114* (17), 8720–8745.
20. Sirés, I.; Brillas, E.; Oturan, M. A.; Rodrigo, M. A.; Panizza, M. Electrochemical Advanced Oxidation Processes: Today and Tomorrow. A Review. *Environ. Sci. Pollut. Res. Int.* **2014**, *21* (14), 8336–8367.
21. Martinez-Huitle, C. A.; Rodrigo, M. A.; Sires, I.; Scialdone, O. Single and Coupled Electrochemical Processes and Reactors for the Abatement of Organic Water Pollutants : A Critical Review. *Chem. Rev.* **2015**, *115* (24), 13362–13407.
22. Moreira, F. C.; Boaventura, R. A. R.; Brillas, E.; Vilar, V. J. P. Electrochemical Advanced Oxidation Processes: A Review on Their Application to Synthetic and Real Wastewaters. *Appl. Catal. B: Environ.* **2017**, *202*, 217–261.
23. Zhao, G.; Wen, T.; Chen, C.; Wang, X. Synthesis of Graphene-based Nanomaterials and Their Application in Energy-related and Environmental-related Areas. *RSC Adv.* **2012**, *2* (25), 9286–9303.
24. Ambrosi, A.; Chua, C. K.; Bonanni, A.; Pumera, M. Electrochemistry of Graphene and Related Materials. *Chem. Rev.* **2014**, *114* (14), 7150–7188.
25. Muñoz, R.; Gómez-Aleixandre, C. Review of CVD Synthesis of Graphene. *Chem. Vap. Depos.* **2013**, *19* (10–12), 297–322.
26. Li, X.; Cai, W.; An, J.; Kim, S.; Nah, J.; Yang, D.; Piner, R.; Velamakanni, A.; Jung, I.; Tutuc, E.; et al. Large Area Synthesis of High Quality and Uniform Graphene Films on Copper Foils. *Science* **2009**, *324* (5932), 1312–1314.
27. Chen, Z.; Ren, W.; Gao, L.; Liu, B.; Pei, S.; Cheng, H.-M. Three-dimensional Flexible and Conductive Interconnected Graphene Networks Grown by Chemical Vapour Deposition. *Nat. Mater.* **2011**, *10* (6), 424–428.

28. Bae, S.; Kim, H.; Lee, Y.; Xu, X.; Park, J.-S.; Zheng, Y.; Balakrishnan, J.; Lei, T.; Ri Kim, H.; Song, Y. Il; et al. Roll-to-roll Production of 30-inch Graphene Films for Transparent Electrodes. *Nat. Nanotechnol.* **2010,** *5* (8), 574–578.
29. Liang, X.; Sperling, B. a.; Calizo, I.; Cheng, G.; Hacker, C. A.; Zhang, Q.; Obeng, Y.; Yan, K.; Peng, H.; Li, Q.; et al. Toward Clean and Crackless Transfer of Graphene. *ACS Nano* **2011,** *5* (11), 9144–9153.
30. Moraes, F. C.; Gorup, L. F.; Rocha, R. S.; Lanza, M. R. V; Pereira, E. C. Photoelectrochemical Removal of 17 beta-estradiol Using a RuO_2-graphene Electrode. *Chemosphere* **2016,** *162,* 99–104.
31. Hummers, W. S.; Offeman, R. E. Preparation of Graphitic Oxide. *J. Am. Chem. Soc.* **1958,** *80* (6), 1339–1339.
32. Marcano, D. C.; Kosynkin, D. V; Berlin, J. M.; Sinitskii, A.; Sun, Z.; Slesarev, A.; Alemany, L. B.; Lu, W.; Tour, J. M. Improved Synthesis of Graphene Oxide. *ACS Nano* **2010,** *4* (8), 4806–4814.
33. Zhang, J.; Yang, H.; Shen, G.; Cheng, P.; Zhang, J.; Guo, S. Reduction of Graphene Oxide via L-ascorbic Acid. *Chem. Commun.* **2010,** *46* (7), 1112–1114.
34. Parvez, K.; Wu, Z. S.; Li, R.; Liu, X.; Graf, R.; Feng, X.; Müllen, K. Exfoliation of Graphite into Graphene in Aqueous Solutions of Inorganic Salts. *J. Am. Chem. Soc.* **2014,** *136* (16), 6083–6091.
35. Parvez, K.; Li, R.; Puniredd, S. R.; Hernandez, Y.; Hinkel, F.; Wang, S.; Feng, X.; Mu, K.; Engineering, C.; Road, D. Electrochemically Exfoliated Graphene as Solution-processable, Highly Conductive Electrodes for Organic Electronics. *ACS Nano* **2013,** *7* (4), 3598–3606.
36. Su, C. Y.; Lu, A. Y.; Xu, Y.; Chen, F. R.; Khlobystov, A. N.; Li, L. J. High-quality Thin Graphene Films from Fast Electrochemical Exfoliation. *ACS Nano* **2011,** *5* (3), 2332–2339.
37. Wu, L.; Li, W.; Li, P.; Liao, S.; Qiu, S.; Chen, M.; Guo, Y.; Li, Q.; Zhu, C.; Liu, L. Powder, Paper and Foam of Few-layer Graphene Prepared in High Yield by Electrochemical Intercalation Exfoliation of Expanded Graphite. *Small* **2014,** *10* (7), 1421–1429.
38. Guo, J.; Zhang, T.; Hu, C.; Fu, L. A Three-dimensional Nitrogen-doped Graphene Structure: A Highly Efficient Carrier of Enzymes for Biosensors. *Nanoscale* **2015,** *7* (4), 1290–1295.
39. Krauss, B.; Lohmann, T.; Chae, D. H.; Haluska, M.; Von Klitzing, K.; Smet, J. H. Laser-induced Disassembly of a Graphene Single Crystal into a Nanocrystalline Network. *Phys. Rev. B: Condens. Matter Mater. Phys.* **2009,** *79* (16), 1–16.
40. Ferrari, A. C.; Robertson, J. Raman Spectroscopy of Amorphous, Nanostructured, Diamond-like Carbon, and Nanodiamond. *Philos. Trans. A. Math. Phys. Eng. Sci.* **2004,** *362* (1824), 2477–2512.
41. Kovtyukhova, N. I.; Ollivier, P. J.; Martin, B. R.; Mallouk, T. E.; Chizhik, S. A.; Buzaneva, E. V.; Gorchinskiy, A. D. Layer-by-layer Assembly of Ultrathin Composite Films from Micron-sized Graphite Oxide Sheets and Polycations. *Chem. Mater.* **1999,** *11* (3), 771–778.
42. Schroder, D. K. In *Semiconductor Material and Device Characterization;* 3rd Ed.; John Wiley and Sons: New York, NY, 2006.
43. Pei, S.; Cheng, H. M. The Reduction of Graphene Oxide. *Carbon* **2012,** *50* (9), 3210–3228.
44. Filik, H.; Çetintaş, G.; Koç, S. N.; Gülce, H.; Boz, İ. Nafion-graphene Composite Film Modified Glassy Carbon Electrode for Voltammetric Determination of P-aminophenol. *Russ. J. Electrochem.* **2014,** *50* (3), 243–252.

45. Wei, B.; Tokash, J. C.; Chen, G.; Hickner, M. A.; Logan, B. E. Development and Evaluation of Carbon and Binder Loading in Low-cost Activated Carbon Cathodes for Air-cathode Microbial Fuel Cells. *RSC Adv.* **2012**, *2* (33), 12751–12758.
46. Liu, L.; Zhao, F.; Liu, J.; Yang, F. Preparation of Highly Conductive Cathodic Membrane with Graphene (Oxide)/PPy and the Membrane Antifouling Property in Filtrating Yeast Suspensions in EMBR. *J. Memb. Sci.* **2013**, *437*, 99–107.
47. Zhou, M.; Yu, Q.; Lei, L. The Preparation and Characterization of a graphite–PTFE Cathode System for the Decolorization of C. I. Acid Red 2. *Dyes Pigment.* **2008**, *77* (1), 129–136.
48. Pokpas, K.; Zbeda, S.; Jahed, N.; Mohamed, N.; Baker, P. G.; Iwuoha, E. I. Nafion-graphene Nanocomposite in Situ Plated Bismuth-film Electrodeon Pencil Graphite Substrates for the Determination of Trace Metals by Anodic Stripping Voltammetry. *Int. J. Electrochem. Sci.* **2014**, *9* (2), 5092–5115.
49. Gibney, E. 2D or Not 2D. *Nature* **2015**, *522*, 274–276.
50. Le, T. X. H.; Bechelany, M.; Champavert, J.; Cretin, M. Support Information: A Highly Active Based Graphene Cathode for the Electro-Fenton Reaction. *RSC Adv.* **2015**, *5* (53), 42536–42539.
51. Lv, Z.; Chen, Y.; Wei, H.; Li, F.; Hu, Y.; Wei, C.; Feng, C. One-step Electrosynthesis of Polypyrrole/Graphene Oxide Composites for Microbial Fuel Cell Application. *Electrochim. Acta* **2013**, *111*, 366–373.
52. Rafiee, J.; Mi, X.; Gullapalli, H.; Thomas, A. V; Yavari, F.; Shi, Y.; Ajayan, P. M.; Koratkar, N. A. Wetting Transparency of Graphene. *Nat. Mater.* **2012**, *11* (3), 217–222.
53. Chen, Z.; Ren, W.; Gao, L.; Liu, B.; Pei, S.; Cheng, H.-M. Three-dimensional Flexible and Conductive Interconnected Graphene Networks Grown by Chemical Vapour Deposition. *Nat. Mater.* **2011**, *10* (6), 424–428.
54. Hass, J.; Varchon, F.; Millán-Otoya, J. E.; Sprinkle, M.; Sharma, N.; de Heer, W. A.; Berger, C.; First, P. N.; Magaud, L.; Conrad, E. H. Why Multilayer Graphene on 4H–SiC(0001) Behaves Like a Single Sheet of Graphene. *Phys. Rev. Lett.* **2008**, *100* (12), 125504.
55. Morcos, I.; Yeager, E. Kinetic Studies of the Oxygen—Peroxide Couple on Pyrolytic Graphite. *Electrochim. Acta* **1970**, *15*, 953–975.
56. Yeager, E. Dioxygen Electrocatalysis: Mechanisms in Relation to Catalyst Structure. *J. Mol. Catal.* **1986**, *38*, 5–25.
57. Bard, A. J.; Faulkner, L. R. In *Electrochemical Methods: Fundamentals and Applications*; John Wiley and Sons: New York, NY, 2000.
58. Grewal, Y. S.; Shiddiky, M. J. A.; Gray, S. A.; Weigel, K. M.; Cangelosi, G. A.; Trau, M. Label-free Electrochemical Detection of an Entamoeba Histolytica Antigen Using Cell-free Yeast-scFv Probes. *Chem. Commun.* **2013**, *49* (15), 1551–1553.
59. Yu, X.; Zhou, M.; Ren, G.; Ma, L. A Novel Dual Gas Diffusion Electrodes System for Efficient Hydrogen Peroxide Generation Used in Electro-fenton. *Chem. Eng. J.* **2015**, *263*, 92–100.
60. Zhou, L.; Zhou, M.; Hu, Z.; Bi, Z.; Serrano, K. G. Chemically Modified Graphite Felt as an Efficient Cathode in Electro-Fenton for P-nitrophenol Degradation. *Electrochim. Acta* **2014**, *140*, 376–383.
61. Ma, L.; Zhou, M.; Ren, G.; Yang, W.; Liang, L. A Highly Energy-efficient Flow-through Electro-Fenton Process for Organic Pollutants Degradation. *Electrochim. Acta* **2016**, *200*, 222–230.

62. Hu, J.; Sun, J.; Yan, J.; Lv, K.; Zhong, C.; Deng, K.; Li, J. A Novel Efficient Electrode Material: Activated Carbon Fibers Grafted by Ordered Mesoporous Carbon. *Electrochem. Commun.* **2013**, *28*, 67–70.
63. Hu, Z.; Zhou, M.; Zhou, L.; Li, Y.; Zhang, C. Effect of Matrix on the Electrochemical Characteristics of TiO$_2$ Nanotube Array-based PbO$_2$ Electrode for Pollutant Degradation. *Environ. Sci. Pollut. Res. Int.* **2014**, *21* (14), 8476–8484.
64. Ding, Z.; Hu, X.; Morales, V. L.; Gao, B. Filtration and Transport of Heavy Metals in Graphene Oxide Enabled Sand Columns. *Chem. Eng. J.* **2014**, *257*, 248–252.
65. Zhou, M.; Yu, Q.; Lei, L.; Barton, G. Electro-Fenton Method for the Removal of Methyl Red in an Efficient Electrochemical System. *Sep. Purif. Technol.* **2007**, *57* (2), 380–387.
66. Zhou, L.; Zhou, M.; Zhang, C.; Jiang, Y.; Bi, Z.; Yang, J. Electro-Fenton Degradation of P-nitrophenol Using the Anodized Graphite Felts. *Chem. Eng. J.* **2013**, *233*, 185–192.
67. Yu, F.; Zhou, M.; Zhou, L.; Peng, R. A Novel Electro-Fenton Process with H$_2$O$_2$ Generation in a Rotating Disk Reactor for Organic Pollutant Degradation. *Environ. Sci. Technol. Lett.* **2014**, *1*, 320–324.
68. Zhang, H.; Li, H.; Deng, C.; Zhao, B.; Yang, J. Electrocatalysis of Oxygen Reduction Reaction on Carbon Nanotubes Modified by Graphitization and Amination. *ECS Electrochem. Lett.* **2015**, *4* (8), H33–H37.
69. Frysz, C. A.; Shui, X.; Chung, D. D. L. Electrochemical Behavior of Porous Carbons. *Carbon N. Y.* **1997**, *35* (7), 893–916.
70. Hu, X.; Wu, Y.; Li, H.; Zhang, Z. Adsorption and Activation of O$_2$ on Nitrogen-doped Carbon Nanotubes. *J. Phys. Chem. C* **2010**, *114* (21), 9603–9607.
71. Sedlak, D. L. In *Water 4.0*; Yale University Press: New Haven, CT, 2014.
72. Oturan, M. A.; Aaron, J.-J. Advanced Oxidation Processes in Water/Wastewater Treatment: Principles and Applications. A Review. *Crit. Rev. Environ. Sci. Technol.* **2014**, *44* (23), 2577–2641.
73. Latimer, W. M. Oxidation Potentials. *Soil Sci.* **1952**, *74* (4), 333.
74. Oturan, M. A. An Ecologically Effective Water Treatment Technique Using Electrochemically Generated Hydroxyl Radicals for in Situ Destruction of Organic Pollutants: Application to Herbicide 2,4-D. *J. Appl. Electrochem.* **2000**, *30* (4), 475–482.
75. Sirés, I.; Garrido, J. A.; Rodríguez, R. M.; Brillas, E.; Oturan, N.; Oturan, M. A. Catalytic Behavior of the Fe^{3+}/Fe^{2+} System in the Electro-Fenton Degradation of the Antimicrobial Chlorophene. *Appl. Catal. B Environ.* **2007**, *72* (3–4), 382–394.
76. Mousset, E.; Huguenot, D.; Van Hullebusch, E. D.; Oturan, N.; Guibaud, G.; Esposito, G.; Oturan, M. A. Impact of Electrochemical Treatment of Soil Washing Solution on PAH Degradation Efficiency and Soil Respirometry. *Environ. Pollut.* **2016**, *211*, 354–362.
77. Huguenot, D.; Mousset, E.; van Hullebusch, E. D.; Oturan, M. A. Combination of Surfactant Enhanced Soil Washing and Electro-Fenton Process for the Treatment of Soils Contaminated by Petroleum Hydrocarbons. *J. Environ. Manage.* **2015**, *153*, 40–47.
78. Mousset, E.; Oturan, N.; van Hullebusch, E. D.; Guibaud, G.; Esposito, G.; Oturan, M. A. Influence of Solubilizing Agents (Cyclodextrin or Surfactant) on Phenanthrene Degradation by Electro-Fenton Process—Study of Soil Washing Recycling Possibilities and Environmental Impact. *Water Res.* **2014**, *48* (1), 306–316.
79. Mousset, E.; Oturan, N.; van Hullebusch, E. D.; Guibaud, G.; Esposito, G.; Oturan, M. A. Treatment of Synthetic Soil Washing Solutions Containing Phenanthrene and Cyclodextrin by Electro-oxidation. Influence of Anode Materials on Toxicity Removal and Biodegradability Enhancement. *Appl. Catal. B Environ.* **2014**, *160–161*, 666–675.

80. Mousset, E.; Wang, Z.; Lefebvre, O. Electro-Fenton for Control and Removal of Micropollutants—Process Optimization and Energy Efficiency. *Water Sci. Technol.* **2016,** *74* (2), 2068–2074.
81. Oturan, N.; Brillas, E.; Oturan, M. A. Unprecedented Total Mineralization of Atrazine and Cyanuric Acid by Anodic Oxidation and Electro-Fenton with a Boron-doped Diamond Anode. *Environ. Chem. Lett.* **2012,** *10* (2), 165–170.
82. Trellu, C.; Péchaud, Y.; Oturan, N.; Mousset, E.; Huguenot, D.; van Hullebusch, E. D.; Esposito, G.; Oturan, M. A. Comparative Study on the Removal of Humic Acids from Drinking Water by Anodic Oxidation and Electro-Fenton Processes: Mineralization Efficiency and Modelling. *Appl. Catal. B: Environ.* **2016,** *194*, 32–41.
83. Miao, J.; Zhu, H.; Tang, Y.; Chen, Y.; Wan, P. Graphite Felt Electrochemically Modified in H_2SO_4 Solution Used as a Cathode to Produce H_2O_2 for Pre-oxidation of Drinking Water. *Chem. Eng. J.* **2014,** *250*, 312–318.
84. Yu, F.; Zhou, M.; Yu, X. Cost-effective Electro-Fenton Using Modified Graphite Felt That Dramatically Enhanced on H_2O_2 Electro-generation Without External Aeration. *Electrochim. Acta* **2015,** *163*, 182–189.
85. Assumpção, M. H. M. T.; Moraes, A.; De Souza, R. F. B.; Gaubeur, I.; Oliveira, R. T. S.; Antonin, V. S.; Malpass, G. R. P.; Rocha, R. S.; Calegaro, M. L.; Lanza, M. R. V.; et al. Low Content Cerium Oxide Nanoparticles on Carbon for Hydrogen Peroxide Electrosynthesis. *Appl. Catal. A Gen.* **2012,** *411–412*, 1–6.
86. Zhang, G.; Yang, F.; Gao, M.; Fang, X.; Liu, L. Electro-Fenton Degradation of Azo Dye Using Polypyrrole/Anthraquinonedisulphonate Composite Film Modified Graphite Cathode in Acidic Aqueous Solutions. *Electrochim. Acta* **2008,** *53* (16), 5155–5161.
87. Wang, Z. X.; Li, G.; Yang, F.; Chen, Y. L.; Gao, P. Electro-Fenton Degradation of Cellulose Using Graphite/PTFE Electrodes Modified by 2-Ethylanthraquinone. *Carbohydr. Polym.* **2011,** *86* (4), 1807–1813.
88. Zhang, G.; Zhou, Y.; Yang, F. FeOOH-Catalyzed Heterogeneous Electro-Fenton System upon Anthraquinone @ Graphene Nanohybrid Cathode in a Divided Electrolytic Cell : Catholyte-regulated Catalytic Oxidation Performance and Mechanism. *J. Electrochem. Soc.* **2015,** *162* (6), H357–H365.
89. Mousset, E.; Frunzo, L.; Esposito, G.; van Hullebusch, E. D.; Oturan, N.; Oturan, M. A. A Complete Phenol Oxidation Pathway Obtained During Electro-Fenton Treatment and Validated by a Kinetic Model Study. *Appl. Catal. B: Environ.* **2016,** *180*, 189–198.
90. Pimentel, M.; Oturan, N.; Dezotti, M.; Oturan, M. A. Phenol Degradation by Advanced Electrochemical Oxidation Process Electro-Fenton Using a Carbon Felt Cathode. *Appl. Catal. B: Environ.* **2008,** *83* (1–2), 140–149.
91. Oturan, M. A.; Pimentel, M.; Oturan, N.; Sirés, I. Reaction Sequence for the Mineralization of the Short-chain Carboxylic Acids Usually Formed upon Cleavage of Aromatics During Electrochemical Fenton Treatment. *Electrochim. Acta* **2008,** *54* (2), 173–182.

INDEX

A
Adsorption, 164–165
Amineterminated butadiene-acrylonitrile, 77

B
Barium titanate, 281–282
Bench tests
  results of, 247
Betalain dyes, 294
Bionanocomposites
  chitosan–HNT, 228–229
  PBS–HNT, 227
  PLA–HNT, 224–226
  PVOH–HNT, 229
  starch–HNT, 227–228
Biopolymers, 219
  studied, 220
Block copolymers, 78–81
  geometries produced by, D
  micro and nanostructures of epoxy and, 85–94
    AB di-block copolymer micelles, 87
    AFM image of PS, 92
    photo cross-linking nanostructured blocks, 87
    poly A/poly B blocks copolymers, 88–90
    star-block copolymers micellar structures, 91
    thermosetting blends, 94

C
Carboxyl-terminated butadiene-acrylonitrile (CTBN), 79
Cashew nut shell liquid (CNSL), 49, 135, 136
  experimental details, 50–52, 136
    FFT plot, 52
    FRF function, 51
    hammer modal analysis, impact, 51
    modal analysis, impact, 50
    postprocessing of the vibration response, 50–51
  mounting and failure of, 137
  results and discussion, 52–53
    crack properties, 138
    damping percentages, 53
    load, displacement plots of, 137
    natural frequencies, 52
    Q factors, 53
    tensile test results, 137
Cellulose and nanocellulose, 197
  acid hydrolysis, 200–201
  alkaline solution, 199–200
  application of, 201
  bleaching treatment, 200
  extraction of, 199–201
  lignocellulosic biomass, 197
  nanofibers, 199
  structure of, 197, 198
Chitosan–polyvinyl alcohol blend (CS–PVOH), 228–229
Clays
  classification of, 216–217
Clean technologies, 203
Copper (Cu), 308
  electroreduction, 309

D
Diglycidyl ether of bisphenol S (DGEBS), 72
Dual-beam thermal lens (TL) technique, 3
Dye-sensitized solar cell (DSSC), 293
  experimental
    characterization, 296–297
    dye sensitization and cell assembly, 296
    synthesis of ZnO nanostructures, 295–296
  natural pigments, 294
  results and discussion, 297–303

comparison of, 301
current density against voltage, 300
SEM images of ZnO, 298
XRD patterns of different seeds, 299
ZnO nanostructures, growth of, 299

E

Elastic solids
　mechanical properties of, 23, 24
Electrochemical advanced oxidation processes (EAOPs), 321
Electroreduction
　using Cu as catalysts, 309
Epoxy
　AFM image of, B, C, G
　AFM phase, B, C
Epoxy resin, 72, 113
　AFM phase, 74
　dynamic mechanical analysis of cured DGEBS, 73
　fracture energy *versus* CSR content, 76
　glass transition temperature, 75
　tan δ curve, 72
　tensile strength *versus* core–shell rubber particle, 75
　TGA thermograms of DGEBS, 73
　thermal properties of cured DGEBS, 73

F

Feedstock polymer materials, 239–241
　thermally sprayed polymers and polymer composites, 239–240
Ferroelectric material, 275
　piezoelectric and pyroelectric, relationship between, 277
Flame spraying of polymers, 235
Flory–Huggins theory, 84

G

Graphene, 159, 160–161, 255–256
　characterization of
　　AFM image of, 163
　properties of
　　electronic, 257
　　optical, 257
　　thermal, 257–258
　remediation methods

　　adsorption, 164–165
　storage and energy conversion devices
　　electrochemical supercapacitors, 265–266
　　fuel cells, 266–268
　　lithium-ion batteries, 261–265
　structure of, 256
　synthesis of
　　bottom-up approach, 162–163
　　photocatalysis, 165–166
　　top-down approach, 161–162
　synthesis techniques
　　ex situ, 258
　　in situ, 259–260
　toxicity of, 179–181
Graphene-based nanostructured materials
　characterization of
　　purity and structural properties, 327–329
　　sheet resistance, 330
　　surface morphology, 325–327
　manufacture of
　　binder-free, 333
　　methods, 331
　　using binder, 330, 332–333
　performance comparison of, 348, 349
　properties of, 333
　　electrical conductivity, 336–337
　　electroactive surface area, 337–339
　　ORR activity, 340–341
　　surface morphology, 334–335
　　wettability, 335–336
　synthesis of, 322–323
　　bottom-up methods, 323–324
　　steps for applications, 323
　　top-down methods, 324–325
　water/wastewater treatment applications of
　　context and development of EAOPS, 342–343
　　energy requirements, 350–351
　　H2O2 accumulation concentration, 343–344
　　influencing parameters and impacted properties of, 352
　　mineralization efficiency, 346–347
　　pollution degradation efficiency, 345–346

Index 361

stability of electrode, 347, 350
Gun Terco-P
  features of, 238

H

Halloysite bionanocomposites, 215
Halloysite nanotubes (HNTs), 216–218
  bionanocomposites
    applications of, 229–230
    chitosan–HNT, 228–229
    PBS–HNT, 227
    PLA–HNT, 224–226
    PVOH–HNT, 229
    starch–HNT, 227–228
  biopolymers, 219
    studied, 220
  CEC values of, 219
  crystalline structure of, 218
  processing techniques
    electrospinning, 222
    in-situ polymerization, 222
    LBL method, 223–224
    melt processing, 221
    solution mixing, 220–221
  structural unit of, 218–219
Harvesting energy, 273

I

Inverse relaxation (IR), 26–28
  experimental realization of, 31–34
  index
    yarn to fabric flow, 34–35
  inverse creep, 26
  linear polymer molecule, 28
  linear viscoelastic models, 28–31
    Kelvin (Voigt) model, 29
    Maxwell model, 29
  occurrence of, 26
  viscoelastic properties, 26–27

L

Layer-by-layer deposition method (LBL method), 223–224

M

Metachromasy, 56
  absorption spectrum
    MB–NaHep, A
    TB–NaHep, A
  biological activity of, 56
  materials and methods
    apparatus, 57
    reagents, 57
  methods
    alcohols and urea, reversal, 58
    stoichiometry of polymer–dye complex, 57–58
    surfactants and electrolytes, reversal, 58
    thermodynamic parameters, 58
  physicochemical properties, 56
  results and discussion, 59–68
    absorption spectra of MB, 59, 60
    alcohols and urea, reversal, 62–63
    determination of stoichiometry, 60–61
    interaction parameters, 64–67
    methylene blue, structure of, 67
    structure of dye, effect of, 67–68
    surfactants and electrolytes, reversal, 64
    toluidine blue, structure of, 68
Metal nanoparticles, 128, 141
  application of green chemistry, 144
  bottom-up approach, 142
  chemical methods, 142–143
  copper nanoparticles, 144–145
  experimental, 154–155
    materials required, 146
    synthesis techniques for, 147
  green approach, 143
  Lycurgus cup, 142
  properties, 142
  results and discussion, 147–151
    FTIR peaks, 150–151
    PL emission spectra of, 157
    room temperature photoluminescence emission spectra, 156
    SEM images of, 156
    TEM image of, 148
    UV–vis spectra of, 148–149
    XRD pattern for, 155
  top-down approaches, 142
  traditional, conventional methods, 146

N

Nanocrystalline ZnO thin film, 37
Nanomaterials, 322

P

Piezoelectric materials, 273–274
  applications
    biomedical, 288
    energy harvesters, 289–290
    flexible electronics, 289
    tactile sensors, 289
  classification of
    ceramics, 280
    hard, 279
    lead-free ceramics, 281–282
    PZT, 280–281
    quartz and Rochelle salt, 279–280
  commercial applications, 275
  energy harvesting, 287–288
  ferroelectric and pyroelectric, relationship between, 277
  naturally occurring, 275
  piezoelectric property, 275
  theoretical formulation, 278
Piezoelectricity, 274
  poling
    corona, 286
    electrode, 286
  polymer nanocomposites, 286–287
  polymers, in, 282–283
    polyamide 11 (nylon 11), 285–286
    polyvinylidene cyanide-vinyl acetate, 285
    PVDF, 283–284
    PVDF-TrFE, 284–285
Poly(ethylene oxide) block based copolymer system, 101–106
  AFM phase image, 103
  connectivity of PEO–PEP, 104
  morphology and behavior of, 102
  TEM image, 103
  Tg of noncrystalline material, 104–105
Poly(ethylene oxide)–poly(ethylene-*alt*-propylene) (PEO–PEP), 102
Polyamide 11 (nylon 11), 285–286
Polymer blend, 82–84
  binodal curve, 83
  Flory–Huggins theory, 84
  immiscible, 82–84
  spinodal curve, 83
  tie-line, 84

Polymer electrolyte membrane fuel celltype reactor
  electrochemical reduction of CO_2, 307
  electrochemical CO_2 conversion techniques, 308
  experimental
    cell configuration, 311
    electrochemical reduction, 311
    reactant solutions, 311
  results and discussion
    anode reactant, effect of, 313
    cathode reactant, effect of, 314
    electrochemical cell, 312–313
    temperature, effect of, 316–317
    time, effect of, 315
Polymer flame-spraying gun
  design of, 236–239
    coating properties, 238
    fuel mixture air–propane, 236–237
    temperature of various combustion mixtures, 237
    Terco-P, 238
Polypropylene (PP) hybrid composites, 15–17
  crystallization characteristics, 15–16
  experiment
    DSC measurements, 17
    materials, 17
    sample preparation, 17
    thermogravimetric analysis, 17–18
  results and discussions
    crystallization parameters, 18
    DSC measurements, 18
    kinetic parameters, 20–21
    thermogravimetric analysis, 19–21
  thermal properties of, 15, 16
Polyvinyl alcohol (PVA)-based thin films, 127, 128–229
  experimental setup, 130
  materials and methods, 129
  metal nanoparticles, 128
  optical filter parameters
    bandwidth, 128
    blocking range, 128
    center wavelength, 128
    optical density, 128, 129
  optical properties, 127
  results and discussion, 130–132

absorption spectrum of erythrosine films, 131
blocking capability of Ag-doped, 132
OD plot for Ag nanoparticle-doped, 132
transmittance of Ag-doped, 132
transmittance spectrum of erythrosine films, 131
Polyvinyldene cyanide-vinyl acetate, 285
Polyvinyldene fluoride (PVDF), 283–284
 molecular structure of, 284
Polyvinyldene fluoride-trifluoroethylene (PVDF-TrFE), 284–285
PS (polystyrene)-*alt*-PEO (poly(ethylene oxide))
 AFM image of, E, F
Pyroelectric material, 276
 coefficient, 276
 piezoelectric and ferroelectric, relationship between, 277

R

Rh6g, 8–9
 dye concentration
  fluorescence behavior of, 10
  variation of diffusivity with, 9

S

SIS block copolymer system, 106–113
 epoxidation of polyisoprene block, 107
 inclusion of CNT, 111
 morphological transformation of, 110
 phase diagram of, 109
 possible mechanisms, 108
 PSgMWCNT, mechanism of dispersion, 113
 schematic representation, 106
 tapping mode-AFM phase image, 109, 112
 TEM microstructure, 111
 uncured and cured, 110
Stress relaxation, 25–26
Styrene–butadiene–styrene (SBS) tri-block copolymer, 95
Styrene–isoprene–styrene (SIS), 106
Sustainable industrial processes, 212

T

Textile wet-processing, 203, 204

aim of study, 205–206
decomposition of hydrogen peroxide, 205
materials, 206
methods, 206–208
 chemical pre-treatment and bio-pretreatment of Lycra, 207
pretreatment, 204
results and discussion
 absorbency, 208
 bursting strength, 210
 effluent test results, 211
 fabric test results, 209
 pH of fabric, 208
 resources used of, 212
 weight, 208
 whiteness and yellowness index, 210–212
sustainable economy, 205
Thermal lens (TL)., 4, 12

W

Water, 159–160
 desalination, 178–179
 inorganic pollutants, 166–167
  arsenic, 167–170
  cadmium, 173
  chromium, 171–173
  copper, 175
  ions, 177–178
  lead, 174–175
  mercury, 170–171
  radionuclides, 176–177

Z

Zinc oxide (ZnO) nanostructures, 4–5, 153–154, 282
 electrical characterization, 45–46
  electrical parameters of, 45
 experimental details, 38–40
  deposition parameters, 40
  preparation of sample, 6–7
  pulsing time of precursor gases, 39
  setup, 5–6
 forms, 38
 optical characterization, 43–45
  $(\alpha h\nu)^2$ versus $h\nu$ plot, 44
  absorbance spectra of, 43
  photoluminescence spectra, 44

PL spectra of, 45
transmission spectra of, 43
results and discussion, 40
  Rh6g, 8–9
  TL, 7–8
  variation of thermal diffusivity with, 11
  variation of TL signal amplitude, 12
scanning electron microscopic images of, 7

structural characterization, 40–42
  comparison of XRD pattern of, 41
  compositional analysis of, 42
  EDX spectrum of, 42
  SEM image of, 42
  XRD of, 41
thin film, 37
transparency and conductivity, 38